Renewable Energy:
Research, Development and Policies

snova
New York

Renewable Energy: Research, Development and Policies

Integrated Energy Systems: Design, Control and Operation
Bikram Das (Editor)
Abanishwar Chakraborti (Editor)
Arvind Kumar Jain (Editor)
Subhadeep Bhattacharjee (Editor)
2023. ISBN: 979-8-89113-206-1 (Softcover)
2023. ISBN: 979-8-89113-227-6 (eBook)

Photovoltaic Systems: Advances in Research and Applications
Sudip Mandal, PhD (Editor)
Pijush Dutta, PhD (Editor)
2023. ISBN: 979-8-89113-102-6 (eBook)

The Fundamentals of Thermal Analysis
Mamdouh El Haj Assad, PhD (Editor)
Ali Khosravi, PhD (Editor)
Mehran Hashemian, PhD (Editor)
2023. ISBN: 979-8-88697-759-2 (Hardcover)
2023. ISBN: 979-8-88697-875-9 (eBook)

Solar Collectors and Systems
Mamdouh El Haj Assad, DSc (Editor)
Mohammad Alhuyi Nazari, PhD (Editor)
2023. ISBN: 979-8-88697-774-5 (Softcover)
2023. ISBN: 979-8-88697-860-5 (eBook)

The Future of Solar Power
Professor Dr. Hussain H. Al-Kayiem (Editor)
2023. ISBN: 979-8-88697-709-7 (eBook)

More information about this series can be found at
https://novapublishers.com/product-category/series/renewable-energy-research-development-and-policies/

Mehmet Landrey
Editor

Wind Energy

Assessment, Developments and Technology

snova
New York

Library of Congress Cataloging-in-Publication Data

ISBN: 979-8-89113-528-4

Published by Nova Science Publishers, Inc. † New York

Contents

Preface

Wind energy is one of the fastest-growing sources of renewable energy. The Biden administration's decarbonization goals (i.e., carbon-pollution-free electricity by 2035 and net-zero economy by 2050) will require at least a threefold increase in the U.S. wind energy deployment rate.

Airborne wind energy (AWE) is "the conversion of wind energy into electricity using tethered flying devices" Chapter 1 provides an analyses and assessment of AWE systems. Wind-solar-storage hybrid power plants represent a significant and growing share of new proposed projects. Chapters 2 and 3 identify areas that are particularly suited for wind-solar hybrid power plant development. Chapter 4 examines the U.S. land-based and offshore wind energy supply chain, including raw materials, processed materials, components, sub-components, and recycling. Chapter 5 explores how material needs for wind energy might change under two U.S. wind deployment scenarios: Current Policies and High Deployment. Chapters 6 and 7 describe the vision and purpose for AWAKEN. AWAKEN is an international, multi-institutional wind energy field campaign designed to answer the most pressing science questions about the impacts of wind turbine wake effects.

Chapter 1

Airborne Wind Energy*

Jochem Weber
Melinda Marquis
Aubryn Cooperman
Caroline Draxl
Rob Hammond
Jason Jonkman
Alexsandra Lemke
Anthony Lopez
Rafael Mudafort
Mike Optis
Owen Roberts
and Matt Shields
National Renewable Energy Laboratory

Acknowledgments

The National Renewable Energy Laboratory would like to acknowledge all the workshop participants and individuals who shared their expertise in airborne wind energy with the authors of this report. This report was sponsored by the U.S. Department of Energy's Wind Energy Technologies Office.

* This is an edited, reformatted and augmented version of National Renewable Energy Laboratory, operated by Alliance for Sustainable Energy, LLC, for the U.S. Department of Energy (DOE), Technical Report, NREL/TP-5000-79992, dated August 2021.

In: Wind Energy
Editor: Mehmet Landrey
ISBN: 979-8-89113-528-4

List of Acronyms

AEP	annual energy production
ATB	Annual Technology Baseline
AWE	airborne wind energy
CapEx	capital expenditure(s)
DOD	U.S. Department of Defense
DOE	U.S. Department of Energy
FAA	Federal Aviation Administration
FCR	fixed charge rate
GW	gigawatt
IEA	International Energy Agency
km	kilometer
kW	kilowatt
LCOE	levelized cost of energy
LLJ	low-level jet
m	meter
m/s	meters per second
MW	megawatt
MWh	megawatt-hour
NCF	net capacity factor
NREL	National Renewable Energy Laboratory
O&M	operation and maintenance
OpEx	operational expenditure(s)
R&D	research and development
TPL	technology performance level
TRL	technology readiness level
WRF	Weather Research and Forecasting

Executive Summary

Airborne wind energy (AWE) is "the conversion of wind energy into electricity using tethered flying devices" (Schmehl 2020.) Pursuit of AWE and airborne wind energy systems began in 1980 (Loyd 1980). Interest and investment in AWE have grown substantially in the last decade, with approximately 70 active research entities including over 20 technology developers globally. This report describes technical analyses of various aspects of AWE and insight gained from dedicated outreach[1] provided to the U.S. Department of Energy's Wind Energy Technologies Office to underpin its response to the congressional request in the Energy Act of 2020 for a report on the "potential for, and technical viability of, airborne wind energy systems to provide a significant source of energy in the United States."

A wide variety of AWE technology concepts, operational principles, and designs have been under development primarily over the last two decades and in Europe and the United States. Developments target a diverse range of applications and markets and are currently at low-to-intermediate technology readiness levels. Generally, they have not yet achieved market-entry requirements; particularly, with respect to reliability and techno-economics, although developers have started delivering early-adopter market pilot projects (Brabeck 2021) and are offering preorders (Peschel 2021) that are both in the 100-kilowatt (kW) unit capacity range. AWE technology developments targeting land-based and offshore competitive grid markets have achieved higher levels of technical sophistication. However, reliable, large-scale megawatt-class technology market rollout may require more than 10 years of dedicated research and development, depending on the level of support and effort provided. Overall, the technology development has not converged toward a single preferred archetype, and the design space has not been fully explored. AWE technology is fundamentally different from traditional wind technology, specifically wind energy produced by wind turbines that have towers, hereafter referred to as "traditional wind," in nearly every aspect and over the entire life cycle. AWE can thus be considered as a separate renewable energy branch, and its success will hinge on the technology's ability to generate power in a cost-effective and reliable manner over the long term.

The ability of airborne wind energy to be cost competitive with other generation sources at a commercial scale by 2030, the time frame of interest in the congressional request, depends primarily on the achievable power rating, maximum tether length, and capacity density.

Airborne wind power plants across the United States comprising 5-megawatt (MW) devices may require capital expenditures of under $1,000/kW to produce the same levelized cost of energy as traditional wind turbines that are anticipated to have capital costs of around $1,200/kW for commercial

---

[1] Other contributors to this report include a literature review, 2-day workshop with more than 90 participants, 14 meetings with technology developers, and more than 20 individual stakeholder meetings, with more than 50 experts from industry and academia.

deployment in 2030. Airborne wind power plants with smaller, 500-kW devices may have higher capacity factors than the reference 5-MW traditional technology in broad regions of the United States. When coupled with potential capital expenditure reductions, AWE has the potential for low-levelized-cost-of-energy devices; however, commercial-scale deployment may be limited because of an estimated capacity density between 0.4 and 4 MW/square kilometer (km^2), which are typically lower than the average 3 MW/km^2 of traditional wind. The ability of airborne wind energy technology to achieve these performance metrics is currently unknown because of a lack of cost and operational data, which represents a significant source of uncertainty for the viability of the technology at a commercial scale.

The wind energy resource potential relevant to AWE (primarily targeting heights between 200 and 800 meters [m]) in the United States is significant. AWE's land-based technical potential for the conterminous United States varies drastically with AWE system design—between 420 and 34,573 gigawatts (GW) and 1,615 and 92,469 terawatt-hours. For comparison in 2018, the total U.S. consumption of electricity was 4,222.5 terawatt-hours. Positive wind shear above 200 m was not found consistently in our analyses, and selected sites and wind profiles appeared to be rather flat, with more evidence of positive shear on land than offshore. Using power curve projections of assumed mature, fully developed, and commercially operated AWE technology based on generic models and an overview of pilot experiment experience, we determined significant energy output at selected sites, showing annual energy production per installed capacity like traditional wind. The exploitation of high-altitude (200 m and higher) winds is strongly dependent on and influenced by technology-archetype-driven design constraints, such as the experience or avoidance of high-tether drag and power-to-weight ratios.

The technical airborne wind energy potential in the Unites States considering social, environmental, and licensing constraints is substantial. The analysis is critically dependent on per-unit installed capacity and tether length, which is directly related to the setback assumptions employed. For example, the rigid-wing, 500-kW AWE device with a 227-m tether length could yield 9,029 GW of technical potential, which is comparable to the 7,827 GW of technical potential for traditional land-based-wind technology. The absolute potential of airborne and traditional wind should not be interpreted as additive, as there is significant spatial overlap between the two estimates. The available and targeted wind resources of both traditional wind and airborne wind energy technology are vast, many times greater than the U.S. demand for electricity

(Lopez et al. 2021; Musial et al. 2016). Technical potential for AWE and traditional wind occur in similar areas spatially and further investigation into where AWE may represent an increase in technical potential is recommended. The technical potential estimates are based on hypothetical, fully commercialized AWE technology, which are compared against traditional wind systems expected to be operational in the year 2030. To supplement the findings in this document, further research could investigate opportunities for AWE that might not be feasible for traditional wind and vice versa. The statements on resource potential and technical potential from the land-based considerations can be extrapolated to offshore environments.

A variety of markets have been identified for AWE technologies, ranging from early-adopter applications to microgrids, distributed wind, and offshore fixed or floating. The research and development challenges of AWE are similar in complexity to those in other precommercial renewable energy technologies, wherein a variety of technological solutions are under development and convergence has not been reached and would require a similar magnitude of dedicated investment to harvest the significant technical potential.

Based on the extensive stakeholder input gathered in the interviews and workshop mentioned earlier, and on our analyses and assessment of AWE systems, resource potential, techno-economic potential, technical potential, and commercialization, we believe it is worth considering a 10-year research plan.

The following research, development, demonstration, and commercialization recommendations outline a conceptual 10-year program to examine and validate the technical and economic viability of AWE technologies.

Near-Term – Years 1 to 4: Initiate Fundamental Research and Craft a Plan for Commercialization

- Develop a national AWE research, development, and flight research facility
- Facilitate collaboration among stakeholders to address airspace restrictions, radar impacts, device lighting, avian impact studies, and social perception

- Develop complete sets of functional requirements of AWE systems for diverse applications and markets to fully capture the problem and opportunity statements
- Support national and regional cost and feasibility studies
- Develop methods and tools to identify potential technology showstoppers and critical areas of attention and drive innovation
- Initiate fundamental research on critical issues regarding aerodynamics, controls, materials, siting, cost and feasibility drivers, environmental issues, reliability, safety, flight optimization, and social acceptance
- Attract global expertise and technology developers through accelerator programs
- Develop open-source, multifidelity simulation capabilities that model AWE systems
- Establish reference models for various AWE technologies and sizes
- Support hardware development of next-generation industry AWE systems
- Perform policy scenario development and evaluation to identify and facilitate the most effective research, development, demonstration, and deployment strategies along with related assistance and support mechanisms
- Coordinate aspects of the AWE research development agenda with the new International Energy Agency Wind task on AWE.

Midterm – Years 3 to 10: Evaluation Campaigns and Research to Accelerate Technology Development

- Foster industry engagement by funding industry research and development, supporting the use of new simulation capabilities, and providing a research facility site with space, instrumentation, and expert personnel to get flying time on prototypes and precommercial systems
- Establish an international working group to develop international design requirements to ensure the integrity of the promising AWE technologies.

Long Term – Years 6 to 10: Technology Advancement and Fundamental Research

- Select the most promising concepts for long-term reliability evaluation at the national research facility to prove system durability
- Fund development of demonstration projects in commercial early-adopter markets, like remote communities, agriculture, microgrids, and progressive utilities
- Continue fundamental research on aerodynamics, controls, materials, siting, cost and feasibility drivers, environmental issues, reliability, safety, flight optimization, and social acceptance.

1. Introduction

The concept of airborne wind energy (AWE) conversion—whereby tethered flying devices are used to convert wind energy into electricity—was first investigated and originally formulated through theoretical modeling (Loyd 1980). Since then, a wide variety of technology concepts, operational principles, and designs have been considered and technology development efforts have strongly increased in intensity over the last two decades. The main developments have been primarily in Europe, across a multitude of technology developers (Bechtle and Zilmann 2021; Cherubini et al. 2015) and in the United States, with the flagship development by Makani (Echeverri 2020) that ended in early 2020 when Alphabet discontinued support.

These five topics provided a framework for the research performed for this report:

- *Technology assessment and upscaling* addresses airborne wind energy technology systems across technology archetypes including concept of operations, design space, flight, structure, power conversion chain, takeoff, and landing and considers the state of the art as well as challenges and opportunities for upscaling the various systems to deliver significant energy.
- *Techno-economic analysis and markets* considers the cost, logistics, and supply-chain considerations for a future, commercial-scale AWE system to identify market share opportunities. The relative benefits and drawbacks of a mature AWE system are compared with expected

characteristics of traditional land-based and offshore wind systems for megawatt-scale machines and hundred-megawatt-scale projects. Comparing AWE to traditional wind and offshore wind systems requires defining a representative full-scale AWE system, which draws from the technical potential and technology assessment analyses within this project, as well as from the existing AWE literature and conversations with industry experts.

- *Resource potential and energy output* estimates the wind resource across the United States based on existing literature, over land and offshore, and at altitudes viable for AWE. In addition, this topic considers whether the AWE resource is overlapping or complementary to traditional wind, whether AWE would impact climate, and the estimated impact of wakes from AWE devices, which ultimately provide an estimated potential energy output.
- *Technical potential, social and environmental impacts, and permitting* explores the siting opportunities and challenges for AWE regarding social and environmental considerations. In addition, this topic explores the land-use requirements of different AWE technologies These siting considerations led to the development of possible siting restrictions and land-use/ocean requirements. Assumptions developed in this area, combined with the resource potential and technology assessments, were used to estimate the overall technical potential of airborne wind energy systems effectively, as well as the quantity (megawatts [MW]) and quality (megawatt-hours [MWh]) of available wind resource for AWE development.
- *Research, development, demonstration, and commercialization needs* addresses potential pathways for commercializing AWE, from demonstration to scaled concepts, and for both land-based and offshore AWE applications. This topic covers the potential overlap of AWE and traditional wind turbines, and it identifies opportunities for developing AWE technologies in locations where traditional wind turbines may not apply. It also covers recommendations for future research, potential approaches for AWE technology development and demonstration, and commercialization approaches.

2. Methodology

This section briefly describes the way in which this report addresses the congressional request regarding focus and scope, the consideration of AWE in relation to traditional wind energy, and the approach taken to develop the report.

2.1. Addressing the Congressional Request: Focus and Scope

This report describes technical analyses of various aspects of AWE provided to the U.S. Department of Energy (DOE) Wind Energy Technologies Office to underpin DOE's response to the congressional request for a report on AWE in the Energy Act of 2020, which states:

> "(i) IN GENERAL. — Not later than 180 days after the date of the enactment of this Act, the Secretary shall submit to the Committee on Science, Space, and Technology of the House of Representatives and the Committee on Energy and Natural Resources of the Senate a report on the potential for, and technical viability of, airborne wind energy systems to provide a significant source of energy in the United States.
>
> (ii) CONTENTS. — The report under paragraph (1) shall include a summary of research, development, demonstration, and commercialization needs, including an estimate of Federal funding requirements, to further examine and validate the technical and economic viability of airborne wind energy concepts over the 10-year period beginning on the date of the enactment of this Act."

To support DOE's response to the congressional request for a report on AWE, the National Renewable Energy Laboratory (NREL) performed basic analyses, primarily in the areas of resource potential and energy output, technical potential based on the resource potential and estimates of siting constraints, and techno-economic analysis based on the technological potential and technology assessments. Expectations of demonstration requirements and paths to commercialization are also provided.

Thus, it is important that the assessments aim to identify the knowns and the known unknowns, and a high-level understanding of both, and point to the unknowns regarding the challenges and opportunities.

Given the level of research and technology developments underway globally, the diversity of considered and possible AWE system archetypes, and the specific characteristics of this technology field, the findings of this report may not be final. They do, however, provide an increased understanding of the potential that AWE can contribute to the U.S. energy mix, its growth, and its significant transformation.

2.2. Relationship to Traditional Wind Energy

By using wind as its energy source, AWE may appear, at first glance, to be in direct competition with traditional wind energy technology (defined as wind turbines built on towers. To assess if this assumption is valid or could reduce the scope of the value proposition associated with AWE, this study aims to address the following questions when reflecting upon the congressional request:

Is AWE in direct competition with traditional wind energy? Influencing factors may include:

- Desired access to identical and mutually exclusive installation locations
- Significant direct or secondary technology cost implications, such as raw material consumption
- Strongly different cost implications on foundations and support structures (such as in offshore floating wind), respectively, or other reasons to be identified.

Is AWE complementary to traditional wind energy? Influencing factors may include:

- Temporal characteristics of the wind resource
- Capacity-factor-related characteristics of either technology
- Possibilities of co-location, with levels of system integration, or other reasons to be identified.

Is AWE independent of and thus supplementary to traditional wind energy? Influencing factors may include:

- Capability to access different wind resources. Reasons may be associated with location, level of resource intensity, and inherent access to resource at different heights above ground
- Type of markets and applications targeted (e.g., microgrid, distributed wind, deployment speed, plant operational duration, adaptability, visual impact, social benefit)
- Independent supply chains or other reasons to be identified.

2.3. Report Development Approach

To begin the study, NREL performed a literature review and reviewed the publications. The team led and participated in a 2-day workshop that included presentations, a panel discussion, and breakout group discussions covering the following topics, as described in Section 3:

- Resource potential and energy output
- Technical potential, social and environmental impacts, and permitting
- Techno-economic analysis and markets
- Technology assessment and upscaling
- Research and development (R&D), demonstration, and commercialization needs.

In addition, the NREL team performed baseline analyses in resource potential, technical potential, and techno-economic potential. Furthermore, we conducted individual meetings with experts from industry and academia, including AWE technology developers to gather information on the range of technology types under development and the sector status and needs.

Figure 1 illustrates the fundamental approach we took to develop the report. The core topics of resource, markets, technology, and R&D activities are common threads and considered in all phases of the report development, from outreach and knowledge acquisition to scientific assessment and recommendations.

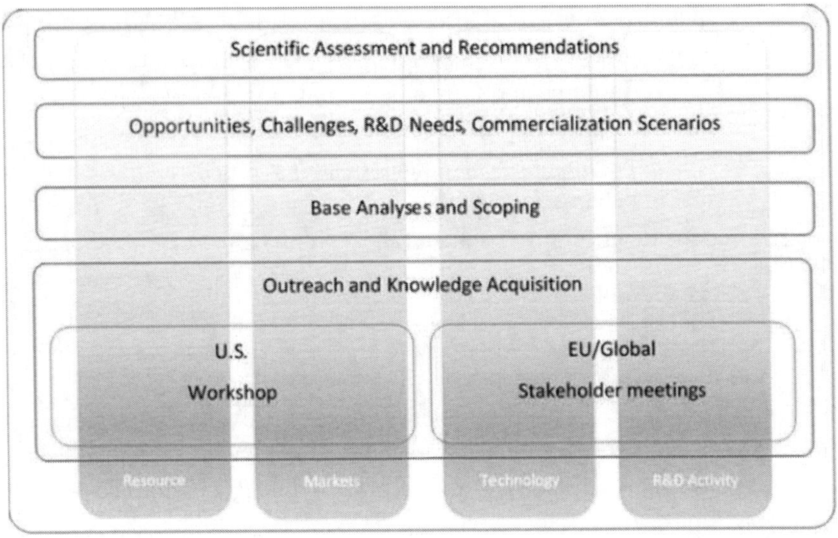

Note: EU is the European Union.

Figure 1. Report development approach, starting at the bottom and moving up.

3. Outreach and Knowledge Acquisition

3.1. U.S. Workshop

The workshop (Weber et al. 2021) focused on U.S. stakeholders in AWE, with approximately 100 experts from AWE technology developers, operators, engineering or consulting firms, governmental organizations, national laboratories, and universities. The event began with an explanation from the Wind Energy Technologies Office regarding the context and purpose of the workshop, including the report to Congress noted earlier. Next, three presentations provided overviews of the AWE workshop held in September 2020 by the International Energy Agency (IEA) Wind Task 11 (Hayek 2020), European AWE R&D efforts, and U.S. R&D efforts, respectively. In addition, five panelists discussed a variety of AWE aspects including markets, systems, technology deployment status challenges, opportunities, and needs. The second day was devoted to two sessions of five parallel breakout groups to discuss the five themes described in Section 3. Outcomes from each of the breakout group discussions were summarized and published in the workshop proceedings (Weber et al. 2021).

3.2. European Union/Global Stakeholder Meetings

Separate from the workshop, we reached out to over 50 stakeholders and met with 14 different technology developers (four from the United States and 10 from the European Union [EU]) including Abound, Ampyx Power, Enerkite, eWindSolutions, KiteGen, KiteKraft, Kitemill, KitePower, kPower, Makani (former), Saipem, Skypull, Skysails, and Windlift.

With a focus on gathering input for the congressional request, the main agenda covered:

- Key characteristics of each developer's technology
- Targeted markets and intended rollout
- Technology development status regarding technological readiness level and techno-economic performance
- Status and challenge of the sector
- R&D needs such as capabilities, tools, and infrastructure
- Suitable support types and mechanisms.

Additionally, we engaged with a small number of academic research entities in the United States and Europe including the Delft University of Technology, NC State University, and University of Delaware.

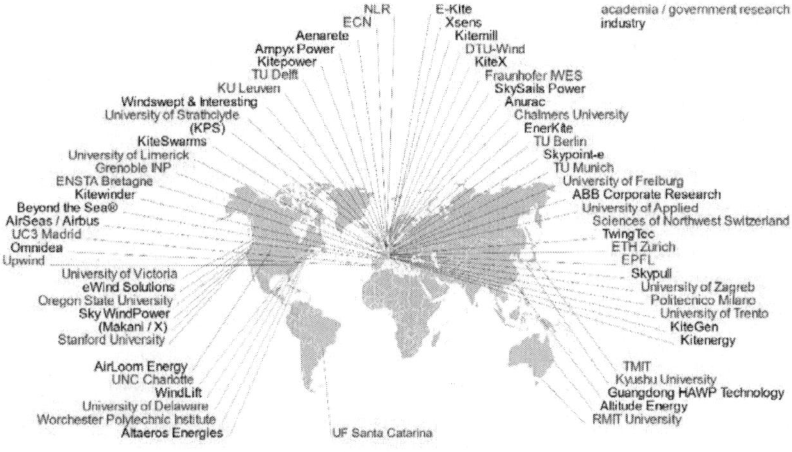

Figure 2. Global R&D landscape.
Image courtesy of Roland Schmehl.

Currently, the global AWE research and development community spans about 70 players from industry, academia, and governmental research entities, as shown in Figure 2. An updated map of AWE stakeholders based in the United States is shown in Figure 3.

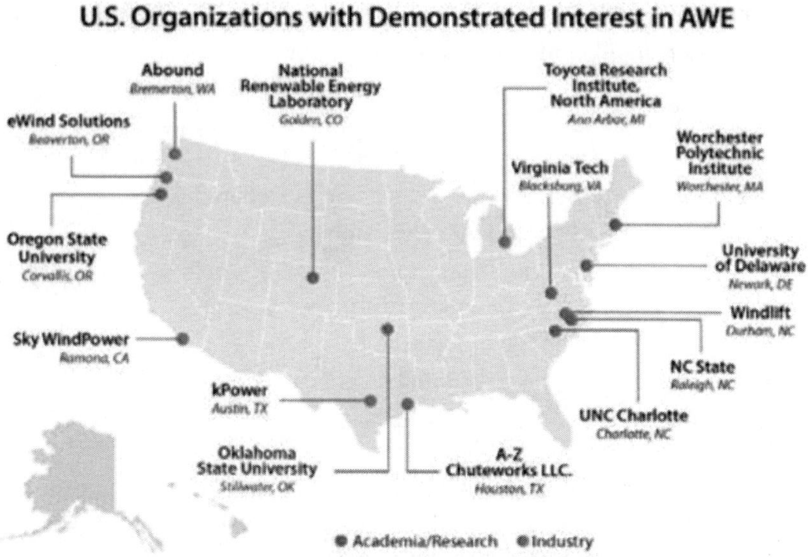

Note: for additional entries, please contact the authors to update.

Figure 3. AWE U.S. stakeholder map.
Image created by John Frenzl, NREL.

3.3. Literature Review

NREL performed literature reviews on AWE. They used the Zotero online reference management tool to hold 238 publications that the team used. Further, publications for each of the five subtopics listed in the executive summary, as well as publications about AWE more generally, were collected and used by the team. This extensive literature overview was valuable to gain insight and understanding into the field; however, only a portion of key publications is directly referenced in this report.

4. Technology Assessment and Upscaling

In this section, we provide a high-level perspective on the status of AWE technology development and its potential to become a viable and commercially deployed contributor to renewable energy. The emphasis here will be on the sector, rather than individual technologies. Next, we present an overview of technology archetypes and the design space followed by technology assessment criteria and status of technology with respect to techno-economic potential and technology readiness. Consideration on upscaling, reliability, autonomous flight, and specific AWE characteristics are offered before overall conclusions are drawn. This work is based on the U.S. Airborne Wind Energy Workshop, several EU and global stakeholder and industry meetings, literature study, and analysis.

4.1. Key Findings

Key findings in AWE technology assessment and upscaling include the following:

- Airborne wind energy technology developers are pursuing a wide variety of functional and operational concepts and design approaches. Overall, the sector has not converged to a single concept approach, which is normal when developing new, complex technology and at this precommercial stage of the industry. Based on available information, we conclude that the design space has not been fully explored or examined sufficiently.
- Over the last two decades of intensified technology development globally, some design approaches have been identified as less promising and are no longer worth pursuing, such as aerostats. Several technology developments can be categorized around a set of similar concepts with varying embodiments and realization of key functional requirements.
- The prevailing technology development branches focus on fundamental working principles and design choices, such as:
    - Fly-gen, with electricity generation in the air aboard the airborne subsystem, or ground-gen, with electricity generation on the ground

- Rigid-wing or soft-wing designs and variations thereof
- Methods of takeoff and landing.
- The AWE technology development sector is targeting a wide range of markets, from remote microgrids and distributed wind to large-scale, land-based, fixed, and floating offshore applications with differing requirements regarding automation, reliability, durability, scale, and economics. Specifically:
  - Some technology developments targeting early-adopter markets, such as island communities, have partly reached intermediate technology readiness with one developer having shipped its first commercially sold units to pilot projects this year (Brabeck 2021).
  - Individual technology developments targeting large-scale power markets have partly reached high levels of sophistication yet are further away from potential market entry regarding technology readiness and meeting demanding techno-economic requirements.
- Overall, the AWE sector needs to advance significantly with respect to:
  - Automation in all operational modes and design conditions
  - Reliability and durability supporting system availability
  - Techno-economic performance at scale.

The efforts outlined in the previous bullets will require large dedicated additional effort and time and can be supported through valuable learning and maturation in early-adopter market operations.

4.2. Background in AWE Technology and Upscaling

This section provides information on the background and context of assessing AWE technology and factors in scaling up the technology.

4.2.1. Research Questions and Challenges

The following key questions are critical in the advancement of AWE technology development:

- How can an airborne wind energy system be assessed against sectorwide, market-specific functional requirements to achieve greater technology convergence, and how can AWE progress on time-, cost-, and risk-effective technology development trajectories toward market entry?
- How can the required levels of long-term continuous operation, covering all operational modes, be achieved to gain significant operational experience and trigger the onset of learning curves with respect to system reliability and economics?
- How can airborne wind energy technology best be scaled up to achieve significant contributions to the U.S. energy system? What is the role and importance of a) upscaling AWE unit size and capacity, b) upscaling by airborne unit number and topological density, and c) achieving higher degrees of system integration of multiple airborne units?

4.2.2. Methods and Tools Needed

For AWE to advance, the following efforts need to be fulfilled:

- Combining a) experimental research and operation, both exploratory as well as long-term operation covering all control states and modes of operation with b) numerical simulation for design and optimization along with subsystem and holistic integrated system technology assessments
- Enabling fast-turnaround prototype characterization and long-term pilot project operation and validation infrastructure in dedicated research facilities and early caretaking pilot projects in combination with extensive knowledge sharing across the industry
- Developing internationally agreed-upon reference systems to serve as a common basis for comparison, along with achieving consensus on key performance indicators and establishing methods that allow for complete system assessment methods.

4.3. Technology Archetypes, Design Space, and Classification

AWE systems can be classified into two basic approaches: fly-gen and ground-gen, defined here (shown in Figure 4 in an offshore setting):

- *Fly-gen* AWE systems fly onboard turbine-generator units that are connected to an airframe along figure-eight or circular crosswind trajectories defined by the tether length. Electricity is generated onboard the kite. The onboard-generated electrical power is transmitted to the ground station through a conductor in the tether. The system is designed to achieve high relative velocity on the flying wind turbines.
- *Ground-gen* AWE systems generate electricity by flying in figure-eight or circular crosswind trajectories while spiraling upward and reeling out the tether that connects the kite to a winch-generator unit on the ground station. Electricity is generated on the ground. Once the kite reaches its maximum operational altitude, it is retracted, and the tether is reeled in. The system is designed to achieve high lift on the wing for maximum mechanical power transmitted through the tether.

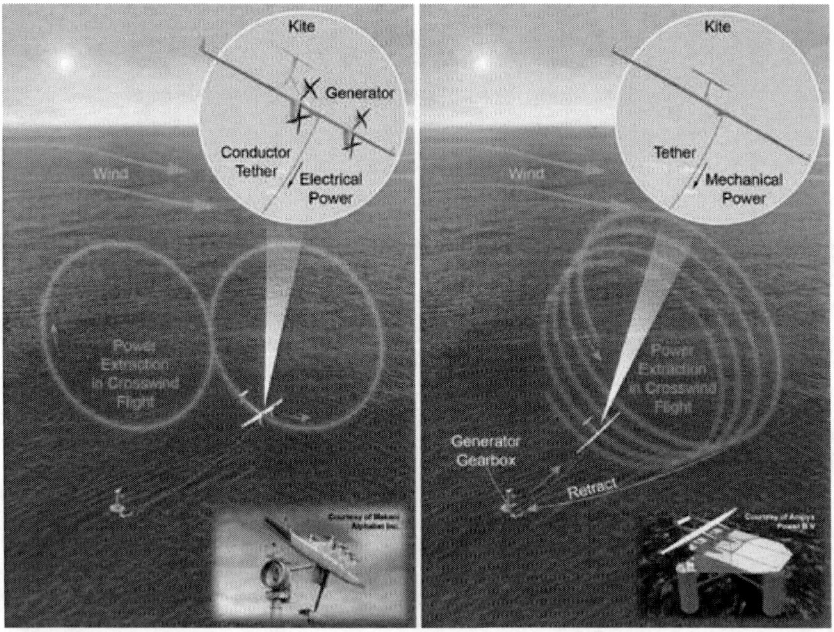

Figure 4. Operational principles of fly-gen (left) and ground-gen (right) crosswind AWE systems in an offshore setting. *Images created by Besiki Kazaishvili, NREL.*

The range of designs and approaches used to implement AWE systems extends well beyond the simplified description of these two fundamental concepts. Figure 5 shows AWE prototypes in flight—a subset of the technologies that are currently under development.

Figure 5. AWE system prototypes currently under development (in flight).
Images based on Schmehl (2018).

Because AWE is aiming for deployment in a variety of markets and applications, many technology approaches, concepts of operation, and design are under consideration.

Figure 6 classifies AWE systems based on a sector overview from 2018 (Schmehl 2018). The first level of categories identifies where electrical generation takes place:

- Ground-gen: generation occurs at a ground station that is either fixed or mobile
- Fly-gen: generation occurs onboard the airborne system.

The second level of categories identifies groups with respect to flight operations:

- Crosswind: a flight path in crosswind with a large directional component that is perpendicular to the wind direction to increase relative velocity on the kite and swept harvesting area. The crosswind flight path may be implemented through flying reciprocating patterns, such as figure eights, or returning flight patterns, such as circles.

- Tether-aligned: a flight path that aligns with the tether direction, and which potentially reduces both the relevance to tether drag and the opportunity for area sweep.
- Rotational: a rotor-type system with a circular flight path of aerodynamically active kite elements.

Colors in Figure 6 indicate several additional design choices. Wings are categorized by their structural design as flexible (like parasail kites) or rigid (like traditional wind turbine blades or fixed-wing aircraft). Concepts are further classified with respect to takeoff and landing operations, distinguishing between vertical takeoff and landing and horizontal takeoff and landing. AWE concepts that are lighter than air make up an additional category.

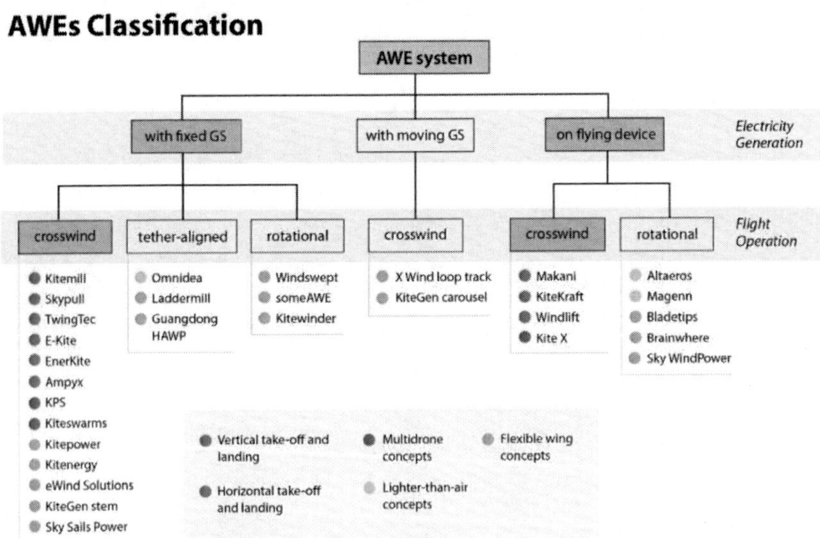

Note: GS = ground station.

Figure 6. Classification of traditional AWE technology archetypes investigated based on Faggiani and Schmehl (2018), Echeverri et al. (2020), Eijkelhof et al. (2020), and Vimalakanthan et al. (2018). *Image created by John Frenzl, NREL.*

Over the last two decades of global technology development, some AWE design approaches, such as lighter-than-air systems, have been identified as less promising and are no longer widely pursued.

Instead, the prevailing technology development branches are focusing on fundamental working principles such as fly-gen or ground-gen, and the fundamental flight dynamics, wing structure and wing designs, takeoff and landing methods. There are complexities in the air, on the ground, and various methods of scaling through unit size, number of units, spatial distribution, combined multisystem arrangements, and cluster density.

Airborne wind energy R&D efforts have not converged to a single archetype and operational concept. The lack of convergence reflects the relative immaturity of the technology and may indicate uncertainty regarding the future success of the sector. At the same time, the diversity of design concepts provides ample opportunity for further improvement, as the sector is not locked into a confined design space.

Table 1. Conceptual Design Attributes and Variants to Determine the Design Space and Possible Combinations of Concept and Design Options (includes input from Brabeck [2021], Kruijff [2021], Peschel [2021], Harklau [2021], Ippolito [2021], and Isensee [2021])

Concept Attributes	Variants
Electrical generation location	In flight, ground-based fixed, ground-based mobile, other
Flight path	Stationary, in tether direction, crosswind with unidirectional rotation, crosswind with bidirectional rotation, other
Wing structure	Soft single layer, soft ram air multicell, soft with inflatable support, soft with highly nonlinear elastic reinforcement, semirigid, segmented rigid, rigid, multiwing, other
Takeoff	Vertical, horizontal, use of main aerodynamic system, use of ancillary embedded aerodynamic system, centrifugal, catapult, lift system, support system, fan, other
Landing	Vertical, horizontal, use of main aerodynamic system, use of auxiliary-embedded, dedicated aerodynamic system, other
Flight region	Low (up to 300 meters [m]), medium (300 m to 600 m), high (above 600 m), combined, other
Control	Diverse active control surfaces: elevator, elevon, flap, wing deformation; tension control; part passive, other
Energy-conversion stages	Direct generation, mechanical, hydraulics, other
Tether topology	Single, bridle, dual, multiple, other
Kite topology	Single, twin, rotor, staggered, combined (identical, varied), multiple, network architectures, other
Farm integration	Independent unity, staggered height levels, farm control, other
Reference	Fixed, moving, single, multiple, absolute, relative, other
Other attributes	Known, to be defined, other

Even though a variety of concepts and related design implementations are being considered, it is also evident that the design space has not been explored fully enough to draw decisive conclusions on the probability of success of individual concepts within different market segments. To illustrate this, the classification criteria employed in Figure 6 have been extended in Table 1 by increasing the number of variants, adding new concept classification attributes, and identifying related design variants. This preliminary selection of technology concept attributes and design options is incomplete yet illuminates the size of the potential design space.

4.4. Upscaling

One of the critical questions around the technological advancement of AWE is how to scale up, as stated in a recent publication: "The big issue right now in airborne wind energy is scaling" (Anderson 2019). In this context, scaling ultimately relates to the installed capacity. There are inherent scientific and engineering challenges associated with upscaling AWE technology, and these challenges vary depending on the application: fly-gen or ground-gen systems, soft or rigid wings, and overall system integration. Relevant physics and engineering relate to aspects such as tether drag, weight, tension, and fatigue; conductivity; wing loads; flight path; speed; and accelerations (Weber 2019).

It is important to note that the requirements for upscaling installed capacity depend on the target market and application, and that upscaling may be measured in different ways. Three relevant metrics are:

- Installed capacity per unit
- Installed capacity per infrastructure
- Installed capacity per land or sea area.

Installed capacity per unit (or airborne device) is particularly relevant for AWE developers targeting utility-scale floating offshore wind farm deployment (Echeverri et al. 2020; Kruijff 2021). Maximizing the capacity in megawatts per device is critical to achieving economic viability because of the high per-unit cost of the balance of plant—including the floating substructure, mooring system, and subsea electrical cable—as well as offshore operation and maintenance (O&M). For floating offshore AWE applications, engineering design scaling analysis based on levelized cost of energy (LCOE) results in optimal device scaling between 7.5 and 15 MW per unit (Kruijff 2021). Successful upscaling of the capacity per unit will depend on fundamental physical similarity considerations for the power-to-mass ratio and most importantly, on determining achievable values through practical engineering design options (Santos 2021).

Concepts for upscaling of installed capacity per critical infrastructure focus on maximizing capacity at a single ground station, floating platform, and other balance-of-plant subsystems. Options under study or development include multikite system configurations using a common tether, and parallel single-layer or two-layer multikite layouts (De Schutter et al. 2019a, 2019b; Kiteswarms 2021). While increasing the complexity of control in operation and transitional states, such system architectures may also help reduce tether drag.

Upscaling of installed capacity per occupied area of land or sea may be implemented by staggering operational heights, thereby using AWE systems with a different tether length and inclination angles, and developing plantwide control strategies (Leuthold et al. 2017, 2018). Other concepts with higher levels of system integration have been proposed, such as rotary-torsion-based, multiwing, multilayer AWE systems (Beaupoil 2019; Read 2018; Tulloch et al. 2019) or branched tethers and large multitether kite networks (Santos 2021). Multikite system integration via a carousel-type system has also been proposed (Ippolito 2021).

In summary, upscaling is critical for grid-market applications and may be implemented by upscaling the unit capacity and through scaling by number of units or degree of system integration.

4.5. Reliability and Autonomous Operations

The need to advance and demonstrate system reliability and continuous long-term autonomous operation and control has been identified across the AWE sector (Fagiano 2019; Vermillion et al. 2021). As with many functional requirements, autonomous operation is required across all markets; however, at different levels of sophistication. The requirement is particularly acute for offshore AWE systems, which are highly capital- and maintenance-cost-intensive and demand the highest degree of reliability and automation to facilitate certification and long-term validation. While advancing though the precommercial development stages, developers concentrate on specific and narrow test objectives to demonstrate and validate equipment functionality and autonomous operation algorithms. Some systems require wind speeds of 3 to 5 meters per second (m/s) close to the surface for takeoff, whereas most AWE systems use a variety of active launch mechanisms (see Table 1).

Individual flight times vary between minutes and hours and systems have performed several hundreds of operational flights (Kruijff 2021; Brabeck 2021; Harklau 2021). Individual developers have reported continuous operation over several days (Brabeck 2021; Bormann 2021). Although some developers have demonstrated autonomous flight and other critical automated functions including takeoff and landing of reduced-scale prototypes (Echeverri 2020; Brabeck 2021), achieving reliable, fully automated, long-term operation represents the most significant challenge for the industry.

Critical subsystems for functional, reliability, and durability testing and validation include power conversion units (Echeverri 2020), back-to-back tether spooling systems (Kruijff 2021), and structural maximum load and fatigue testing (Bormann 2021). SkySails Group developed, tested, validated, and commercially operated an airborne wind ship propulsion system with mechanical pulling power levels of up to 2 MW and reported a total of over 3,000 hours of operation since the first demonstration at sea over 10 years ago (Brabeck 2021). This system does not include energy conversion to electricity; thus, it does not represent an integrated AWE system, but demonstrates the functionality of a key airborne subsystem in automated flight up to commercial readiness.

It is important to mention that fully autonomous operation is not a key requirement in many mini and microgrid applications (Zywietz 2019). In specific applications in which part-manual operation or system reset support is available—such as military operations, industrial microgrids, and remote installations—the requirement for autonomy may be somewhat softened, and criteria such as deployment transportation may dominate (Creighton 2021; Peschel 2021). Such applications could serve as early-adopter markets, providing opportunities for commercial operation of AWE systems that initiate learning curves for improving system reliability, durability, and efficiency. The electrification of significant energy markets, such as agriculture, could potentially be supported in single or hybrid energy source microgrids with AWE systems (Bormann 2021; Schaefer 2021) soon, with the first commercial 100-kilowatt (kW) systems being sold to pilot projects this year (Brabeck 2021).

4.6. AWE Technology Characteristics and Relationship to Traditional Wind Energy

Reflecting on the questions posed earlier on whether AWE technology can be regarded as (a) in direct competition with, (b) complementary to, or (c) independent of and thus additional to traditional wind, the following observations can be made with respect to AWE system characteristics:

- *Temporal, local wind resource correlation.* The temporal correlation of the wind resource targeted by traditional wind and by airborne wind technology in each location is not shown to be anticorrelated in a consistent manner (Bechtle 2019) and not of a complementary nature within the typical operational heights above ground.
- *Spatial resource access in height above ground.* Traditional wind and airborne wind energy technology largely target and access the wind resource through airflows at various heights and across different height ranges, with potential overlap depending on technology concept. Most AWE technology developers target heights up to 800 m above ground (Brabeck 2021), with some developers targeting greater heights up to 2,000 m above ground (Ippolito 2021) and beyond (Calverley 2021).
- *Spatial resource access in a location.* Traditional wind and airborne wind energy technology may target and access different wind resource locations, which is driven by a variety of factors including minimal viable resource requirements (Schmehl 2021), transportation and deployment requirements (Creighton 2021), site access, farm design, and terrain complexity.
- *Construction materials and processing.* Traditional wind and airborne wind energy technology partly require a different raw material resource at significantly different amounts per installed capacity, with AWE requiring significantly less raw material (Schmehl 2021) estimates at around 20% of what is required for traditional wind (Bormann 2021).
- *Supply chain.* Traditional wind and airborne wind energy technology require and access different and largely independent supply chains, fabrication, and manufacturing processes. For example, foundations are not required to transfer significant torque load for AWE systems (Peschel 2021).

- *Installation and deployment.* Traditional wind and airborne wind energy technology have very different installation and deployment requirements. AWE offers faster deployment and installation with significantly lower installation equipment requirements (Bormann 2021).
- *Workforce.* Traditional wind and airborne wind energy technology have different workforce requirements. While both require manufacturing employees, airborne wind is expected to require and provide a higher degree of O&M employment opportunities because of replacement requirements for key components, such as the tether and soft kite (Peschel 2021).
- *Environmental impact and adaptability.* Because of the lack of operational experience, the environmental impact of airborne wind energy requires thorough investigation and is not well understood. The nature of traditional and AWE systems in operation and their control needs and options may result in different environmental interactions and adaptation, including the interruption of operation of a traditional system and grounding of an AWE system.
- *Financial modeling and risk profiles.* Potentially low capital investment and high component and subsystem replacement and maintenance costs during the operational life span of AWE technology requires diverse financing approaches, capital expenditure (CapEx) payback, and investment risk profiles. These circumstances may also allow shorter economically viable periods of operation than those of traditional wind technology in some specific markets such as hybrid microgrids (Zywietz 2019).
- *Upgrades and technology adjustments.* Replacing components and subsystems (e.g., tether, soft kite) in AWE technology over the installation's life span is an operational cost burden. At the same time, it offers the possibility of upgrading the airborne subsystems as the technology advances. It also allows adaptation to site conditions, and to increase the capacity factor for a given ground station through kite adjustment and selection.
- *Flexibility and mobility of assets.* The potentially low weight per installed capacity and modest deployment effort may allow redeployment of AWE technology assets at different operational sites. Fast deployment, mobility and swift relocation of assets are

highly valued properties and satisfy imperative system requirements in some markets (Creighton 2021).

4.7. AWE Technology Development – Sector Status

This section provides a high-level perspective of the AWE sector based on two technology development progress metrics: technology readiness level (TRL) and technology performance level (TPL). "TRL at its most basic definition describes the maturity of a given technology relative to its development cycle" (Sanchez 2011) and is considered at material, component, device, and system levels. To assess AWE systems, we draw from DOE's TRL definition (Sanchez 2011) as well as TRL definitions specific to wind energy (De Rose et al. 2017) and ocean energy (Fitzgerald et al. 2012; De Rose et al. 2017), a sector with similar technological diversity. TPL (Weber 2012; Weber et. al. 2013) describes the potential for techno-economic performance in ocean energy during precommercial technology development and similarly applies to AWE. Although TRL defines the degree of readiness toward becoming a commercially available product, TPL reflects the potential of the product to perform economically in the market. The TPL assessment of a technology is based on a holistic set of cost and performance drivers relating to capital expenditures, operational expenditures, power conversion, availability, and acceptability. The joint consideration of TRL and TPL provides a useful platform for determining the precommercial technology development status and identifying efficient technology development trajectories with respect to cost, time, and risk (Weber et al. 2017). This approach has also proven to be a valuable framework in the AWE technology sector (Weber 2019; Hayek 2020; Weber et al. 2021; Marsh 2021; Zerweckh 2021).

The TRL/TPL framework is used to assess and represent the progress and status of technology development in the AWE sector as a whole and to identify challenges and opportunities to successfully enter different markets and related applications. Figure 7 provides an overview of this high-level assessment and perspective. While individual technology development examples are included for clarification, the aim is to identify trends and segments of the sector.

Overall, evaluating technology readiness and techno-economic performance potential is a rough estimation, and is exclusively based on high-level, incomplete public domain information. It is important to note that within a specific AWE system development, different degrees of readiness and techno-economic performance can be observed across the system levels of component, subsystem, and integrated system and may vary considerably across the various performance attributes. The visual representation in Figure 7 does not specify individual developers and technology developments may be identified at a lower or higher development status when subjected to more in-depth assessments. Building on the knowledge and observations gained from direct stakeholder meetings and in the 2021 AWE workshop (Weber et al. 2021), we identified three main trends and technology development status groups across the global AWE sector and are shown in Figure 7.

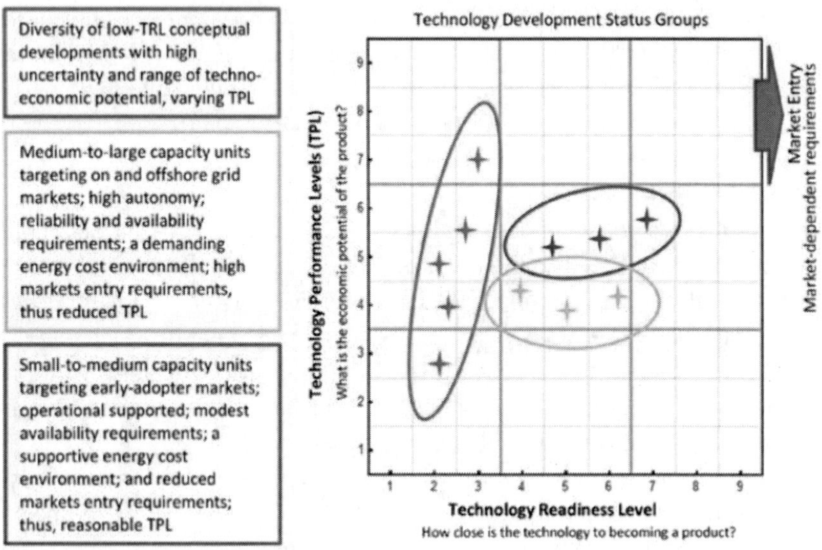

Figure 2. Technology readiness levels (TRLs) and technology performance levels (TPLs) of the main technology development status groups (left) across the global AWE sector.

Figure 7 combines technology developments that target different market segments, ranging from niche markets and early-adopter markets to large-scale, land-based and offshore grid markets. It is important to note that market entry to these markets is subject to significantly diverse market entry requirements, especially in relation to fully autonomous operation, reliability, capacity, scale, and economics. Thus, technological readiness and techno-economic performance potential is expressed in relative terms to the targeted and absolute market entry requirements, as follows:

- *Early stage* (green). A variety of technology developments are at the conceptual stage, with low-to-medium TRLs and high uncertainty and are spread over the range of TPLs. The attributes driving techno-economic potential vary considerably. Some show high potential to satisfy market requirements with individual design concepts resolving challenges that present other systems with significant trade-offs. One example is the developments of stacked multikite systems (De Schutter et al. 2019a, 2019b). Such systems can resolve the trade-off between large tether length and associated tether drag, which is typically found in single kite developments. In some multikite systems, the long common main tether is not subject to fast lateral motion when transmitting power to the ground station, whereas individual kites with short tether branches operate at high velocities in cross-flight. Multikite configurations are considered both for fly-gen and ground-gen. While conceptually offering high installed capacity and power density per deployment space, such systems pose complex swarm control challenges.
- *Targeting large impact* (yellow). Some developers targeting land-based and offshore grid markets are preparing medium- and large-capacity units. Challenges to market entry include meeting requirements for operational autonomy, reliability, and availability in a demanding, low-energy cost environment. Some of these systems have reached higher levels of technical sophistication and detail as the technologies are refined. Technology developments are advanced and derisk prototype demonstrations through subsystem bench characterization and validation (Kruijff 2021; Echeverri et al. 2020) that are supported by custom numerical simulation and control optimization tools. Because of the demanding market entry requirements, these systems are clustered at intermediate TPLs.
- *Targeting early adoption* (blue). A considerable portion of AWE technology developers are targeting early-adopter markets in high-energy cost environments that may have a high level of O&M support with somewhat less stringent automation or availability requirements, such as extended operations support to include launch preparations (Peschel 2021). Examples include microgrids, remote industrial plants, hybrid systems, military, agriculture, and others. Current devices vary from a subscale prototype to unit sizes in the 100- to 200-kW range. First developments are entering early-adopter pilot

projects (Peschel 2021) and selling first units (Brabeck 2021) to strategic clients and island communities, partly in collaboration with energy companies. The reduced market entry requirements lead to these developments reaching medium to medium/high TPLs.

A consistent allocation of AWE technology types, such as fly-gen and ground-gen or rigid- or soft-wing designs, to market types or the development groups described earlier is not suggested here. However, note that most developers are pursuing ground-gen technology concepts with a variety of takeoff and landing approaches. Technologies with a reduced level of complexity in the airborne subsystem appear to have advanced closer to the targeted market requirements and have demonstrated higher operational duration and time (Brabeck 2021).

Most technology developments first aim for market entry into an early-adopter market typically associated with an installed capacity per unit of well below 500 kW. A small number of developments have aimed (Echeverri et al. 2020), or are aiming (Kruijff 2021), for market entry into competitive grid markets with a large installed capacity per unit and in offshore applications.

While early-adopter markets may serve as niche or potentially growing end markets, it is important to note that these markets may also facilitate technological evolution in commercial operating environments toward higher capacity and significantly more competitive grid markets if supported through intensive ongoing R&D activities.

4.8. Summary of Technology Assessment and Upscaling

AWE technology is fundamentally different from traditional wind technology based on fundamental principles and through all lifecycle stages, including engineering design, procurement, financing, manufacturing, deployment, installation, operations, maintenance, upgrading, decommissioning, circular economic, environmental, and societal integration and could be regarded as a unique addition to the renewable energy sector. The available natural, potential technical, and targeted U.S. wind resources of both traditional wind and airborne wind energy technology are vast, and more than twice the U.S. energy demand (Section 7, Lopez et al. 2021; Musial et al. 2016), with a significant spatial overlap between the two estimates. However, because of this large available resource and characteristic features of the technologies, a separate deployment could be possible to provide a renewable energy contribution to the national electric energy demand and renewable energy transition when and if AWE is commercially available.

A wide variety of technology concepts, operational principles, and designs have been under development over the last two decades, originating from early theoretical modeling work in 1980 (Loyd 1980). Current technology readiness levels vary from low to medium, and in individual cases, medium-high (up to TRL 7, [Brabeck 2021]). The potential for airborne wind energy systems to achieve high techno-economic performance for market entry and to satisfy customer requirements varies significantly over technology concepts, depending on the requirements of their target markets and applications. Globally, a first developer is now entering pilot projects with sales of 100- to 200-kW units supporting remote island communities (Brabeck 2021) and another is offering preorders of mobile 100-kW AWE systems (Peschel 2021). Both technologies are ground-gen devices with flexible wing designs.

Technology developments targeting land-based and offshore competitive grid markets may have achieved higher levels of technical sophistication yet are further away from fulfilling the demanding market requirements. AWE technologies may be able to offer competitive economic performance based on the relatively low material requirements. However, achieving this performance depends on reliability, durability, survivability, and fully autonomous operations, some of which still require significant advancements to enter markets.

Early-adopter and special application market entries are starting to emerge and in the near-term for up to 3 years, whereas reliable, large-scale multimegawatt class technology for power to the grid may require 10 years for market rollout, depending on support and effort. Component, subsystem, and integrated system testing, validation, and operation will play a significant role in shaping technology development, increasing operational experience, and establishing reliability. The sector may benefit from the growth of enabling technology in related sectors such as sensors, controls, drone technologies, material science, engineering modeling, and high-performance computing.

Because of the uncertainty of technological success and the rather low global R&D investments to date and especially in the United States (with an approximate $13-million direct investment from federal entities), AWE technology development requires significantly more support to increase knowledge, evidence, and validation to understand the impact potential that AWE has for the U.S. electricity sector. In what form, and to what degree the AWE sector will provide a significant contribution to the U.S. renewable supply hinges on the rate of technology development and whether it can overcome the challenges associated with achieving commercial readiness, reliability, and techno-economic performance. The current initiation of pilot projects (Brabeck 2021) in the 100-kW class may trigger the onset of technological evolution of AWE systems through commercial operational experience gathered in diverse early-adopter markets. Higher-capacity multimegawatt AWE systems for larger land-based and offshore grid markets may evolve from early-adopter markets, in parallel with or in addition to considerable precommercial research, development, and validation activities.

Both the risk of failure as well as the risk of lost opportunity need to be considered, and further research and analysis is needed to better understand and quantify these risks – as well as the opportunities – to overcome the challenges and realize the technical potential of AWE.

5. Techno-Economic Analysis and Markets

In this section, we focus on AWE systems for grid-connected, commercial-scale applications and identify relevant techno-economic challenges and opportunities for these systems. The analysis centers on land-based, multimegawatt devices that could potentially be deployed by 2030.

5.1. Key Findings

Key findings in AWE techno-economic analysis and markets are:

- If airborne wind energy can achieve ambitious performance metrics (such as multimegawatt devices requiring tether lengths on the order of 200-300 meters [m]), then it has the potential to produce a lower LCOE in 2030 than land-based traditional wind because of lower capital costs; however, there has been little-to-no experimental validation of these performance goals to date. The industry needs to demonstrate the ability to achieve these metrics for AWE to be cost competitive with traditional wind.
- Site-specific capacity factors and uncertainty in operational expenses leads to a wide range of breakeven CapEx values, clustered between $500/kW and $1,500/kW for a 5-MW rigid-wing AWE device and

between \$1,000/kW and \$3,000/kW for a 500-kW rigid-wing device. Breakeven CapEx represents the highest acceptable CapEx for the airborne wind system to match the levelized cost of energy of commercially available traditional wind turbines in 2030 (represented by the NREL Annual Technology Baseline (ATB) 5-MW turbine (NREL 2020) and indicates that a 500-kW AWE device shows higher (i.e., more easily achievable) breakeven cost margins than a 5-MW AWE device. The broad range of values indicates that there are regions where site conditions are better suited for airborne wind than traditional wind. Improved estimates of these values (for example, through long-duration research assessment programs) will help clarify target capital costs for AWE systems to be cost competitive with traditional wind.

- The southeastern region of the United States demonstrated consistently high breakeven CapEx values, indicating that this may be a good target region for AWE.
- Smaller devices may face challenges for commercial-scale deployment because of lower capacity densities than traditional wind turbines but may have a good opportunity for deployment in early-adopter markets (e.g., off-grid, remote locations) because of potential capital cost savings. Deployment in these remote markets would require the devices to achieve high reliability to lower O&M demands.
- The range of capacity densities for commercial-scale airborne wind power plants extends to higher capacity densities than those in existing traditional wind power plants (2–19.6 MW/km^2 for a 5-MW rigid-wing kite and 0.6–4.0 MW/km^2 for a 500-kW rigid-wing kite); however, these densities do not account for wake losses or other siting constraints.

5.2. Background in AWE Techno-Economic Assessment and Markets

This section provides information on the background and objectives of our techno-economic assessment and market analysis of AWE.

5.2.1. Research Questions and Challenges

To guide our techno-economic assessment, we asked ourselves the following questions about AWE:

- Can future AWE systems achieve the target performance metrics outlined in this section to reach cost parity with traditional wind? What are the major technological hurdles that need to be addressed?
- Can a real-world AWE power plant achieve the high-capacity densities currently being modeled, or will they be diminished by wake interactions, siting constraints, or permitting considerations? Can technological advances such as plant-level control or stacked kites further increase capacity density?
- Is airborne wind energy best suited for a bulk energy production market where it competes with traditional wind, or for niche markets that take advantage of value streams, such as ease of installation and portability?
- Are there particular regions in which the benefits of AWE present a clear advantage over traditional wind (for example, the higher potential cost margins available in the southeastern United States)? How can an AWE system be optimized to maximize the benefits in these regions?

5.2.2. Methods and Tools Needed

For AWE to advance, the following efforts need to be met:

- Estimation of failure rates and maintenance schedules for commercial AWE systems.
- Improved component- and system-level cost models incorporating physics-based design, life cycle cost analysis, and learning effects from improved supply chains and industry experience.

The AWE industry is currently at a precommercial stage, with a focus on evaluating individual units at small scales. As such, comparing costs or performance between existing AWE systems and commercial-scale traditional wind turbines is not appropriate, as the latter technology has benefited from experiential learning, economies of scale, maturing supply chains, and technological advancement (Junginger and Louwen 2019). In this section, we develop potential topologies for future commercialized AWE systems and assess the relative benefits and drawbacks of a tethered wind turbine compared with a traditional wind turbine for a commercial operation date of 2030.

This analysis assumes that AWE systems continue to grow to megawatt-scale devices that can be clustered in commercial-scale wind power plants. We prescribe hypothetical power curves for 500-kW and 5-MW AWE turbines and solve for approximate system characteristics that would be required to achieve these power ratings. We use these system properties to generate nationwide capacity factors, as described in Section 7, and subsequently find the breakeven capital costs required for systems with these capacity factors to be cost competitive with future traditional wind turbine systems, as described by NREL's ATB (NREL 2020). This analysis focuses primarily on how AWE can be cost competitive with commercial-scale, traditional wind energy. Although several interesting niche or distributed markets exist for AWE systems, these value propositions are based on additional considerations beyond costs and are discussed qualitatively at the end of this section and in the following sections.

Finally, this analysis focuses on land-based deployment. Although the offshore wind energy market is of interest to the AWE community, the time horizons for offshore AWE deployment are expected to be beyond 2030, as the industry focuses on commercializing the technology on land before moving to the harsher offshore environment (Weber et al. 2021).

The analysis used here does not identify a specific technology pathway or preferred technology solution. Instead, the premise is that if the AWE industry can achieve the assumed power performance and plant design characteristics, then an opportunity exists to realize the benefits of the technology and obtain significant market penetration. If these goals are not met (for example, if measured power output falls short of expected production goals as was the case with the Makani M600 prototype (Echeverri et al. 2020), airborne wind energy will face a significantly more challenging path to market.

5.3. Definition of Potential AWE System Characteristics

5.3.1. Power Curves

The power curve of a wind turbine is a critical design specification that shows the power output of the machine over a range of wind speeds. Wind turbines are characterized by their rated power, which is the maximum power a particular device can extract from the prevailing wind. Several power curves have been reported in the literature for existing airborne wind energy prototypes and planned devices (Faggiani and Schmehl 2018; Echeverri et al. 2020; Eijkelhof et al. 2020; Vimalakanthan et al. 2018); however, these power curves are highly specific to their respective designs and do not reach the 5-MW power rating that is expected to be typical of an average traditional wind power plant installed in 2030 (NREL 2020).

To use a generic power curve for this analysis, we specify two power ratings of 5 MW and 500 kW; assume cut-in, cut-out, and rated wind speeds; follow a wind-speed-cubed law between cut-in and rated wind speeds; and assume a linear drop-off in power for wind speeds above rated.

This power curve represents the output electrical power of the system after aerodynamic, mechanical, electrical, and AWE-specific losses (such as cosine and gravity losses) have been applied. These assumptions are based on power curves available in the AWE literature and were refined with input from industry experts during the 2021 Airborne Wind Energy Workshop (Weber et al. 2021). The power decrease after rated power and the relatively low cutout speeds are attributed to the adverse effects of tether drag at higher wind speeds.

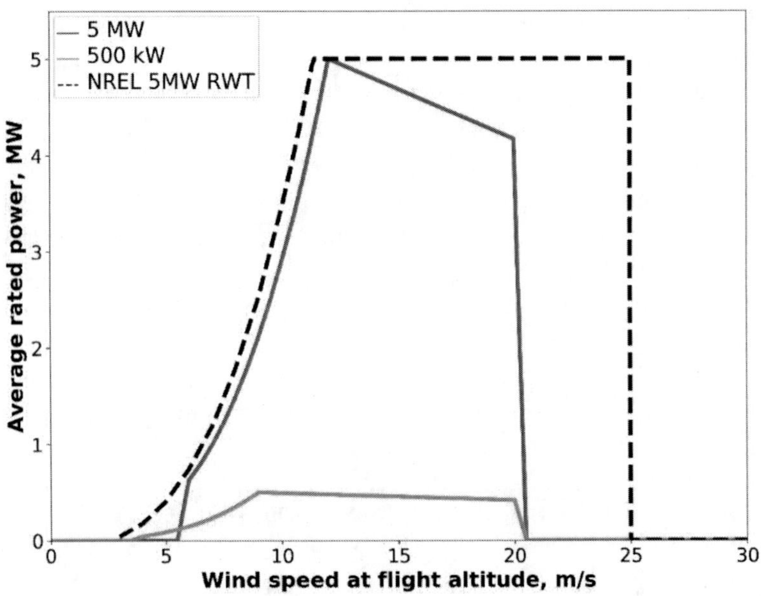

Figure 8. Target power curves for commercial-scale AWE systems in 2030. The NREL 5-MW reference wind turbine (RWT; traditional) power curve is also shown.

Although the AWE industry has developed prototypes on the 100-kW scale, achieving a 5-MW device by 2030 will require significant technological advancement from the current state of the art; however, industry experts suggest that this is an ambitious but potentially achievable goal, and that AWE devices with a rated power on the order of 1 MW are likely necessary for airborne wind energy to be viable in a commercial setting relative to traditional wind (Weber et al. 2021). Industry experts agreed that higher power ratings for AWE, equivalent to the 15-MW traditional wind turbines expected to be deployed offshore by 2030, are not achievable in the next decade (Weber et al. 2021).

5.3.2. Geometric and Flight Characteristics

The size and flight path of an AWE device will dictate its ability to achieve its intended power performance; furthermore, the length of the tether required to conduct flight operations will directly impact the siting requirements and constraints for an airborne wind power plant (Faggiani and Schmehl 2018). In this analysis, we take a simplified approach to estimate the wing size and flight radius required to generate the power curves specified in Figure 8. A more detailed analysis that considers mass scaling, controller design, and the time-varying space requirements of the power generation cycle is outside the scope of this study, which limits the ability of the current method to directly evaluate the required tether length and flight path that a particular device can achieve. We address this uncertainty with a sensitivity analysis for short and long tether lengths.

Aerodynamically, AWE devices can be broadly categorized into two types: rigid wings and flexible wings, as described in Section 4.2. For each power curve in Figure 8, we determine the size of both a rigid and a flexible AWE system that would be required to achieve the rated power. The length of the tether that connects the wing to the ground station may remain nominally constant for a fly-gen AWE system in power production mode (Echeverri et al. 2020) or may vary with time as a ground-gen AWE turbine reels in and out during the power generation and retraction phases of the pumping cycle (Schmehl 2018). The maximum tether length will drive setback requirements from existing infrastructure to avoid collisions in the case of a crash. Directly calculating the required tether length for a particular AWE device requires simulating the time-dependent flight performance of the wing while accounting for the loads imparted by the tether as well as the aerodynamics of the wind (Eijkelhof et al. 2020) and is beyond the scope of this study. Instead, we prescribe a short and a long tether length for each wing to evaluate the range of impacts that this parameter can have on deployment potential, as described in Section 7. We derived the range of prescribed tether lengths from input from industry experts during the 2021 Airborne Wind Energy Workshop based on their proposed designs for megawatt-scale AWE devices (Weber et al. 2021).

In summary, a rigid and a flexible wing are designed for each of the two power curves in Figure 8; then, a short and a long tether length are prescribed for each of the power rating/wing type combinations, leading to eight scenarios. For each device, we specified a minimum ground clearance of 30 m. The smaller wing spans and tighter flight circles of rigid wings permit lower average flight altitudes than flexible wings. The resulting geometric and flight characteristics are presented in Table 2.

Table 2. Wing and flight properties for rigid and flexible wings at 5-MW and 500-kW rated power

Power Rating	5 MW				500 kW			
Wing type	Rigid		Flexible		Rigid		Flexible	
Tether length (m)	1,000	300	1,000	700	500	200	600	350
System[2] lift/drag ratio	15	15	5	5	15	15	5	5
Wing area (m^2)	192	192	1,637	1,637	45	45	388	388
Wingspan (m)	43.8	43.8	127.9	127.9	21.3	21.3	62.3	62.3
Average flight altitude (m)	200	200	414	414	200	200	217	217
Flight radius (m)	131	131	384	384	64	64	187	187

5.3.3. Plant Design and Characteristics

A common value proposition outlined in the AWE literature is the potential for tethered wind turbines to achieve higher capacity densities than traditional wind. Capacity density, measured in MW/km^2, refers to the generation capacity within a given amount of land area. For land-based traditional wind turbines, empirical wind farm capacities range from 1–18 MW/km^2 (Harrison-Atlas, Lopez, and Lantz, forthcoming) and offshore wind farm capacity densities range from 3.1 to 18.7 MW/km^2 (Deutsche WindGuard GmbH 2018). Drivers for these capacity densities include the required spacing between wind turbines to reduce downstream wake effects as well as siting constraints (such as terrain complexity) and accessibility or navigation through the wind plant (Musial et al. 2016). Airborne wind energy may have reduced wake interaction effects, as the relatively small surface area of the kite sweeps through a larger airspace volume than traditional wind turbine blades

[2] Including tether drag.

(Faggiani and Schmehl 2018). Theoretical capacity densities between 6 and 32 MW/km^2 have been estimated in the literature (Faggiani and Schmehl 2018; European Commission 2018; De Lellis et al. 2016); however, further research is required to determine how the impact of wakes and other siting constraints affect the actual capacity density of airborne wind power plants (Weber et al. 2021).

We use an approach developed in the AWE literature for determining the capacity density based on the minimum geometric spacing between ground stations (Faggiani and Schmehl 2018). This methodology accounts for the elevation angle, tether length, and flight radius of the devices listed in Table 2 and outputs the minimum unit separation and resulting circular footprint for each device. Scaling the rated power of the device by the footprint produces an estimate for the capacity density of the wind power plant. We modify this approach slightly for the 500-kW AWE system, which we assume is intended to be deployed similarly to traditional distributed wind (i.e., a small number of turbines in off-grid applications), and simply assume that the footprint of each device is a circle with a radius equal to the tether length. Finally, we assign a setback of 1.25 times the tether length to each device to define the required spacing from existing infrastructure. The setback and capacity density results are provided in Table 3 and used for the spatial analysis of Section 7.

Table 3. Setback and capacity density

Power Rating	5 MW				500 kW			
Wing type	Rigid		Flexible		Rigid		Flexible	
Tether length (m)	1,000	300	1,000	700	500	200	600	350
Minimum unit separation (m)	427	285	893	719	N/A	N/A	N/A	N/A
Footprint (km^2)	0.57	0.26	2.50	1.62	0.79	0.13	1.13	0.38
Capacity density (MW/km^2)	8.7	19.6	2.0	3.1	0.6	4.0	0.4	1.3
Setback (m)	1,250	375	1,250	875	625	250	750	438

5.3.4. Breakeven Capital Expenditures

The final step in the techno-economic analysis of potential commercial-scale AWE systems is evaluating the technology costs. Several AWE researchers and developers have produced LCOE values for their designs, typically based on component-level cost estimates (European Commission 2018). Table 4 summarizes such cost estimates in the literature. Using this approach for future technologies is challenging because of the uncertainty around how component prices and supply chains may evolve in the next decade (Junginger and Louwen 2019) and the immaturity of AWE-specific supply chains (current designs borrow heavily from the aviation market) (Weber et al. 2021). Instead, we begin with the performance characteristics derived in Sections 5.3.1–5.3.3 and solve for the capital costs required to match the LCOE that the ATB wind turbines are anticipated to achieve in 2030 (NREL 2020). This results in the breakeven CapEx that would allow airborne wind energy systems to be cost competitive with the projected future costs of traditional wind.

The levelized cost of energy is calculated using:

$$LCOE = \frac{FCR \times CapEx + OpEx}{8{,}760 \times NCF}$$

where FCR is the fixed charge rate that must be collected annually (%/year), CapEx are all capital expenditures including installation and soft costs ($/kW), OpEx are the annual operational expenditures ($/kW/year), and NCF is the net capacity factor (scaled by the 8,760 hours in a year to yield the annual energy production).

The steps to find the breakeven CapEx for airborne wind energy throughout the United States are as follows:

- Set the values of LCOE, OpEx, and FCR equal to the 2030 ATB moderate technology values in the LCOE equation
- Use the range of net capacity factor values for different AWE systems developed in the spatial analysis of Section 7
- Solve the LCOE equation for CapEx at each spatial location in the United States where the technology was deployed. This result represents the breakeven CapEx required for an AWE system to achieve the LCOE of the ATB wind turbine, given similar operational expenses and financing rates.

Table 4. Cost Estimates from literature for AWE systems and traditional wind energy systems in 2030

	System Type	Rated Power	Capital Expenditures (€/kW)	Operational Expenditures (€/kW/year)	Capacity Factor (%)	LCOE (€/MWh)	Source
AWE systems	Flexible pumping kite	100 kW	3,800	110	54%	120	(Faggiani and Schmehl 2018)
	Flexible pumping kite	1.2 MW	599–956	27–54	35%	33–59	(Heilmann 2012)
	Land-based pumping kite	2 MW	1,096		34%–45%		(De Lellis et al. 2016)
	Floating offshore flexible kite	2 MW	1,250				(European Commission 2018)
	Floating offshore rigid wing	2 MW	3,800	110		120	(European Commission 2018)
Traditional systems (reference)	Land-based	5.5 MW	1,048	33	47%	21	(NREL 2020)
	Fixed-bottom offshore	15 MW	2,298	62	47%	43	(NREL 2020)
	Floating offshore	15 MW	2,554	49	51%	40	(NREL 2020)

The interpretation of these results is that an actual AWE system will result in cost savings relative to the ATB reference wind turbine if it achieves a lower CapEx than the breakeven value. In other words, higher values of breakeven CapEx provide more opportunity for cost savings, whereas low values of breakeven CapEx mean that the AWE system in question has a stricter (and possibly unattainable) target cost to compete with traditional wind.

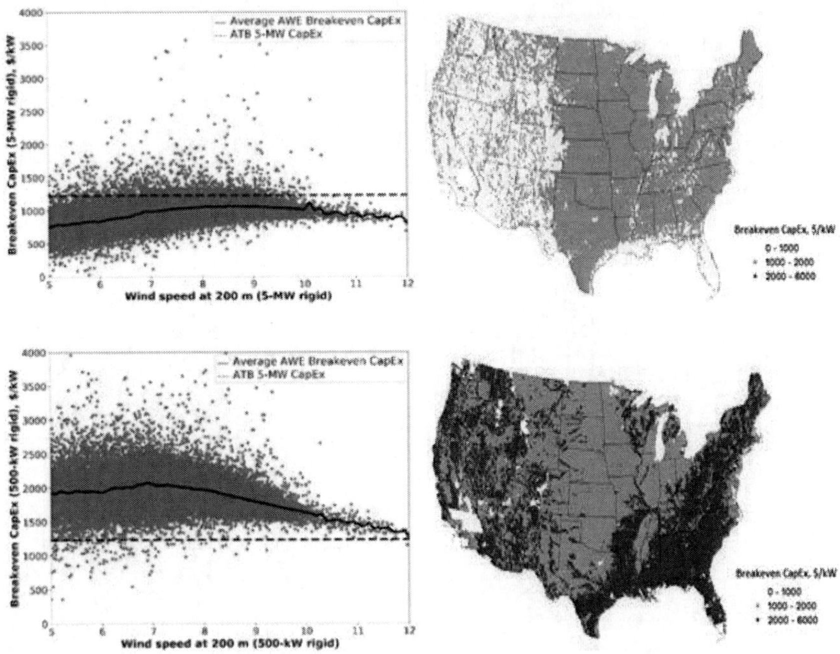

Figure 9. Breakeven CapEx in $/kW for a 5-MW (top) and 500-kW (bottom) rigid-wing AWE system deployed throughout the United States. Lower values of breakeven CapEx indicate a more stringent target for AWE to be cost competitive with the ATB 5-MW reference traditional wind turbine. Spatial maps of the breakeven CapEx data points are shown on the right.

The results for rigid-wing AWE systems rated at 5 MW and 500 kW, both using the shorter tether lengths from Table 2, are shown in Figure 9 as spatial heat maps and as a function of wind speed at each spatial location. As this simplified analysis uses the same power curves for rigid and flexible wings, and the short tether lengths for the different topologies are relatively similar for both wing types, there are no significant differences in the results for rigid and flexible wings at 500 kW. Results for 5-MW flexible-wing AWE systems are not shown as the lower capacity densities and long tether lengths make them less feasible at the commercial scale. The CapEx of the ATB 5-MW traditional wind turbine is also shown for reference.

In Figure 9, the scatter in the results shows that the breakeven CapEx for the 5-MW rigid-wing AWE system varies significantly by location, but it has an average value of around $1,000/kW for wind speeds between 8 and 12 m/s; by comparison, the ATB 5-MW wind turbine has a CapEx value of $1,227/kW. This difference indicates that the 5-MW AWE system is cheaper than the traditional wind system to produce the same cost of energy. The spatially resolved map shows that regions in the central plains are reasonably generous to airborne wind energy, with breakeven values between $1,000/kW and $2,000/kW, which can be attributed to high capacity factors in the region (as shown in Figure 9). However, regions west of the Rocky Mountains tend to require lower values of breakeven CapEx for the AWE system to be competitive (indicating that these are less advantageous markets). These breakeven values are lower than reported CapEx values for conceptual AWE systems (summarized in Table 5), indicating that the AWE industry will have to achieve further cost reductions beyond these project values to be cost competitive with traditional wind.

The 500-kW wing also shows significant spatial dependencies and performs well in the central plains, with breakeven values close to the ATB wind turbine CapEx, although it shows significant advantages in breakeven CapEx in the Southeast and Northeast, where standard traditional wind turbines may underperform because of lower wind speeds (although low-specific-power traditional wind machines may also be competitive in these regions). Again, the higher values of breakeven CapEx mean that, if an AWE system can achieve capital costs below this value, it will provide a cost savings relative to traditional wind at that location. Further, these savings represent a direction for future research efforts to better characterize the marginal value of airborne wind energy relative to traditional wind in these regions while considering sensitivities to key parameters such as flight altitude, power rating, and cost factors.

The results in Figure 10 indicate that, on average, a 5-MW AWE system will require around 18% lower capital costs than a traditional wind turbine to produce the same LCOE because of lower capacity factors at many spatial locations. This cost of energy also assumes that the AWE system can obtain the same financing rates and operational costs as a more mature, traditional wind system; the bankability and OpEx costs of airborne wind energy remain a significant uncertainty facing the industry (Weber et al. 2021). These results indicate that a 500-kW system is more favorable on a per-unit LCOE basis, with breakeven CapEx values typically higher than a 5-MW traditional wind system at a given location; however, using these smaller devices at a commercial scale requires facing additional challenges because of the lower capacity densities than traditional wind, which would demand more land area for a given project capacity. These challenges would be exacerbated for a 500-kW flexible wing system, which has longer tether lengths and lower capacity densities than a 500-kW rigid wing. It is likely that a commercially deployed AWE system would be bracketed by these two scenarios to balance the economies of scale and higher capacity densities of the 5-MW system with the better capacity factors and cost margins of the 500-kW system.

5.3.5. Cost Category Breakdowns

We also provide an approximation of how the breakeven CapEx is distributed among different cost categories based on LCOE models for rigid wings (Echeverri et al. 2020; Kruijff and Ruiterkamp 2018) and flexible wings (Heilmann 2012) in Table 5. As previously mentioned, some care must be taken when assessing these cost categories, as bottom-up LCOE projections for AWE systems are shrouded with uncertainty around the maturation of the supply chain for individual components. It is also important to note that the provided values are the overnight capital costs and do not include the costs to replace key components; these operational costs may be significant, as it is possible that major components such as kites and tethers may need to be replaced every 6–12 months (Weber et al. 2021). A sensitivity analysis around these operational costs is conducted in the following section. The results are then plotted in Figure 10.

Table 5. How the breakeven CapEx is distributed among different cost categories based on LCOE models for rigid wings

Percentage of Overall CapEx		
Wing Type	Rigid	Flexible
Kite	35.7	10.3
Ground station	19.3	15.5
Electrical	16.7	41.8
Tethers	3.0	6.1
Installation	4.2	0.0
Foundation	10.3	26.2
Other	10.8	0.0

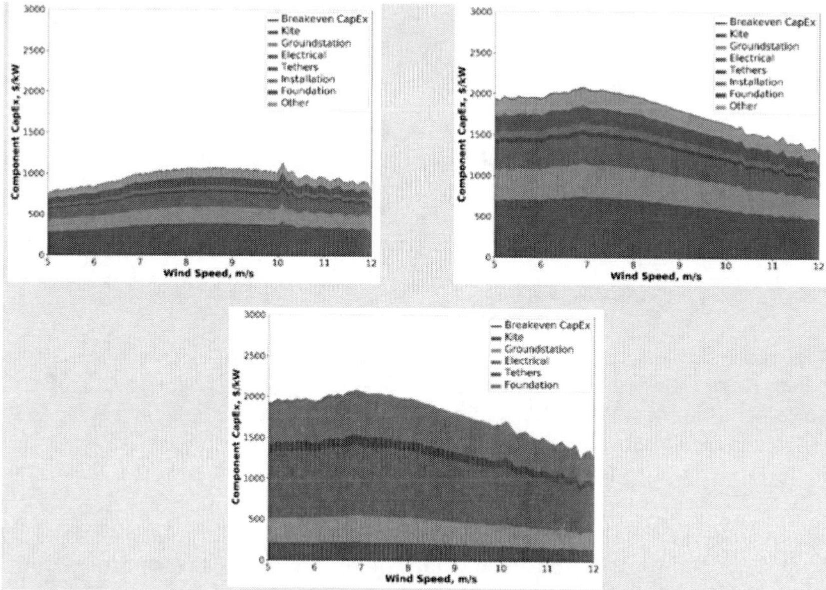

Figure 10. Breakeven CapEx component breakdown for a 5-MW rigid wing (top left), a 500-kW rigid wing (top right), and a 500-kW flexible wing (bottom).

5.3.6. Sensitivity to Operation and Maintenance Costs

The results for the breakeven CapEx use the same values of OpEx as the traditional ATB wind turbines; however, conversations with private developers have reflected a range of both optimistic and conservative assumptions about OpEx for AWE systems. The benefits of maintaining an AWE system relative to a traditional wind turbine are that all service operations can be conducted on the ground without the need for expensive cranes or vessels, and that many components are cheaper than their traditional wind equivalents; however, the drawback of AWE systems is that the high cyclic loading on the kite and tether system may result in more frequent replacement of key components (more than once per year), accruing additional cost and labor. As insufficient evaluation data or predictive models exist to effectively estimate the failure rates and associated OpEx of an AWE system, or even to definitively identify if these costs will be higher or lower than those of traditional wind systems, we also prescribe a range of OpEx values corresponding to 50% and 150% of the ATB value to demonstrate the sensitivity of breakeven CapEx to maintenance costs. These sensitivities are plotted along with the original values of breakeven CapEx against wind speed in Figure 11.

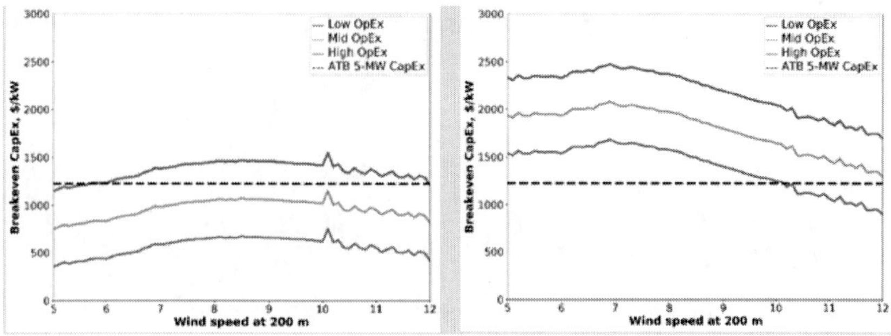

Figure 11. Sensitivity of breakeven CapEx to a range of low, mid, and high OpEx values for a 5-MW rigid wing (left) and a 500-kW rigid wing (right). Mid-OpEx corresponds to the ATB 5-MW traditional wind turbine value of $39.0/kW/yr, low OpEx corresponds to an optimistic scenario of 50% lower costs ($19.5kW/yr), and high OpEx corresponds to a conservative value of 50% higher costs ($58.4/kW/yr).

Increasing the operational costs of a 5-MW airborne wind energy system by 50% decreases the breakeven CapEx to around $500/kW, or around half of the expected CapEx values for multimegawatt AWE systems (Table 5). These low breakeven CapEx values indicate that it will be difficult to compensate for any unforeseen increases in OpEx with lower capital costs; however, if improved accessibility and lower component costs lead to reduced operational costs, the 5-MW AWE system may be able to realize cost savings relative to traditional wind for a wide range of wind speeds. The 500-kW rigid-wing system retains breakeven CapEx values above the ATB 5-MW traditional wind turbine under most scenarios.

5.3.7. Assumptions and Caveats

We made the following assumptions and caveats in the techno-economic assessment:

- The current analysis focused exclusively on land-based deployment. While offshore deployment is a potential target for the airborne wind energy industry, it is expected to take place on a longer time horizon than 2030 (Weber et al. 2021).
- The simplified scaling tools do not capture how mass scales with wing area, which directly impacts the ability of the wing to fly in its optimal turn circles and obtain the desired power rating. As a result, this analysis assumes that the AWE devices can successfully fly with the

flight radii and tether lengths specified in Table 2 without directly modeling their ability to do so.

- The fixed charge rate in the LCOE equation is equal to that of the ATB land-based traditional wind turbine. This implicitly assumes that the less-proven AWE technology has become equally bankable and is as low risk as traditional wind technology by 2030. LCOE is highly sensitive to FCR, and less-advantageous financing terms will significantly increase the cost of energy (Shields, Beiter, and Kleiber 2021).
- The capacity densities of airborne wind energy plants are computed without considering wake losses, siting constraints, or permitting requirements. Instead, they represent the minimum spacing required to avoid collision between neighboring kites. It is currently uncertain if real-world plants would be able to achieve these idealized capacity densities.
- Operational expenses are uncertain, as no long-term deployments currently exist; consequently, there are no data on failure rates and maintenance schedules. The AWE industry does not provide a consensus as to whether these costs will be greater or less than traditional wind systems.

The estimated power curves and flight geometries assume:

- A 50% loss factor applied at all wind speeds because of electrical, pumping, gravity, cosine, and mechanical losses. This factor was derived from what published performance data are available (Echeverri et al. 2020; Luchsinger 2013) and was refined by conversations with industry experts. Losses will vary at different wind speeds and for different wing/tether designs. This assumption limited the study's ability to evaluate the effects of tether drag on performance.

The following power relationship is between the lift-to-drag ratio and wind speed from (Ahrens, Diehl, and Schmehl 2014):

$$P = \frac{2}{27}\rho v_w^3 \left(1 + \left(\frac{L}{D}\right)^2\right)\sqrt{C_L^2 + C_D^2}$$

where ρ is the air density, v_w is the wind speed, L/D is the lift-to-drag ratio, C_L is the lift coefficient, and C_D is the drag coefficient. We make the following assumptions:

- A minimum 30-m ground clearance
- A flight radius of 3 times the wingspan
- An average flight altitude of 200 m or the flight radius plus the ground clearance, whichever is higher.

A detailed investigation of launch/land designs, costs, and trade-offs was outside the scope of this study; however, these systems are critical to the viability and deployment of autonomous and commercial-scale AWE systems.

5.3.8. Technical and Economic Viability of Airborne Wind in the Next 10 Years

The results in this section show a potential pathway for airborne wind energy to be cost competitive with traditional wind at a commercial scale if the technology can achieve the performance and sizing values presented here. Being cost competitive at the commercial scale would require the technology to attain all or some of:

- Power ratings on the order of 1 MW
- Short tether lengths on the order of a few hundred meters
- Tight flight radii of 3–4 times the device wingspan
- High capacity densities of above 10 MW/km^2
- Low capital and/or operational expenditures
- High reliability and low replacement frequencies for major components.

A commercially viable product would not necessarily have to accomplish all of these goals; for example, a 5-MW rigid wing with costs on the same order as a 5-MW traditional wind turbine may be economically attractive if it can achieve the high-power densities provided in Table 5, whereas a 500-kW flexible kite may be advantageous if sufficient land area exists to take advantage of the high capacity factors and low capital costs at the expense of capacity densities lower than traditional wind. The results in Figures 15, 16, and 17 (as well as Section 7) show that these pathways to economic viability are likely to be highly spatially dependent and will depend on the permitting and regulatory feasibility of AWE systems.

The airborne wind literature proposes theoretical designs that have the potential to achieve the techno-economic goals stated earlier, which is reinforced by the perspective of industry practitioners (European Commission 2018; Weber et al. 2021). However, the lack of experimental data from commercial-scale devices means that significant uncertainty exists around the performance, cost, and maintenance specifications. The most advanced prototype of a large-scale rigid-wing kite, the 600-kW Makani M600, significantly underperformed its projected power and flight performance despite having the advantage of a well-funded program (although this result may not be generalizable to the full spectrum of AWE design choices, such as ground-gen or flexible kites) (Echeverri et al. 2020). Experimental validation of the performance of large-scale AWE systems, coupled with realized capital cost reductions, advancing supply chains, and better-understood maintenance strategies and costs, will be critical to determine if airborne wind energy can directly compete with traditional wind on an LCOE basis.

Airborne wind energy faces additional risks to commercialization, as described in European Commission (2018). In addition to the technology risk outlined earlier, major barriers facing the industry include:

- A lack of convergence on optimal system designs, particularly with regards to the kite, control system, and takeoff/landing strategies. The multiple of system designs also introduces a degree of financial risk as investors are faced with a paradox of choice and are uncertain of which technology to support.
- The need to demonstrate safe and reliable autonomous operation over extended deployment periods, including bad weather and multiple-kite systems (particularly for the yet-to-be-evaluated multimegawatt systems).
- The current immaturity of the supply chain and need to leverage economies of scale to reduce costs.
- Building trust and support with investors and external stakeholders.
- Investment risk as developers set their sights on larger machines that may never reach prototype or commercial-scale evaluation.

5.3.9. Viability for Offshore Wind

The value proposition for deploying AWE systems offshore is primarily related to reduced foundation size, which makes up a significant portion of the CapEx of fixed-bottom and floating traditional wind turbines (Stehly, Beiter, and Duffy 2020). The low overturning moments of a tethered system directly connected to the foundation can drastically reduce the size, and therefore the cost, of the support structure (Kruijff and Ruiterkamp 2018). Additional benefits are often cited, such as the strong offshore wind resource and the reduced demand (relative to traditional wind systems) for expensive and scarce installation vessels to construct and service an offshore AWE plant.

Despite these advantages, airborne wind energy faces several significant challenges to obtaining offshore market penetration. A primary concern is the ability of AWE systems to scale up to a nameplate capacity that can compete with the 12–15 MW traditional wind turbines that are expected to dominate the market over the next decade (Musial et al. 2020). These turbines benefit from the low shear and consistently high wind speeds offshore and require fewer devices to achieve a given generation capacity. While wake effects from these large wind turbines do reduce capacity density, current offshore projects being planned in the United States are already selecting suboptimal, increased spacing between turbines to accommodate navigational and transit lanes within the wind plant (Bureau of Ocean Energy Management 2021; Vineyard Wind 2020). If AWE systems are subject to these same spacing constraints, the much lower rated power of the kites makes it unlikely that a project developer would select the technology for new construction (for reference, an AWE plant with 5-MW turbines at 1-nautical-mile-by-1-nautical-mile spacing would have a capacity density of 1.5 MW/km^2, roughly half that of traditional offshore wind). An additional concern is the potential for increased maintenance frequency offshore and the associated risk and safety concerns. As a result, it is likely that AWE systems would have to achieve nameplate capacities comparable to those of offshore traditional wind turbines to realize the benefits of lower costs.

5.3.10. Viability for Niche Markets

Airborne wind energy offers several value propositions other than LCOE that make it attractive for certain niche markets. For example, ease of transport and installation, lack of dependency on traditional fuels, portability, and the potential for low costs make AWE systems (particularly flexible wing systems) attractive for markets such as disaster relief, distributed generation, developing countries, military bases, and highly complex terrain. These markets are discussed in more detail in Section 8.2.

In summary, the AWE industry has advanced significantly in recent years and the progress showed by many developers has garnered enthusiasm within the sector for the future of the technology and its potential to reduce costs relative to traditional wind. This analysis has shown that, if the AWE industry can achieve relatively ambitious technology and performance metrics, it can reduce capital costs and/or LCOE significantly. However, the lack of experimentally validated performance data at the multimegawatt device scale, uncertainty around achievable capacity densities, uncertainty around O&M costs and strategies, and a myriad of risks and barriers to commercialization, coupled with rapidly declining costs and increasing capacities of traditional wind turbines, mean that significant further growth is required for airborne wind energy to be competitive at a commercial scale. If these metrics are not achieved by the industry, additional penetration into niche markets has some promising characteristics, although additional siting challenges would have to be overcome to make significant inroads into the offshore wind market.

6. Resource Potential and Energy Output

Research into airborne wind energy is a multifactor challenge, with the wind resource being one of those factors. In this section, we summarize the (limited) literature that is available on the wind resource for AWE. Many of the early studies estimate the wind speed profile up to the height of the jet stream (~10 km), which is far beyond the 500-m height that is currently assumed to be the practical maximum height for AWE operation. This section analyzes the wind resource using NREL's 20-year Weather Research and Forecasting model (WRF) simulations, compares the wind speeds at altitudes relevant to AWE devices (100–500 m) with the wind resource at hub height for traditional wind turbines (currently 80–160 m), and provides preliminary estimates for annual energy production and capacity factors at selected sites.

6.1. Key Findings

The following are key finding on AWE resource potential and energy output:

- The wind resource at altitudes viable for airborne wind energy depends on the location and changes with time of day and during the year. In general, wind speed increases with height up to 300 m; above 300 m, average wind speed profiles seem to be mostly flat. Any increase in wind speeds with height may be negated by the effects of tether length and elevation angle, as they tend to reduce gains from higher heights (although this does not apply to all archetypes). Gross capacity factors calculated for 5-MW airborne wind energy devices (using the power curves developed in Section 5.3.1) are like gross capacity factors for traditional wind turbines at the sites analyzed.
- Our studies indicate that the windiest spots for traditional wind turbines are also the windiest spots for AWE devices, therefore AWE competes with traditional wind energy for sites if altitude considerations are neglected.

6.2. Introduction to AWE Resource Potential and Energy Output

This section provides AWE resource potential and energy output, the atmospheric conditions that drive AWE system design, as well as recommendations, and needed methods and tools.

6.2.1. Recommendations

The following are recommendations for improving understanding of atmospheric conditions and their relationship to AWE systems.

- Recommendations for future research include gaining a better understanding of turbulence, gusts, fluctuations, and intermittency of the airborne wind energy resource. Instrumental campaigns that can give insights into these research topics are needed.
- Experts recommend evaluating AWE devices regarding high-wind performance, high-wind controls, and safety relating to gusts (e.g., how to dampen out instantaneous power surges).
- Future research should also address climate impacts on the wind resource, wakes, and extreme events.

6.2.2. Methods and Tools Needed

The following are methods and tools to improve understanding of atmospheric conditions related to AWE:

- Instrumental campaigns to quantify the resource and understand turbulence, gusts, and fluctuations of the wind resource at altitudes up to 500 m
- U.S.-wide analysis of high-altitude winds and wind shear using simulations and observations
- Wake modeling
- Climate modeling.

6.3. Airborne Wind Energy Potential Worldwide

According to the Monin-Obukhov theory, wind speed increases with height in the atmospheric boundary layer following a logarithmic law. Above the boundary layer height, where geostrophic winds roar in the absence of the ground's influence, winds are assumed to be faster. Even higher up, the jet stream promises very high winds at altitudes around 10 km, which about a decade ago was deemed promising for airborne wind energy. For cxample, Archer and Caldeira (2009) assessed wind power at altitudes between 500 and 12,000 m above ground worldwide. They noted that high-altitude winds are not steady and strong all the time, and that the altitudes at which winds are the strongest depend on the weather. They concluded that the optimal wind power densities and heights are above 6 km and that the United States benefits from relatively high power densities (>0.5 kW/m^2), with optimal heights above 500 m. In contrast, Miller et al. (2011) state a few years later that jet stream wind power does not have the potential to become a significant source of renewable energy.

Archer et al. (2014) identified optimal locations for airborne wind energy globally. High-altitude wind resources were modeled or analyzed for Ireland (Lunney et al. 2017), the Middle East (Yip et al. 2017), and The Netherlands (The Dutch Offshore Wind Atlas 2021). Much of the airborne wind energy resource has been analyzed in Europe. Looking at 12 locations in Germany, Gambier (2014) found strong wind shear between 200 and 1,000 m at night, whereas during the day the wind shear was small. Airborne Wind Europe publish high-altitude wind energy maps over Europe, based on ERA5 reanalysis data (Bechtle and Zilmann 2021). These maps consider higher wind speeds at lower altitudes and apply the highest wind speed available at a site at any given hour at any altitude up to 500 m. This method of using the highest winds available considers the ability of airborne wind energy systems to dynamically lower their flight altitude to harvest these stronger low-level winds should they occur. Bechtle and Zillmann, the authors of the 2021 article on the Airborne Wind Europe website, claim that "winds over Europe at the best altitude up to 500 m are much stronger and steadier than at a fixed altitude of about 100 m," and that in most places, the mean wind speed at high altitude is at least 1–2 m/s higher than at the 100-m altitude.

Bechtle et al. (2019) analyze the wind resource over Europe up to 500 m and compare it to the wind resource at 100 m, a height typically relevant for traditional wind turbines. Example wind speed profiles show that at times the wind resource is higher at higher altitudes, and at other times the wind resource is higher at lower altitudes. Bechtle et al. also analyzed wind power density. They claim that over most of Europe, the 5th percentile wind power density for variable heights increases by more than 100% compared to 100 m.

6.4. Wind Shear and Airborne Wind Energy Resource in the United States

Since airborne wind energy devices fly in circles or figure-8 configurations that span an area between ~100 and 500 m above ground, wind shear, which is the change in wind speed with altitude, is important to consider. Power output fluctuations related to wind shear can occur when the AWE flight path spans a range of altitudes (Roland Schmehl, personal communication).

Contrary to popular belief, wind speeds are not always higher at 500 m compared to 100 m (Bechtle et al. 2019), which was confirmed by workshop participants (Weber et al. 2021), as well as an analysis conducted at NREL with WRF simulations over Hawaii, the Pacific Northwest, and the mid-Atlantic Coast (the methodology used for California in Optis et al. [2020] was replicated for Hawaii and the Pacific Northwest). Figure 12 (left panels) shows the 20-year average maximum wind speed in the layer between 100 m and 500 m, and (right) the 20-year average difference between 100-m and 500-m wind speeds. It is evident that the wind shear depends on the location. Some locations show a higher wind resource at 500 m than at 100 m, some show the opposite. The reasons and conditions for wind shear patterns are recommended as the subject of future research.

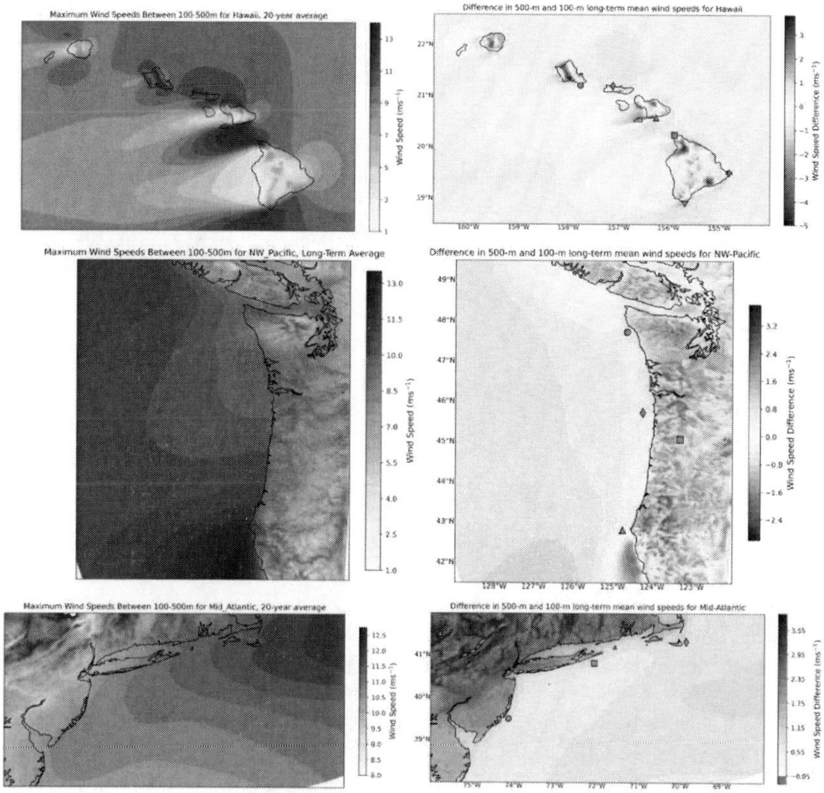

Figure 12. (left) Maximum wind speeds between 100 m and 500 m averaged over 20 years (2000– 2020) and (right) 20-year differences between 500-m and 100-m wind speeds from NREL's WRF simulations (top row) over Hawaii, (middle row) the Pacific Northwest, and (bottom row) the mid-Atlantic Coast. The green symbols denote locations at which wind profiles were extracted and the annual energy production and gross capacity factor calculated.

We further analyzed wind speed profiles at selected locations (symbols in Figure 12, right column). The locations were randomly chosen among areas with promising characteristics for airborne wind energy: offshore and onshore, where the resource is high and shear is positive (i.e., increasing average wind speed with height). We chose more offshore sites for this analysis based on expert interviews, indicating that the most promising locations are offshore. It is evident in Figure 13 that, although wind speeds tend to increase with height at selected locations, this increase is minor. Most of the increase in wind speed happens at altitudes up to 200 m, but above that the wind profiles are rather flat, even at the most promising locations.

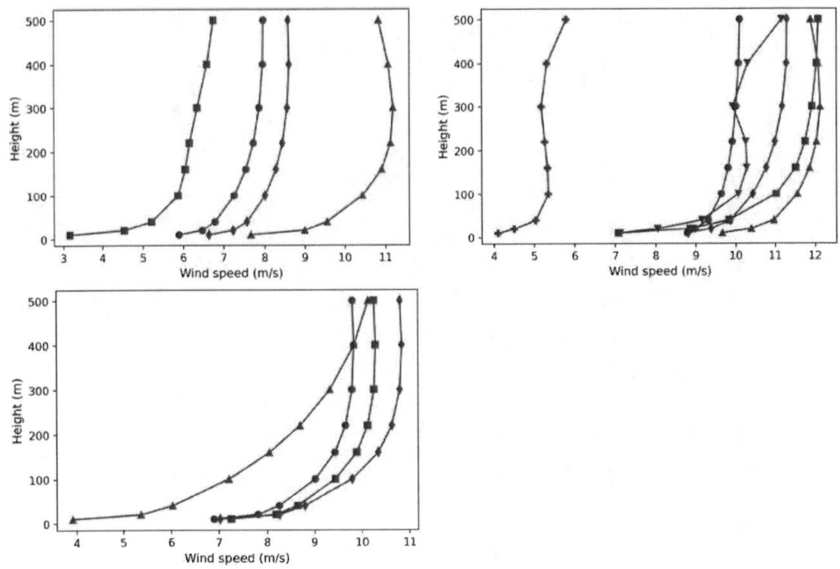

Figure 13. 20-year average wind speed profiles at selected points shown in the difference maps in Figure 10 from NREL's offshore WRF simulations. (Left) over the Pacific Northwest, (right) over Hawaii, and (bottom) over the Atlantic.

We also plotted average wind speed profiles at four locations across the central United States (Figure 14, left), based on WRF simulations at 15-minute intervals for June and July 2018. The locations are denoted in Figure 12 (right), which shows the difference between 500-m and 100-m wind speeds over the modeling domain at a selected time (June 1, 2018, at 0600 Coordinated Universal Time [UTC]) to show an example of the differences in shear across the modeling domain. At that point in time, the shear was extreme, exceeding 12 m/s between 100 m and 500 m. Average shear values across the 2 months are lower, at 3–4 m/s (Figure 14, left). Overall, we see an increase in average wind speeds between 100 m and 300 m, whereas the profiles exhibit little change above ~300 m.

Wind shear was mentioned by experts in the AWE workshop as a topic of importance (Weber et al. 2021). It is unclear whether it is beneficial to pursue higher wind speeds at different heights during high shear. Positive wind shear above 200 m was not found consistently in our analyses, and selected sites and wind profiles appeared to be rather flat, with more evidence of positive shear on land than offshore. More research and field measurements will be required to determine if chasing higher winds will be beneficial.

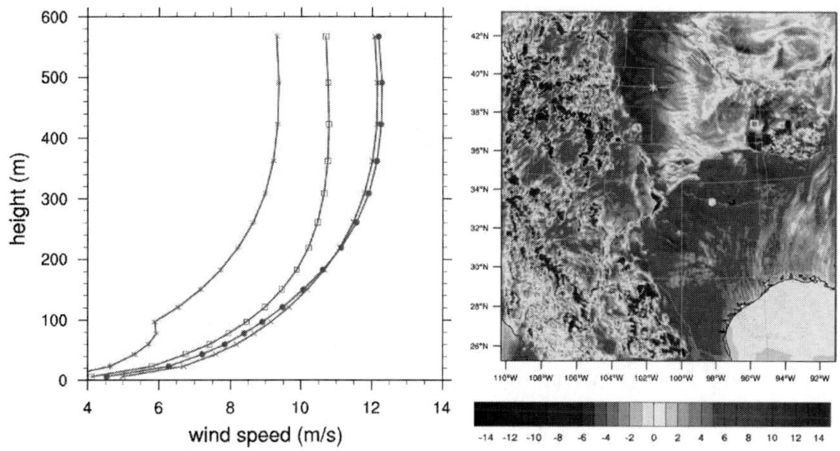

Figure 14. (left) Simulated wind speed profiles at four locations in the WRF modeling domain over the central United States. (right) Difference between wind speeds [m/s] at 500 m and 100 m over the modeling domain on June 1, 2018, at 0600 UTC. The symbols denote locations at which wind speed profiles are taken.

6.5. Wind Resource Variability

Wind speeds are variable in nature. Figure 12 shows that the wind resource between 100 m and 500 m varies spatially. The wind resource also varies temporally. Archer et al. (2014) discuss spatial variability of the wind resource in the U.S. Great Plains, and its variability with regards to changes in altitude and occurrence of the low-level jet (LLJ). They identify the U.S. Great Plains, among other regions, as a location with high wind power densities. The Great Plains LLJ seems to be the major driver. In the summer, it occurs at all hours. It reaches its lowest elevation, which is less than 400 m above ground level, and strongest wind power density (up to 3,000 W/m^2) in the evening and night hours. During the day, the LLJ rises and weakens in the morning and afternoon. The average LLJ in the afternoon is higher and weaker in the evening. In the winter, some locations experience a lower but weaker LLJ in the afternoon than at night. The authors list other appealing properties for AWE in the winter, such as a lower core height (between 300 and 700 m), greater wind speeds (between 15 and 23 m/s), and greater wind power densities (between 2,000 and 5,500 W/m^2) than the summer LLJ.

Here, we analyze the temporal variability of the wind resource for selected locations over the mid-Atlantic, Hawaii, and the Pacific Northwest. Figure 15 (left) shows a histogram of the simulated maximum wind speed differences from one hour to the next over 20 years at a particular location in Hawaii. Maximum wind speeds were used between 80 and 400 m and 120 and 300 m for offshore and onshore locations, respectively. Most changes are within 0.5 m/s, although larger differences up to 5 m/s exist. When these wind speed differences are broken down by months and time of day (Figure 15, right), we see that these differences have a diurnal as well as seasonal cycle. Larger differences can be found in the summer months, for example. Wind speed differences at 100 m show only minor differences at selected sites (not shown).

Likewise, the maximum wind speeds at certain locations change with time of year and time of day (Figure 16–Figure 19). Some sites show maximum wind speeds in the summer (Figure 16 and Figure 18), whereas others show the opposite (Figure 17 and Figure 19). The characteristics of the diurnal cycle vary by site, with some sites maintaining consistent cycles throughout the year, and others experiencing considerable variability (e.g., Figure 19). The discontinuities in Figure 16–Figure 19 arise from diurnal cycles being averaged over each month separately.

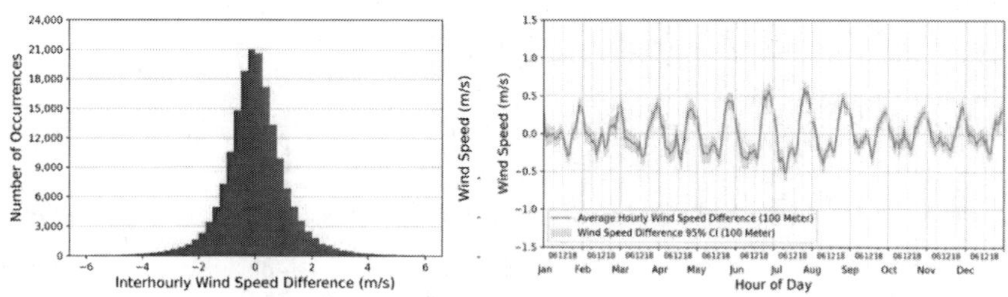

Figure 15. (left) Histogram of simulated wind speed differences from one hour to the next over 20 years at latitude 18.9 and longitude -155.7 (Hawaii, onshore) (right) average hourly wind speed differences as a function of month and time of day [UTC] at latitude 40.8 and longitude -72.1 (mid-Atlantic, offshore).

Maximum wind speeds are on average higher than 100-m wind speeds, and the diurnal cycles differ slightly. The difference between average maximum and 100-m wind speeds varies over the course of a year.

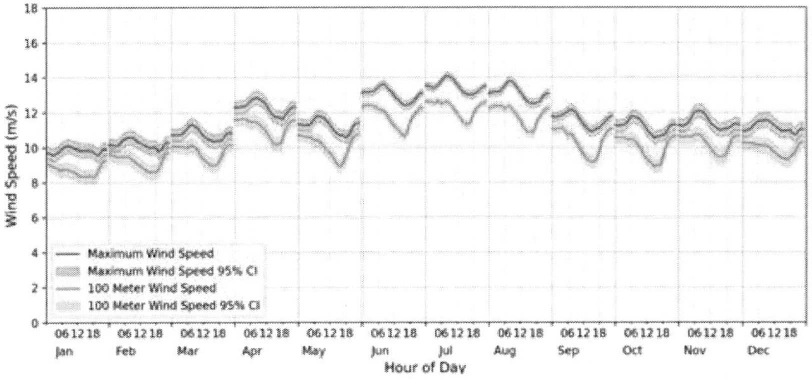

Figure 16. Maximum simulated wind speeds between 80 and 400 m and at 100 m as a function of month and time of day at latitude 21.2 and longitude -157.1 (Hawaii, offshore).

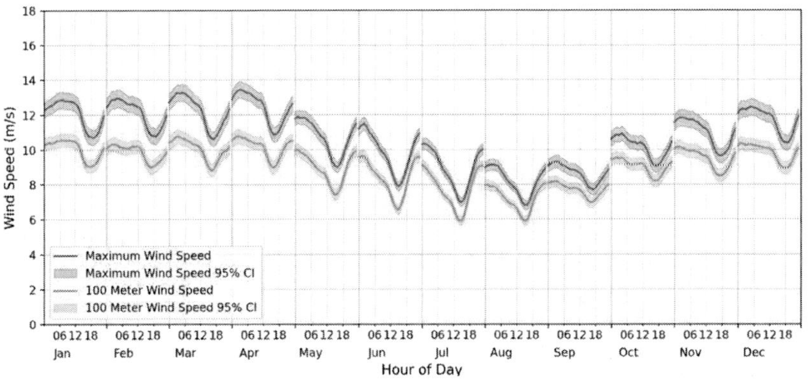

Figure 17. Maximum simulated wind speeds between 80 and 400 m and at 100 m as a function of month and time of day at latitude 39.5 and longitude -74.1 (mid-Atlantic, offshore).

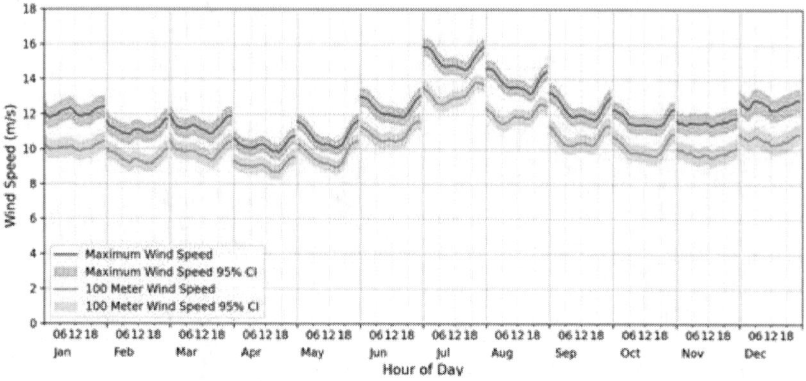

Figure 18. Maximum simulated wind speeds between 80 and 400 m and at 100 m as a function of month and time of day at latitude 42.8 and longitude -124.8 (Pacific Northwest, offshore).

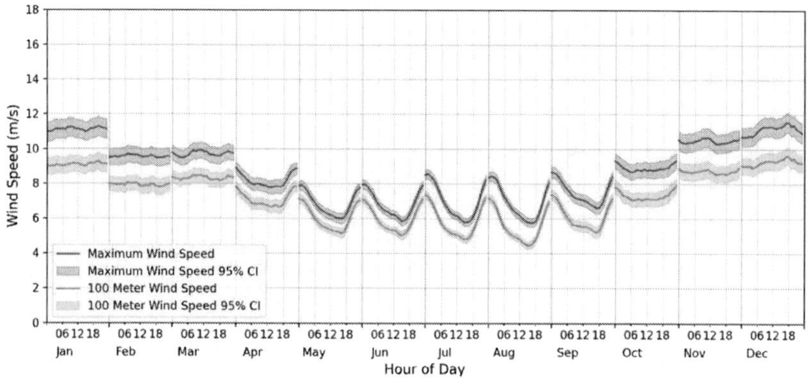

Figure 19. Maximum simulated wind speeds between 80 and 400 m and at 100 m as a function of month and time of day at latitude 47.7 and longitude -124.7 (Pacific Northwest, offshore).

6.6. Annual Energy Production at Selected Sites

The calculation of energy extraction from airborne wind energy devices is different from traditional wind turbines in that the AWE devices do not have a defined swept area. In the literature, only wind speed is frequently used as input to calculate power. The harvesting volume is a consequence of the operation of the system (Roland Schmehl, personal communication).

Airborne wind energy devices have been compared to traditional devices in terms of their applicability of the Betz limit (e.g., De Lellis et al. 2018). De Lellis et al. (2018) showed that, because both traditional and AWE wind turbines harvest power through either torque or drag force, the Betz limit applies. They found that AWE drag power systems can harvest up to 16/27 of the power available in the wind, and AWE lift power systems can harvest up to 4/27. AWE systems seem very ineffective from a Betz perspective, because they sweep a large flow volume from which they capture a comparatively small amount of energy (Roland Schmehl, personal communication; De Lellis et al. 2018). It is important to note that mapping Betz's theory to AWE is difficult because there is no well-defined swept area. Therefore, Roland Schmehl and team came up with the power harvesting factor. This factor uses the wing surface area.

Here, we calculate the annual energy production (AEP) and gross capacity factors using a generic AWE power curve (Section 5.3.1) for location on land and offshore (Figure 20 and Figure 21). For airborne wind energy AEP estimates, we used the maximum power output between 80 and 300 m and 20 and 500 m for rigid and flexible devices, respectively, following the flight altitudes of these kites. We compare the results with AEP for a traditional wind turbine at the same locations, using a 5-MW wind turbine (Jonkman et al. 2009; Figure 8 for both land-based and offshore locations). For the NREL 5-MW wind turbine AEP estimates, we used the maximum power output between 80 and 120 m (the approximate maximum and minimum blade tip heights). The AEP estimates and gross capacity factors vary by location but are similar for rigid and flexible AWE devices as well as traditional wind turbines (Figure 20 and Figure 21; Section 5.3.1).

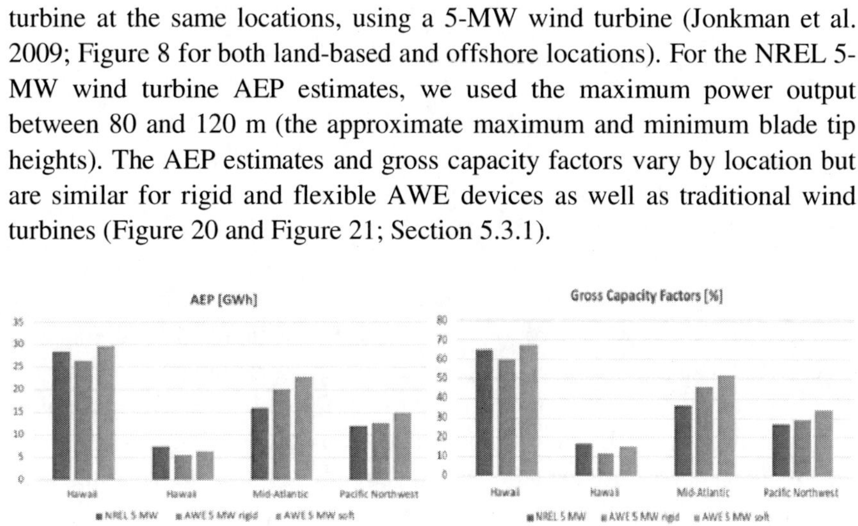

Figure 20. (left) AEP estimates and (right) gross capacity factor (averaged over 20 years) for selected locations over land (Figure 12). The calculations assume 100% availability. (GWh = gigawatt-hour[s]).

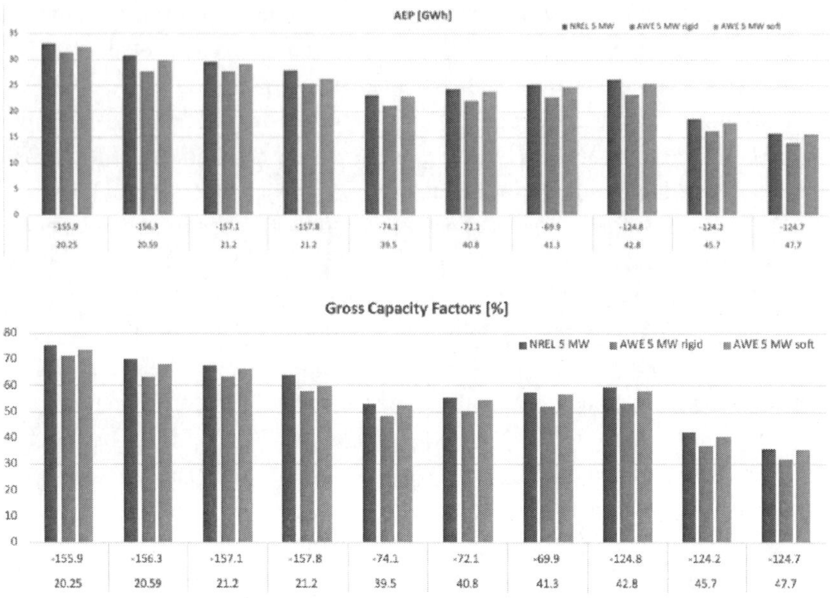

Figure 21. (top) AEP estimates and (bottom) gross capacity factor (averaged over 20 years) for selected locations offshore (indicated by longitude/latitude, see Figure 12). The calculations assume 100% availability.

All calculations assume 100% availability. A 50% loss factor that approximates the effects of aerodynamic, mechanical, electrical, and AWE-specific systems (such as the tether) is included in the AWE power curve and therefore implicitly considered in the energy calculations. This loss factor differs from the calculations for the NREL 5-MW wind turbine, which include electrical losses but neglect other losses (e.g., losses resulting from blade soiling, wake, and so on, representing total losses on the order of 10% of gross production). The power curve for the traditional wind turbines represents well-characterized performance values after decades of commercial technology development and refinement, whereas the ones for AWE are an estimate based on the state of research and technology development (Section 5.3.1). Note also that the assumptions for the AWE power curves might differ depending on the device.

6.7. Further Research Areas

Arent et al. (2011), Li et al. (2018), Miller and Keith (2018), Vautard et al. (2014), and Pryor et al. (2020) investigated the impact of wind power plants and solar farms on the climate. The impact of climate change on wind and solar power generation has also been researched (Craig et al. [2018, 2019] Losada Carreno et al. [2020], Haupt et al. [2016], Karnauskas et al. [2018]).

Areas for future research include investigating the possibility that high-altitude wind power devices could alter the general circulation patterns and have significant effects on global and local climate, and the potential for climate change to alter the wind resource between 100 and 500 m above ground. Archer and Caldeira (2009) predict some changes to temperature, precipitation, and greater sea ice cover worldwide because of deploying AWE devices. They also state that further studies are necessary. Using a climate model, Marvel et al. (2013) find, in a highly theoretical study, that uniformly distributed wind turbines generating the entire global primary power demand of 18 terawatts are unlikely to substantially affect the climate.

Another area of research that is well underway for traditional wind turbines but only nascent in AWE is wake effects: whether AWE devices produce wakes, if they can affect each other in an operational setup, and whether these wakes alter local conditions and the weather. Kaufman-Martin (2021) and Haas and Meyers (2017) have studied wakes for AWE. According to workshop participants, wake losses are expected to be minimal, but experts expressed a need to confirm this assumption, especially for large field deployments (Weber et al. 2021).

Extreme events should be investigated as well. It is assumed that AWE devices could be secured before the arrival of a hurricane, for example, but further research is needed.

6.8. Conclusion on Resource Potential and Energy Output

The wind resource at altitudes viable for airborne wind energy depends on location and experiences daily and annual variations. In general, wind speed increases with height up to 300 m, whereas above 300 m the average wind speed profiles show little change. Any discussion about the wind resource should include tether length and elevation angle, as they tend to significantly reduce energy capture gains from higher heights (based on the generic technology design assumptions used here). Calculated gross capacity factors at selected sites using simulated wind speeds across 20 years (assuming 100% availability, using power curves from Section 5.3.1) are similar.

Our studies indicate that AWE and traditional wind energy are competitors for the best resource locations (i.e., the windiest spots for traditional wind turbines are also the windiest spots for AWE devices). Malz (2020) analyzed wind profiles using ERA5 data and concludes that the annual generation profiles for AWE systems and traditional wind turbines are strongly correlated in time. They further find that "AWE is most valuable to the electricity system if installed at sites with low wind speed within a region. At greater shares of the electricity system, even if AWEs could demonstrate lower costs compared to wind turbines, AWE would merely substitute for them instead of increasing the total share of wind energy in the system."

Recommendations for future research include gaining a better understanding of turbulence, gusts, fluctuations, and intermittency of the airborne wind energy resource. Instrumental campaigns (for example, deployment of lidars) that can give insights into these research topics are needed. Experts recommend evaluating AWE devices with regards to high-wind performance, high-wind controls, and safety relating to gusts (e.g., how to dampen out instantaneous power surges). Echeverri (2020) was recommended as a rich source of ideas in this area. Design atmospheric conditions for AWE systems are another research area, (e.g., do devices need to be able to survive a gust with a 50-year return period?) Future research should also address climate impacts on the wind resource, wakes, and extreme events.

7. Technical Potential, Social and Environmental Considerations, and Permitting

In this section, we identify social, environmental, and regulatory factors that may impact siting decisions for AWE systems and estimate the technical potential for land-based AWE in the United States.

7.1. Key Findings

The following are key findings in the technical potential; social and environmental considerations; and permitting of AWE:

- Tether lengths are an important driver for setbacks (minimum distances from civil infrastructure), and they are likely to vary between AWE designs and sites.
- Land-based technical potential in the conterminous United States varies drastically between AWE designs, ranging from 420 to 34,573 gigawatts (GW) and 1,615 to 92,469 terawatt-hours.
- The U.S. Department of Defense (DOD), Federal Aviation Administration (FAA), and North American Aerospace Defense Command (NORAD) will evaluate AWE siting considerations on a case-by-case basis.
- Evaluating the wind resource at additional heights is needed to assess the potential advantages of higher operational wind speeds.

7.2. Background in AWE Technical Potential, Social and Environmental Considerations, and Permitting

This section provides information on the background and context of assessing the technical potential, social and environmental considerations, and anticipated factors in permitting of AWE.

7.2.1. Research Questions and Challenges

The following are areas for further research related to assessing the technical potential, social and environmental considerations, and anticipated factors in permitting of AWE:

- Further research is needed to understand potential impacts of AWE on wildlife. Specifically, assessing bird and bat risk, as well as potential impacts to prairie grouse species.
- Further research is needed to assess social acceptance of AWE. Key factors include viewshed, lighting, and safety.
- Continued assessment of the potential capacity density of arrays of AWE systems will improve estimates of technical potential and further illuminate possible implications for wildlife and social interactions.
- Assessing AWE offshore technical potential is needed.
- Identifying locations where opportunities for AWE and traditional wind are spatially distinct will illuminate potential sites or markets.

7.2.2. Methods and Tools Needed

The following are methods and tools needed in assessing the technical potential, social and environmental considerations, and anticipated factors in permitting of AWE:

- Assessing wildlife exposure and risk is needed first, and depending on the implications of those assessments, mitigation and avoidance strategies need to be established.
- Wind resource assessment across the range of AWE operational heights is needed. Currently, high-resolution (spatial and temporal) wind resource data for the entire United States is limited to 200 m.

Renewable energy "potential" is often assessed in terms of resource (or geographic), technical, economic, and market potential (Figure 22). Each measure represents additional complexity and input assumptions that leverage similar data and a common analysis flow (Lopez et al. 2021).

Technical potential represents the upper bound of achievable capacity and energy production given social, environmental, and regulatory spatial considerations combined with technology performance and density assumptions. Spatial considerations are represented as spatial constraints to resource potential, capturing already-occupied land (e.g., civil infrastructure, buildings, and so on), safety requirements (e.g., siting ordinances), or locations where development might be challenging because of land management strategies (e.g., national parks, conservation easements, threatened and endangered species habitats).

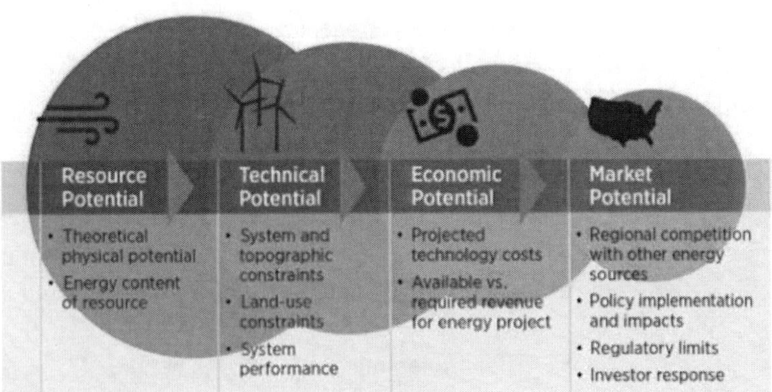

Figure 22. Renewable energy potential. *Image from Brown et al. (2016).*

Critical to assessing technical potential is an understanding that land-use objectives are fluid and evolving over time and space as communities prepare or react to energy development or potential wildlife impacts are identified (Lopez et al. 2021; Wu et al. 2020; Beiter et al. 2016). Further, technology is constantly evolving, leading to increased energy production (Wiser et al. 2020) and changing footprints. Current practices to capture these dynamics involve periodic updates to technical potential in addition to bracketing estimates through scenario analysis.

As described in the following sections, given the novelty of AWE, lack of literature, and lack of existing deployment make siting considerations difficult to anticipate. Furthermore, AWE has operational characteristics that are significantly different than traditional wind technology, making it difficult to leverage or define analogs based on decades of wind, ecological, and social research. To augment what little is known, NREL conducted interviews and held a workshop with leading wind researchers in the ecological and social domains, experts in military and civil radar regulatory requirements, and AWE technology experts.

From a siting perspective, airborne wind energy could provide additional opportunities for deployment in areas that are difficult for traditional wind turbine technology (e.g., steep terrain, or remote locations with limited transportation infrastructure). However, critical questions remain as to the potential environmental and social impacts of AWE. In the following sections, we synthesize high-level research needs and potential impacts gathered through the workshop and interviews. In the final section, we present the technical potential for eight different AWE technologies with plausible siting constraints.

7.3. Environmental and Ecological Siting Considerations

Of critical concern is how wildlife, particularly bird and bat populations, might interact with AWE. Little is known about how wildlife will interact with AWE; however, two studies represent a first attempt to quantify potential interactions. First, a study from Norway (Håland 2018) examined the interaction between AWE and local bird species, providing a single study of potential impacts. In that study, it was found that overall, the potential impact of AWE is low; however, it recommended further studies and exploration of mitigation measures. Second, a study by AMPYX Energy (Schmehl 2018) developed a theoretical model and predicted that their system would cause an average of 2–13 bird kills because of the kite and 11 bird kills because of the tether per year of continuous operation. The author concludes that this is within the range of bird kills at traditional wind facilities (0.6–63 fatalities per year; median value 7).

Given the significant operational differences between AWE and traditional wind turbine technology, few analogs for expected impacts are available. However, existing research into traditional wind turbine technology and interactions with wildlife help to formulate first-order siting considerations and critical research questions for AWE. In addition, research into wildlife interactions with power lines, aircraft, lighthouses, TV towers, and various communications facilities/towers can provide a useful perspective for research on AWE (Håland 2018). Because of limited literature on potential environmental impacts, NREL conducted expert interviews with leading wind and wildlife biologists and ecologists to identify possible concerns and research needs.

A primary concern is better understanding possible lighting requirements for AWE, originating from the FAA. Current lighting recommendations for traditional wind turbines have been shown not to increase collision risk to bats and migrating songbirds (National Wind Coordinating Collaborative 2010). Thus, it can be reasonably assumed that if the FAA allowed similar lighting practices (e.g., momentary flashes 3 seconds in duration) for AWE, attraction of wildlife would be similarly low. Another primary concern is the collision risk associated with the AWE tether or secondary entanglement (e.g., with marine mammals) from a failed operation. (This risk was found to be minimal in the Norwegian study but may depend on context). Further, grassland habitat loss is a concern for prairie grouse species, and they may see increased predation because of any towers or transmission lines associated with development. In addition, consistent with recommended practice for traditional wind, ecologists and biologists recommended avoiding known sensitive or high-abundance locations, including bat roost locations, orographic uplift areas preferred by eagles, and prairie grouse lek sites. Lastly, in interviews and workshop discussions, AWE technology developers expressed the assumption that the flight height of AWE systems was an advantage for avoiding bird and bat species; however, this assumption was refuted by wildlife experts who say that a 200-m to 500-m operation range is within the primary use space and abundance for most species.

Ultimately, to determine the impact of various AWE designs on wildlife, the use of specific sites by individual species needs to be determined, which would enable evaluating exposure by species and finally lead to the ability to assess risk to individual species. Conducting this type of research for the operational range of AWE may require sophisticated radar, near infrared, and thermal videography. These monitoring technologies could be expensive and difficult to implement, especially in offshore environments.

7.4. Social Siting Considerations

Like the environmental siting considerations, little is known regarding the social dynamics and resulting considerations that may unfold for AWE given the lack of literature or deployment experience. To supplement what little is known, NREL conducted expert interviews with leading traditional wind energy social dynamics researchers. There were three primary concerns for AWE from a social perspective: safety concerns, required setbacks from civil infrastructure, and visual impacts. Traditional wind turbines must typically maintain a minimum distance between the turbine tower and buildings, roads, railroads, and transmission lines. Setbacks are established by cities and counties (and some states) and are encouraged as a best management practice by developers. Typical setbacks in the United States are 1.1 times the maximum tip height of the turbine (Lopez et al. 2021), but European countries have established setbacks as long as 1 km (Dalla Longa et al. 2018). As shown in Section 7.5, setback requirements will be the primary driver for the overall potential of AWE. Another point of concern for AWE in common with traditional wind turbines is viewshed impacts to society. Research has shown that for traditional wind, turbines spinning during the day have a greater visual impact than nonsynchronous lighting at night (Firestone et al. 2018). In addition, angling lighting away from populated areas has proven to be an effective strategy to minimize visual disturbance. However, this strategy may not be achievable for AWE devices following time-varying flight paths.

To better determine society's acceptance of AWE, we recommend conducting AWE simulations of lighting for varying sizes and proximity to settlements, assessing people's willingness to pay or be paid for AWE operational characteristics, and assessing deployment in underserved communities to determine their perceptions. Lastly, as is critical to all energy development, community acceptance (or annoyance) is vastly improved when individuals in the proximity of development feel they have a say in the siting and permitting process.

7.5. Department of Defense, Radar, and Federal Aviation Administration Siting Considerations

Siting AWE devices may be more challenging than traditional wind turbines because of the elevations at which AWE devices would function. Close coordination with the FAA and DOD early in the site-selection phase is critical to successfully deploying AWE devices. DOD and NORAD strongly encourage airborne wind energy developers to engage with the DOD Clearinghouse early and often in the site-selection process.

7.5.1. Federal Aviation Administration

The FAA excludes tethered aircraft from regulations that apply to other aircraft, treating AWES purely as obstacles. The FAA concludes that AWE should be studied on a case-by-case basis with respect to the surrounding aviation environment to ensure aviation safety (European Commission 2018). The FAA offers a service for obstruction evaluation and a website[1] that is helpful in performing an initial screening for proposed AWE locations. These circumstances appear to offer a significant advantage over other jurisdictions (e.g., the European Union), where tethered aircraft are considered aircraft.

AWE devices will likely produce a radar signature and noise because of the motion of the AWE. The material of the AWE wing, flight speed, and turning rate will likely impact the signature of the AWE to radar. Guidelines for AWE flight heights near radar installations need to be established in a clear manner to provide AWE developers with clear directives regarding tether lengths and operational heights.

Traditional wind turbine tip heights are typically 500 feet, but tip heights are increasing. Avoidance of radar line of sight at these tip heights is unlikely, as 50–60 nautical miles from a radar installation is required for turbine tips to fall below the horizon. AWE devices may function at even higher flight heights, which will increase their visibility to radar.

In 2014, a memorandum of understanding was signed by DOD, DOE, FAA, and the U.S. Department of the Interior to form the Wind Turbine-Radar Interference Mitigation Working Group[2] and develop the Federal Interagency Wind Turbine Radar Interference Mitigation Strategy.[3] Leveraging these

---

[1] https://oeaaa.faa.gov/oeaaa/external/portal.jsp.

[2] https://windexchange.energy.gov/projects/radar-interference-working-group.

[3] https://www.energy.gov/eere/wind/downloads/federal-interagency-wind-turbine-radar-interference-mitigation-strategy.

efforts by AWE developers may be advantageous for the demonstration and deployment of AWE devices.

7.5.2. U.S. Department of Defense Airspace

The DOD Clearinghouse offers a service at no cost to developers and private parties to assist in siting any structure within DOD airspace or radar viewshed. DOD and NORAD highly recommend coordinating "early and often" with the DOD Clearinghouse on potential evaluation or deployment locations (https://www.acq.osd.mil/dodsc/).

Figure 23 shows DOD training and operation airspace, which covers a large fraction of the continental and offshore United States. Many existing wind farms have been approved by DOD within these flight paths and DOD stressed that these flight paths are not exclusions to AWE but that coordination with DOD is highly recommended. The center of each flight path is the most valuable and frequently flown portion, which should be considered in the initial screening process of AWE sites.

For offshore AWE deployment, the elevations used by DOD are typically higher nearshore and lower farther offshore. Thus, siting large AWE devices farther offshore may require more consideration for DOD airspace. An additional consideration offshore relates to the acoustic monitoring of substructures, which—if used—could lead to additional DOD security concerns around surveillance.

Figure 23. Low-level military flight paths. Image from the Military Aviation and Installation, Assurance Siting Clearing House.

DOD will require review of all components and countries of origin because of security and surveillance concerns. DOD may require the ability to shut down both airborne and traditional wind turbines in case of a national emergency or training exercises. Traditional wind turbines can stop rotor rotation in less than 1 minute, whereas AWE devices may require more time to return to the ground depending on tether length, wind speed, and position of the AWE within its cycle. Coordination with DOD around the length of time required by AWE devices to be grounded is critical to siting those devices.

DOD suggested that AWE research facility sites within or around existing traditional wind turbine farms may be advantageous, as those farms have already been evaluated by DOD and additional radar impact by AWE devices may be minimal. Close coordination with DOD on these opportunities is highly recommended, depending on the impact to radar of the existing wind farm and the proposed AWE location, tether length, and flight path.

7.6. Technical Potential Results

Notwithstanding the list of concerns listed in the previous sections, AWE has a large technical potential within the contiguous United States. To estimate technical potential for land-based AWE, we used key assumptions defined in Section 5. Those assumptions include the capacity density (MW/km^2), power curve, wind speed at a 200-m height (the maximum height available in the wind data set used for land-based wind, discussed in Section 4), and tether length (meters) to determine approximate setback requirements from civil infrastructure. A notable limitation to this analysis is the use of wind speeds at 200 m for all AWE technologies regardless of tether length and operational height. This limitation is because of the lack of wind speed data above 200 m for the contiguous United States in the Wind Integration National Dataset Toolkit (WIND Toolkit) data. In addition, we also used plausible siting constraints based on traditional wind (Lopez et al. 2021) to restrict AWE deployment in unlikely or infeasible locations (e.g., legally, or administratively protected lands like national parks). A full list of exclusions used is provided in Table 6.

To estimate AWE technical potential, we use the Renewable Energy Potential model (Maclaurin et al. 2019), that assesses land availability on a 90-m spatial grid and leverages the WIND Toolkit (Draxl et al. 2015) for estimating generation potential given a power curve.

Table 6. Siting restrictions for AWE technical potential estimation

Siting Exclusion Category	Exclusion Details
Infrastructure	
Setbacks to transmission right of ways, railroads, roads, building structures	Setback = 1.25x tether length = (1,250 m, 875 m, 750 m, 625 m, 438 m, 375 m, 250 m)
Urban areas and airports	Excluded, airports w/10-km setback
Radar	4-km NEXRAD, 9-km short-range Radar/long-range radar
Regulatory	
Protected public lands and conservation easements	Excluded
Physical	
Slope >25%	Excluded
Mountainous landforms and high (>9,000 ft) elevation	Excluded
Water and wetlands (with a 305-m buffer)	Excluded

Tether lengths are an important driver for setbacks, and they are likely to vary between AWE designs and sites. Figure 24 demonstrates the importance of tether length and its influence on land availability when considering setbacks from civil infrastructure. Exploring a broad range of tether lengths was recommended by workshop participants and is included in this assessment.

Figure 25 shows developable areas across the contiguous United States considering all exclusions presented in Table 3 and using a 750-m setback requirement. Most of the technical potential exists in the western United States, where population and civil infrastructure are less dense, providing opportunities for AWE deployment, although these areas are also likely farther from the transmission infrastructure needed for grid interconnection. While implications for AWE deployment in forested areas were not explicitly discussed with environmental experts, it was discussed with workshop participants who expressed the view that operation within forested areas would be difficult and would likely require some forest clearing. Forest exclusions were not applied in this analysis, but they do merit further research and consideration and are likely to be an additional large driver of estimated technical potential.

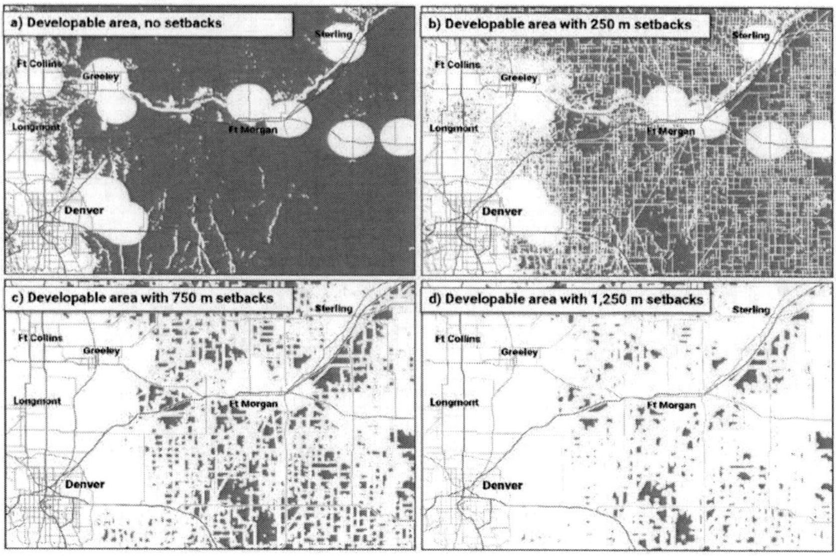

Figure 24. Four maps demonstrating the influences of tether length and setback on developable areas. Blue represents developable areas for AWE: a) all exclusions except setbacks, b) all exclusions plus a 250-m setback requirement, c) all exclusions plus a 750-m setback, and d) all exclusions plus a 1,250-m setback requirement.

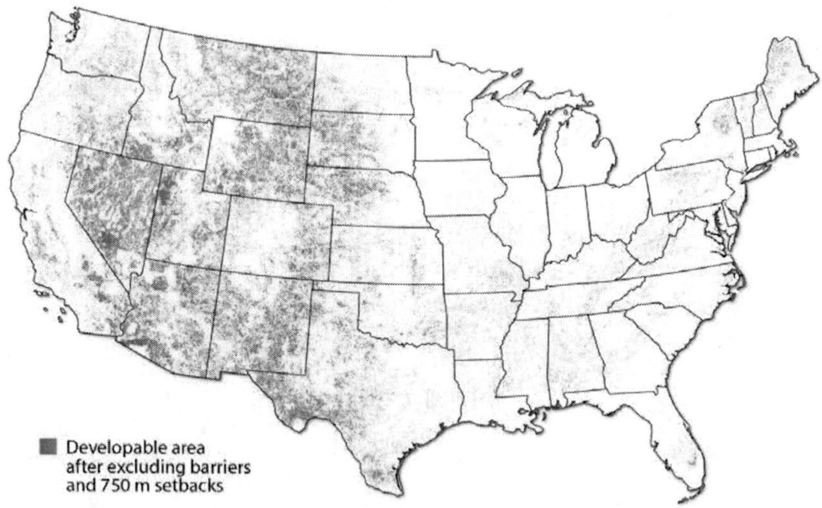

Figure 25. Developable areas for AWE considering all exclusions and a 750-m setback requirement from civil infrastructure.

Table 7. Technical potential results for modeled AWE technologies with varying tether lengths

Technology	Setback From Civil Infrastructure (m)	Capacity Density (MW/ km^2)	Available Land Area (km^2)	Nationwide Mean Capacity Factor (std)	GW	Terawatt-hours
Traditional Wind – ATB Moderate, Reference Siting Regime	228	3	2,616,687	0.36 (0.09)	7,827	24,679
Traditional Wind – ATB Moderate, Open-Access Siting Regime	0	3	5,058,384	0.36 (0.09)	15,175	47,717
Traditional Wind – ATB Moderate, Limited Access Siting Regime	622	3	760,188	0.36 (0.09)	2,280	7,332
Rigid-wing 500 kW (short tether)	250	3.98	2,269,240	0.48 (0.11)	9,029	36,993
Rigid-wing 500 kW (long tether)	625	0.64	1,142,870	0.48 (0.12)	727	2,829
Rigid-wing 5 MW (short tether)	375	19.6	1,764,210	0.32 (0.1)	34,573	92,469
Rigid-wing 5 MW (long tether)	1,250	8.72	530,285	0.3 (0.1)	4,623	11,209
Flexible-wing 500 kW (short tether)	438	1.3	1,567,757	0.48 (0.11)	2,036	8,125
Flexible-wing 500 kW (long tether)	750	0.44	951,804	0.47 (0.12)	420	1,615
Flexible-wing 5 MW (short tether)	875	3.08	811,621	0.31 (0.1)	2,499	6,219
Flexible-wing 5 MW (long tether)	1,250	2	530,285	0.3 (0.1)	1,059	2,568

Technical potential results for AWE are presented in Table 7. Traditional wind technical potential from Lopez et al. (2021) is included for comparison purposes. Three levels of access scenarios are presented for comparison. Each scenario uses the same wind turbine and capacity density, but siting exclusions vary based on plausible siting restrictiveness. For AWE, all exclusions except setbacks are held constant across technology scenarios, revealing the large implications for tether length and its influence on technical potential for AWE.

While the assumed capacity density for traditional wind from Lopez et al. (2021) is 3 MW/km^2, empirical wind farm capacities range from 1–18 MW/km^2 (Harrison-Atlas, Lopez, and Lantz, forthcoming) and offshore wind farm capacity densities range from 3.1 to 18.7 MW/km^2 (Deutsche WindGuard GmbH 2018). These data illustrate the empirical variation of capacity densities that are observed for installed wind farms, which are influenced by resource, technology, power prices, siting constraints, specific power, and other factors. Thus, the wide range of technical potential capacities estimated here for AWE devices is plausible depending on the influences of the aforementioned factors.

While the national summaries provide a quantification of national potential, they obscure regional trends and opportunities. Maps presented in Figure 24 reveal these regional trends and opportunities for different AWE technologies. Flexible-wing technologies have relatively higher capacity factors than rigid-wind technologies, albeit the highest capacity factors are spatially coincident with the traditional wind belt (Great Plains and midwestern United States). However, because of a combination of longer tether lengths and lower capacity densities, there are fewer opportunities for development. Conversely, the rigid-wing technologies have lower relative capacity factors and higher capacity densities, which implies that larger quantities of rigid-wing technologies could be deployed, albeit with lower energy output per unit of land area.

While these results present a first attempt at estimating the technical potential of AWE, further research into the ecological, social, and technical constraints and considerations is needed to refine these estimates and provide a bracketed assessment of AWE's technical potential. A recent report published by the United States Geological Survey (Remington et al. 2021) on the state of America's sagebrush ecosystem revealed a need for broad conservation strategies.

Coincidentally, the research here shows a vast amount of AWE opportunities in sagebrush habitat that may need to be revisited. Lastly, participants in the AWE workshop said their current focus is on land-based applications; however, they do have a long-term goal of deploying offshore. Through simple extrapolation of traditional offshore wind technical potential (Musial et al. 2016), and by using an assumed capacity density of 1.43 MW/km^2, which considers 1-nautical-mile spacing requirements for vessel navigation, we estimate the offshore AWE technical potential to be roughly 1,293 GW for a 5-MW system. Through this simple extrapolation, we find that AWE offshore potential is large; however, the unique challenges of deploying offshore require more rigorous assessment to determine the true potential and we recommend conducting further study of offshore environmental and social considerations for AWE.

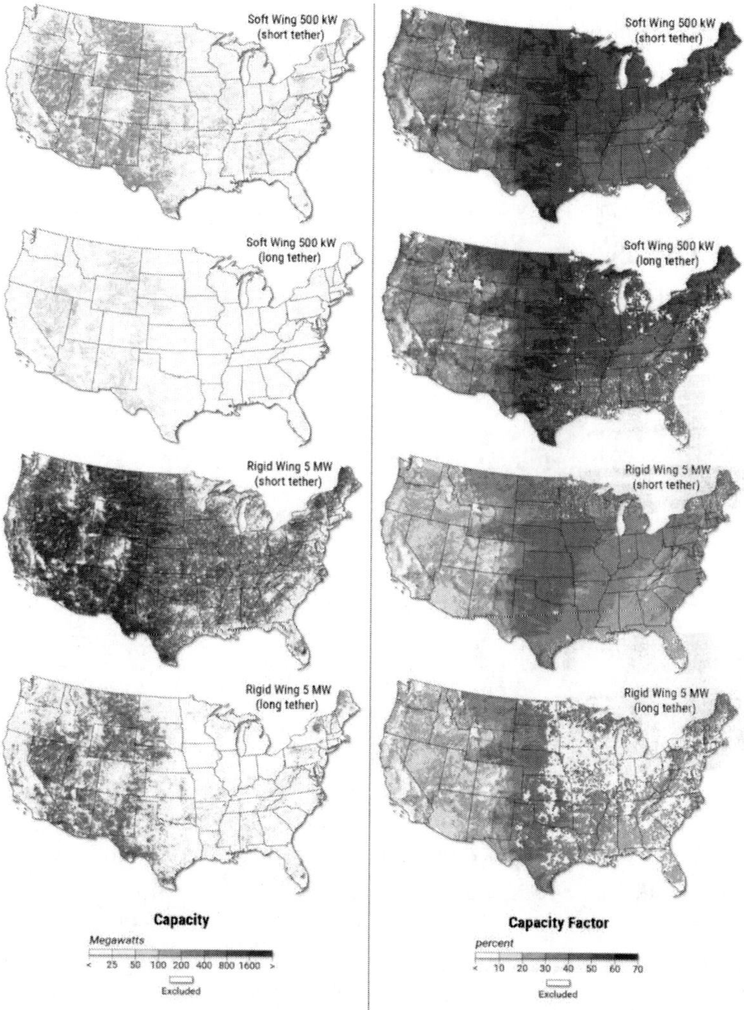

Figure 26. Capacity (left) and capacity factor (right) maps for select AWE technologies.
Image created by Billy Roberts, NREL.

8. Demonstration, Commercialization, and Road Map

This section discusses requirements for demonstration and commercialization of AWE devices and presents a road map for future research, development, and deployment.

8.1. Key Findings

The following are key findings in demonstration and a road map for commercialization of AWE:

- Many companies target small device sizes and niche markets to gain operational hours, which reduces investors' perception of risks and enables production economies of scale and upscaling to reach larger markets.
- Research facility sites that can host both small- and large-scale devices will allow for critical operational hours to develop understanding of turbulence and the atmosphere, design codes, controls, and materials.
- A gap exists between prototype and commercial deployment.
- Funding methods are needed.

8.2. Background in AWE Demonstration and Commercialization

This section provides information on the background and context of assessing AWE demonstration and commercialization needs, including key research questions and challenges and methods and tools we identified.

8.2.1. Research Questions and Challenges

The following are key research questions and challenges we identified to advance AWE demonstration and commercialization:

- Future research should focus on gaining operational hours, evaluation, and development of funding-support structures from prototype stage to commercial deployment at various device scales and sizes.

- There is a need to characterize technology operational parameters to enable development of policies and procedures for permitting AWE devices at the community, state, and national levels, with engagement of FAA and DOD, among other stakeholders.
- Research on technical potential, siting, and environmental issues is critical to determining markets for all stages of deployment.
- What markets exist where traditional wind cannot be deployed? (e.g., markets saturated by solar but with a limited nighttime wind resource).
- What events may trigger a rollout of this technology? Extreme weather, hurricanes, or military operations could encourage a rollout.
- What funding methods may assist in creating the AWE industry?
- How can AWE reduce perceived risks associated with a new technology? If the LCOE of AWE is equal to the LCOE of traditional wind, developers will not choose a less mature technology with more risk.
- What additional factors or metrics can influence the viability of AWE? AWE levelized cost of energy will have to be lower than that of traditional wind for it to be attractive to developers, or other qualities will have to be valued, such as material usage and recyclability.

8.2.2. Methods and Tools Needed

The following are key methods and tools we identified as needed for AWE demonstration and commercialization:

- A road map for developing an AWE industry not only for technical progress but also to advance deployment
- Analysis of national siting, device size, and market potential as technology matures
- Developing a plan and road map for government funding to support AWE companies financially and enable advancement along the learning curve as risk increases
- Programs such as the Competitiveness Improvement Project or design competitions like the Collegiate Wind Competition may also assist in accelerating the industry.

AWE technology is at an early stage of development, with few prototypes deployed and a low number of total operational hours. Demonstration and operation of AWE devices at various sizes are critical to developing understanding of these complex systems and scaling relationships. Developing and securing locations suitable for AWE demonstration are critical to the advancement of AWE. Research locations suitable for a wide range of device sizes help evaluate larger devices suitable for land-based and offshore wind applications. A research facility large enough to allow an array of devices to operate would also enable research on device wake interaction and grid integration. AWE devices may function at elevations above traditional wind turbines and will operate in significantly different areas of the atmosphere, which may impact airspace, wildlife, and radar in ways not yet understood. Evaluating and demonstrating AWE devices to develop best practices for commercial deployment of devices is critical to a future AWE industry.

8.3. Market Development

The Airborne Wind Europe "Policies for Airborne Wind Energy" 2018 report (Petrick 2018) details potential policy and support pathways to enable the AWE industry to reach commercialization. This report is an excellent reference to U.S. policymakers, private investors, local communities, and AWE companies that communicate a need for and potential pathways to developing a commercial AWE industry. As technology concepts advance from prototype evaluations to demonstrations, a "valley of death" gap may exist in which required investment to upscale increases while perceived risk by investors is also high (Petrick 2018; see Figure 27).

Many AWE companies are nearing or within the demonstration phase and need policy, infrastructure, and funding support to reach commercial viability.

Demonstrating small-scale systems may be necessary to gain operational hours and reduce technical risk to utilities and investors. Some companies may elect to upscale devices for demonstration, which can accelerate commercialization of the concept but increase the capital requirements for upscaling, thereby increasing the risk of perceived failures. While small-scale systems may not be the intended final market, they must be commercialized to ensure an early revenue stream and show product potential to investors and utilities. In interviews, AWE companies and developers state that a research facility site would be very beneficial, as space is especially hard to find in Europe. A research facility site suitable for upscaled offshore devices on land would also reduce deployment cost, risk, and time. A land-based research facility site would accelerate timelines for developing siting and permitting policies and procedures by engaging stakeholders such as local, state, and national policymakers including DOD and FAA. A research facility site could also aid in conducting research around environmental impacts and shaping policy.

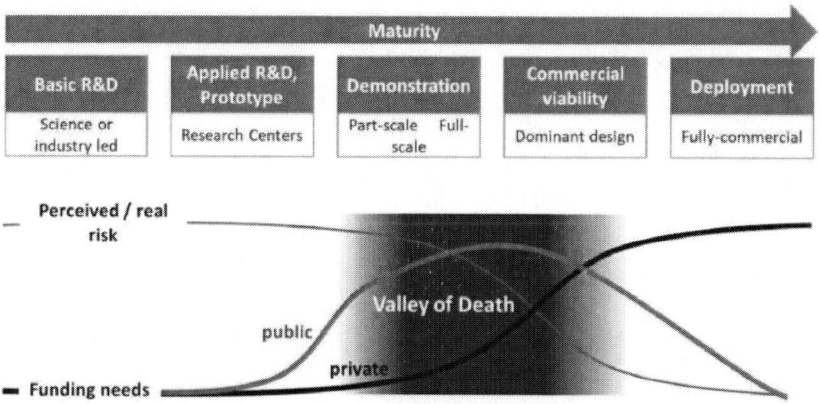

Figure 27. Funding needs increase and risks (perceived and real) decrease as prototype evaluation advances to demonstrations, but a "valley of death" gap exists, wherein funding needs and perceived risk by investors is also high.
Image from Petrick (2018).

Markets identified for initial demonstration and early commercialization include the following:

- Military bases/operations
- Distributed generation and agriculture
- Island communities
- High-energy-cost communities (diesel grids)
- Communities susceptible to extreme winds such as hurricanes
- Sites with highly complex terrain
- International communities without access to reliable power.

The markets identified here may represent high profit potential but typically high risk to operation and deployment with low total market volumes. Many companies identify these markets with 100- to 500-kW devices as an opportunity to gain operational experience and climb the learning curve while aiming toward an end goal of megawatt-scale land-based and offshore devices.

8.3.1. Government and Stakeholder Commercialization Support

Governments can greatly accelerate technology learning not only through financial support but also physical and social investment and learning. Many of the potential barriers to commercial deployment of AWE come from not only the technical learning required but also the ability to access markets. These considerations include the ability to site AWE devices, and a large contribution to the AWE industry can come from establishing policy and regulations.

Specifically:

- Research covering the wind resource and profile at elevation, turbulence, and development of design code simulation and validation is critical to accelerating the AWE industry. Research facility sites are key to gaining operational hours for AWE devices. Governments can assist in identifying and permitting research facility sites to enable R&D of design software.
- DOD and FAA permitting and procedure development is a large task that cannot be achieved by a single company or group of companies. Governmental assistance in coordinating and developing policies and procedures for siting AWE devices would be of great value to the AWE industry.
- Standards regarding design and public safety are needed to gain public acceptance of AWE devices. Developing standards with federal, state, and local governments in concert with component and system design safety standards is critical to enabling deployment of AWE devices.
- Environmental research and policy creation can be enabled by governments through research facility sites and funding of research in coordination with the U.S. Fish and Wildlife Service and others.
- Precommercial demonstrations in coordination with local communities and authorities willing to host demonstrations are a critical step to reducing investors' perceived risk and enabling deployment and technology upscaling. Identifying communities interested in precommercial demonstrations is recommended to show the AWE industry pathways toward commercialization.

- Grid integration standards may be critical to AWE adoption depending on the device generation concept. Ground-gen systems require significant power consumption to reel in the aircraft and dedicated energy storage systems at the device level may become a significant system capital and operational cost.

8.4. Commercialization Scenarios

Most renewable energy technologies have relied on incentives or support from governments. Solar photovoltaics, land-based wind, and offshore wind have all been supported by direct funding and revenue support. As technologies progress through the basic and applied R&D stages toward demonstration, "push" policies, which provide investment or financial support, may be employed to encourage this research and learning. Once technologies reach demonstration and near commercial viability, "pull" policies, which provide revenue support, may be employed to enable the technology to reach a commercial, large-scale deployment.

> "Just as subsidies like the production tax credit in the U.S. were essential to get traditional wind off the ground, substantial policy backing will likely be the only way airborne wind could see widespread adoption at competitive prices." (Shifman 2019)

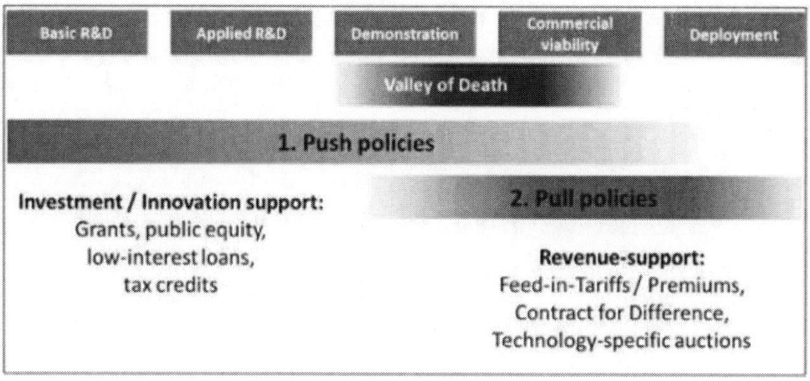

Figure 28. The sequence and types of financial support mechanisms used to support a potential AWE industry. *Image from Zillmann (2020).*

Developing a plan for stages of financial support is critical to enabling a commercial self-supporting AWE industry. Coordination between local, state, and federal governments is recommended to avoid gaps in technology evolution and learning (see Figure 28).

9. Research and Technology Development Needs

The AWE industry is currently limited by a lack of readily available resources for simulating and characterizing the physical response, power performance, structural loads, and cost of existing and future systems. Without these resources, designs must include safety factors that result in oversizing components and driving up costs while adding significant obstacles in the effort to identify the most effective design space. Here, we outline the types of challenges in the AWE industry and propose solutions through simulation software and physical characterization infrastructure.

9.1. Key Findings

The following are key findings in our assessment of infrastructure and technology required by the AWE industry to conduct R&D activities:

- Knowledge sharing via open-source simulation software is key for the industry to mature.
- Types of software applications required for AWE development include:
  - Design exploration (e.g., conceptual design, system architecture, systems engineering)
  - Detailed design (e.g., aerodynamic, structures, control systems, offshore support platforms)
  - Highly resolving (e.g., understanding fundamental physics, calibrating/validating lower-fidelity tools)
  - Plant layout and optimization
  - Technoeconomic analysis
  - Certification

- An AWE-specific simulation environment would comprise a suite of tools of multiple fidelities and applications such as design, analysis, and cost modeling.
- Workforce development is critical to support these needs; specifically targeting the following areas:
    - Domain experts in aircraft, rotorcraft, autonomous vehicles, controls, optimization, and power systems
    - Experts in systems engineering and analytical, engineering, and high-fidelity models
    - Computational scientists with scientific software development expertise.

9.2. Background in AWE Technology Development Needs

This section provides background and contextual information on the infrastructure and technology required by the AWE industry to conduct R&D activities, including key research questions and challenges:

- Reference models and data sets are needed for each system architecture and power generation scale, including:
    - Coordinating with the recently submitted IEA Wind Task proposal that covers reference model development
    - Coordinating with field campaigns of AWE technologies using publicly available measurement data to validate modeling tools
    - Developing industrywide design standards for AWE technologies
- New physics models are needed to simulate the system dynamics and interactions with the atmosphere and environment
- The ability to leverage next-generation computing technologies is needed, such as:
    - Support for high-performance computing resources ([GPU], heterogenous hardware)
    - Integration with optimization, artificial intelligence, and machine-learning ecosystems.

9.3. Simulation Software Environment

Developing a comprehensive simulation environment for addressing AWE design and modeling problems is a complex and major effort. The traditional wind energy industry and its collection of simulation tools are well-suited to serve as a model for how a cohesive, open-source software suite can stimulate technological innovation and encourage research collaboration. The industry challenges addressed by a simulation environment are design exploration (e.g., conceptual design, system architecture, systems engineering), detailed design (e.g., aerodynamic, structures, control systems, offshore support platforms), highly resolving (e.g., understanding fundamental physics, calibrating/validating lower-fidelity tools), plant layout and optimization, technoeconomic analysis, and certification. In all use cases, adequate software will save the industry both money and time, as well as position it to solve existing and future energy challenges.

Knowledge sharing through open-source software is important for the industry to mature. One or a handful of key research institutions must lead the effort to develop the infrastructure with which the industry will refine for their technology-specific requirements. It is critically important for the open-source simulation environment to have a flexible and open license that enables ongoing R&D without the need for each industry member to develop and validate their own unique, proprietary tools.

Simulation software can be divided by fidelity of the underlying models in a range from low to high. All fidelity levels are critical to a comprehensive simulation environment. In addition to their importance to AWE developers, the full range of models are often used in research where higher-fidelity models (e.g., computational fluid dynamics, finite-element modeling, and so on) inform reduced-order models and provide validation for engineering (midfidelity) and analytical (low fidelity) models.

Generally, low-fidelity models yield results in a very short amount of time, but the accuracy or applicability is dictated by the underlying assumptions of the model. These models can be used for design exploration, especially when coupled with optimization systems and data-driven approaches for traversing a design space. Low-fidelity models are often analytical, steady-state, first-order approximations with a runtime for a single data point on the order of a fraction of a second. These types of models enable system developers to explore how the topology of a concept impacts performance and enable technology trade-offs. Other types of low-fidelity models could predict the impact of a kite's wake on other kites downwind and assess the impact on power production; such models would provide important tools for farm layout design, controls design, and cost modeling.

For detailed design, engineering models provide higher-order mathematics and physics at a higher computational cost. These midfidelity models typically simulate nonlinear, time-marching dynamics of a system in near-real time; that is, 1 second of simulation time is computed in 1 second of clock time. Midfidelity models would provide the industry with full individual kite dynamics (e.g., translations, rotations, accelerations), as well as aeroelastic analysis, one-dimensional structural analysis, controls modeling, and farm-scale kite-kite interaction. These tools would integrate with in-depth design tools like systems for optimization, loads analysis, and controls co-design, wherein the controller design is incorporated into the mechanical design at an early stage. Linearization of the underlying nonlinear time-domain engineering models is also important to understand the system response, with applications including eigenanalysis (to identify system natural frequencies, mode shapes, and damping), controls design, and stability analysis. KiteFAST, developed at NREL under contract with Makani, is an example of a midfidelity simulation tool for land-based- and offshore-tethered fly-gen AWE systems and was developed to integrate with OpenFAST, a widely adopted aeroelastic simulation environment for traditional wind turbines.

High-fidelity simulation software provides users with fully resolved physics at a high computational cost. These simulations are often time-consuming to configure and may require days of runtime for a single simulation. However, the result is a time-marching simulation with high-order models that can simulate an AWE system that fully resolves three-dimensional aerodynamics and three-dimensional structural analysis, as well as detailed component-component interaction (i.e., wing wake effects on the horizontal stabilizer for a rigid kite) and farm-scale kite-kite interaction. This level of fidelity is generally used for fundamental research and for calibrating and validating lower-order models.

9.3.1. New Research Areas

New areas of research are required to develop a comprehensive simulation environment for AWE systems. Specifically, reference models detailing targeted technology architectures, physics models for simulating system dynamics, and cost models to determine economic viability are required. Models exist for the traditional wind energy industry and can serve as a basis for the additional research required to accurately describe AWE systems.

The requirements embedded in AWE systems combine the physical performance of aircraft, reliability of autonomous vehicles, and cost performance of typical energy production systems. These extreme requirements will yield designs that push the limits of existing numerical models, requiring additional research for accurate simulation. At the farm scale, new analytical wake models should be developed to capture kite and tether impacts. Further, atmospheric turbulence models should be customized for the operating altitudes of AWE systems. Vortex-based aerodynamic engineering models for aircraft application, coupled with beam-type structural models for kite component-component interaction, will be required to predict critical performance characteristics of individual systems, such as flight qualities and stall behavior. For flexible wing systems, additional research is needed to develop coupled aerostructural models capable of simulating highly elastic components like a fabric wing deforming under dynamic loads and controlled by elastic cables.

Reference models capable of operating in a design condition along with fully capable flight controllers, tether models, tether controllers, and ground station conditions are required. Each AWE system architecture brings unique dynamics, cost, and performance, requiring a full reference model. Furthermore, architecture designs are not expected to scale geometrically, requiring additional reference models for each target power generation scale. The development of this set of reference models will establish baseline technology solutions from which the industry can generate more advanced AWE concepts.

However, a co-dependency exists in that reference models and data sets are required to complete software validation, but the simulation software is required to design AWE systems. For rigid-wing, fly-gen architectures, the open-source Makani models (Oktoberkite and M600) and test data are a major resource for establishing a reference model. For other architectures, an iterative approach to expanding the design space while developing and validating software tools should be employed. Dynamically stable systems with simple components, such as straight beams and well-defined airfoils, can provide the basis for early numerical model validation. Additionally, model validation should be conducted in coordination with industry partners engaging in field campaigns for existing designs.

Controller software in the reference models is an area requiring significant additional research in the industry. It may be feasible to establish a simple controller for ground-gen systems, but the crosswind flight of rigid-wing fly-gen systems present an extremely complex problem. During the NREL effort to develop the KiteFAST software, a significant portion of time was spent characterizing controller behavior and configuring the simulation system so that it could simulate a full crosswind loop without manual input. As this "hands-off" simulation mode was not achieved by the project conclusion, the simulation environment was not fully validated.

The computational landscape for simulation environments continues to change dramatically. Software that supports high-performance computing hardware and heterogeneous hardware systems is increasingly common in the present day and should be anticipated to be the norm within the next 5 years. There is also a trend of integration of very complex algorithms, such as optimization, machine learning, and artificial intelligence, into scientific computing workflows. These tools are becoming increasingly ubiquitous, and it should be expected that the next generation of engineers and researchers will choose software that integrates with these frameworks.

Developing these tools will require integrating aerospace industry expertise with established research fields in the wind energy industry. It will require domain experts in aircraft, rotorcraft, and autonomous vehicle dynamics and controls to establish the high-level requirements for the physical simulation capabilities. Researchers experienced in wind power plant and component design and optimization should inform the specific use cases to target. The development and implementation of the physics models will require experts in analytical wake modeling, systems engineering, and high-fidelity numerical modeling, along with computational scientists with scientific software development expertise. Additional time will be required to validate the numerical models and implementations as the AWE technology landscape changes and new designs are introduced and characterized.

9.4. Physical Characterization and Validation

Close coordination between theoretical modeling, numerical simulation, and physical characterization and validation is required in any science and engineering discipline, especially as new research areas and modeling domains are explored (as described in the previous section). Physical characterization and validation equipment and setups require detailed specification to address the research questions. Example physical domains include:

- Materials science: tensile strength, fatigue, conductivity, specific weight, and multifunctionality
- Aerodynamics: wing flow, turbine flow, load assumption, tether drag, and flow-induced vibration
- Mechanical: structural integrity, deformation, oscillation, and system dynamics
- Electrodynamics: generator efficiency and conductor losses
- Thermodynamics: heat transfer in conductor and generator and environmental conditions
- Control: sensors, data acquisition, control units, and actuators.

At this point, only a high-level exemplary theme specification of physical setups and equipment for exploration, characterization, and validation needs can be provided at different system levels, such as the following material and components, subsystems, and integrated systems:

- Material and components: tether, soft kite fabric, connectors, force introduction: strength, fatigue, wear, elasticity
- Subsystems: control systems, wing structure, base station, tether reel and spooling mechanism, and back-to-back winch setups
- Integrated systems: on-ground base station and on-ground flight unit system validation; in-flight system validation on a dedicated, fully licensed site, suitable for different archetypes, operational concepts; scales; flight dynamics for all states and modes; launch; and landing.

9.5. Procedures, Guidelines, and Standards

Alongside theoretical modeling, numerical simulation, and physical characterization and validation, the development and adaptation of relevant procedures, design, and validation guidelines is critical to preparing processes and systems toward standards and certification. Further details can be found in the IEA Wind AWE task proposal (Petrick et al. 2021) and include the following list of themes:

- Failure mode effect analysis
- Design basis and load cases for structural design
- Risk management framework

- Technology design space definition and archetype classification
- Technology readiness level definition specific to AWE
- Technology performance level definition for AWE specific to key markets: grid, onshore, offshore, early adopter (such as remote and microgrid); see analogous, market-specific technology assessment framework
- Technology assessment frameworks
- Development of a full set of functional requirements for key markets
- Stakeholder outreach development or outreach material, identification of stakeholder needs
- Life cycle assessments.

Conclusions and Recommendations

Based on the insights gained from a literature review, internal analysis, domestic and global outreach through the U.S. AWE workshop, numerous industry interviews, and an overall assessment, the following conclusions are drawn about the potential for and technical viability of AWE to provide a significant source of energy in the United States:

- The AWE resource technical potential is similar in magnitude to the traditional wind potential in the United States; two or more times the total electricity consumption in the United States; and therefore, AWE could provide a significant contribution to the U.S. renewable energy supply.[4]
- AWE technologies are fundamentally new and different from traditional wind turbines because they have different modes of operation and design characteristics, with unique manufacturer supply chains, transport, and installation methods, as well as different operation, maintenance, and upgrade procedures.
- Because AWE is significantly different from traditional wind energy, it requires different pathways for developing policies and procedures for permitting airspace, grid integration, environmental and wildlife

[4] Assuming capacity densities of 3.98 MW/km^2, 500-kW short tether (250 m) of AWE and 3 MW/km^2, 5 MW, 120- m hub height for traditional wind. Technical potential for AWE and traditional wind occur in similar areas spatially and further investigation into where AWE may represent an increase in technical potential is recommended.

impacts, and social perception, including viewshed, noise, and general social acceptance.

- The development of AWE technologies faces unique challenges, and their resolution through R&D requires novel analysis methods, tools, equipment, facilities, and a workforce with new skills.
- Existing AWE system designs remain diverse, and the technology has not converged on configurations that satisfy the requirements of various markets. Several designs under development show promise, and the overall design space has not been fully explored.
- The path to an optimal AWE system design, manufacture, and operation is being investigated by universities and small, entrepreneurs in the United States, whereas AWE is a funded research program in the European Union (European Commission 2018) where most AWE advancements have occurred.
- The R&D challenges of AWE are similar in complexity to the challenges of marine and hydrokinetic technologies of DOE's Water Power Program, and commercializing AWE would likely require a similar or larger level of effort to make progress and rapidly accelerate toward commercial deployment and achieve significant impact.
- The technical complexity of developing successful AWE technologies has similar challenges to developing autonomous flight vehicles and may represent additional challenges related to tethered flight; nonetheless, autonomous flight vehicles are progressing toward commercialization and represent an enabling technology sector for AWE.

The following research, development, demonstration, and commercialization recommendations outline a 10-year program to examine and validate the technical and economic viability of AWE technologies.

Years 1 to 4: Initiate Fundamental Research and Craft a Plan for Commercialization

The following are steps to be taken to advance fundamental research and develop a commercialization plan for AWE:

- Develop a national AWE research, development, and flight research facility that is permitted, flight-certified, fully instrumented (including measurements of inflow turbulence at flight altitude) and staffed for evaluating industry-developed and national laboratory technologies. This research facility covers system characterization and validation at component, subsystem, and system levels, and on bench, rig, and in flight.
- Facilitate collaboration between stakeholders, permitting agencies, and researchers to develop policies and procedures to address airspace restrictions, radar impacts, device lighting, avian impact studies, and social perception to enable development and commercialization of AWE technologies in alignment with existing co-located industries and communities.
- Develop complete sets of functional requirements of AWE systems for the diverse applications and markets to fully capture the problem statements. These are to cover all life cycle stages; the complete range of stakeholder requirements and interactions with ecology, economy, and society; and are expressed in detailed technical specifications, cost, and performance goals; to successfully commercialize AWE technologies.
- Support national and regional cost and feasibility studies to evaluate key cost drivers; market potential; supply chain and workforce development; risk; installation and operational strategies; economic impact; social acceptance; and environmental issues of AWE systems.
- Develop holistic technology assessment methods and tools to identify potential technology showstoppers and critical areas of attention; quantify technology performance and readiness at all system levels and development stages; compare technologies; and drive the technology development along the most cost-, time-, and risk-effective development trajectories.
- Initiate fundamental research on critical issues on aerodynamics; controls; materials; siting; cost; performance and feasibility drivers; environmental issues; reliability; safety; system dynamics; controls; flight optimization; and social acceptance.
- Attract global expertise and technology developers through accelerator programs via operational and evaluation campaigns and fly-offs, as well as tool development and simulation competitions.

- Develop open-source, multifidelity simulation capabilities that model AWE systems in a suite of numerical tools, with a focus on design exploration, detailed design, and highly resolving AWE technology solutions with respect to physical response and dynamics, cost, and environmental and social impact. The open-source framework will provide a means of distributing fundamental knowledge throughout the industry, as well as allow for individual members to enhance proprietary tools through validation and cross-model comparison.
- Establish reference models (including controllers) for various AWE technologies and sizes to enable demonstration and verification of the simulation capability and to □□□establish baseline technology solutions from which to develop more advanced AWE designs.
- Support hardware development of next-generation AWE systems to advance the technology and improve the designs through proven inventive techniques.
- Perform scenario development, evaluation, and refinement of the most effective support and assistance mechanisms for technology development, market entry, commercialization, deployment, and rollout.
- Coordinate aspects of the AWE R&D agenda with the new IEA Wind Task on AWE, where appropriate, and align with international standards and guidelines.

Years 3 to 10: Evaluation Campaigns and Research to Accelerate Technology Development

The following are steps to be taken to accelerate AWE technology development:

- Engage industry by funding industry designs, supporting the use of the new simulation capabilities, and providing a research facility site with space, instrumentation, and expert personnel to get flying time on prototypes and precommercial systems.
- Analyze, document, and disseminate data from evaluation campaigns to validate simulation models and foster sector collaboration while protecting individual participants' intellectual property.

- Hold an AWE program review within 5 years, based on industry engagement, evaluation campaign results, and advanced simulation results. The review results will be used to chart the future course of the R&D program and the development of an out-year program plan for years 6 to 10.
- Establish an international working group—e.g., through the International Electrotechnical Commission—to develop international design requirements to ensure the integrity of promising AWE technologies. The purpose of these design requirements is to establish an appropriate level of safety against damage from all anticipated hazards during the planned lifetime of each AWE unit. Design requirements go together with the needs of the simulation capabilities and evaluation campaigns.

Years 6 to 10: Technology Advancement and Fundamental Research

To advance AWE technology, the following steps need to be taken:

- In accordance with the 6- to 10-year program plan, select the most promising concepts for long-term reliability evaluation at the national research facility to prove system durability. This phase of the program would require funding multiunit advanced systems for reliability evaluation.
- Fund development of demonstration projects in real early-adopter markets, such as remote communities, agriculture, microgrids, and progressive utilities, to gather operational data on market issues and required improvements.
- Continue fundamental research in support of industry on aerodynamics; controls; materials; siting; cost and feasibility drivers; environmental issues; reliability; safety; flight optimization; and social acceptance.

References

Ahrens, U., M. Diehl, R. Schmehl. 2014. *Airborne Wind Energy*. Springer.
Anderson, M. 2019. "Ready Flyer One: Airborne Wind Energy Simulations Guide the Leap to Satisfying Global Energy Demand." *IEEE Spectrum: Technology, Engineering, and*

Science News. https://spectrum.ieee.org/energywise/energy/renewables/ready-flyer-one-airborne-wind-energy-simulations-guide-the-leap-to-satisfying-global-energy-demand.

Archer, C. L., L. Delle Monache, D. L. Rife. 2014. "Airborne wind energy: Optimal locations and variability." *Renewable Energy* 64, 180–186. https://doi.org/10.1016/j.renene.2013.10.044.

Archer, C., K. Caldeira. 2009. "Global Assessment of High-Altitude Wind Power." *Energies*. https://doi.org/10.3390/en20200307.

Arent, D. J., A. Wise, R. Gelman. 2011. "The status and prospects of renewable energy for combating global warming." *Energy Economics*, Special Issue on The Economics of Technologies to Combat Global Warming 33, 584–593. https://doi.org/10.1016/j.eneco.2010.11.003.

Beaupoil, C. 2019. "Practical Experiences with a Torsion Based Rigid Blade Rotary Airborne Wind Energy System with Ground Based Power Generation." Airborne Wind Energy Conference 2019. https://repository.tudelft.nl/islandora/object/uuid:c50f7eb6-f116-4158-9e9b-aa1b05092ab2?collection=research.

Bechtle, P., and U. Zilmann. 2021. "High-Altitude Wind Energy Map Published." Airborne Wind Europe.https://airbornewindeurope.org/resources/high-altitude-wind-energy-map-published-2/. Accessed May 4, 2021.

Bechtle, P., M. Schelbergen, R. Schmehl, U. Zillmann, S. Watson. 2019. "Airborne wind energy resource analysis." *Renewable Energy* 141, 1103–1116. https://doi.org/10.1016/j.renene.2019.03.118.

Beiter, P., W. Musial, A. Smith, L. Kilcher, R. Damiani, M. Maness, S. Sirnivas, T. Stehly, V. Gevorgian, M. Mooney, G. Scott. 2016. A Spatial-Economic Cost-Reduction Pathway Analysis for U.S. Offshore Wind Energy Development from 2015-2030 (Technical Report). Golden, CO: National Renewable Energy Laboratory (NREL). NREL/TP-6A20-66579. https://www.nrel.gov/docs/fy16osti/66579.pdf.

Bormann, A., 2021. Private communication in stakeholder outreach with Enerkite.

Brabeck, S., 2021. Private communication in stakeholder outreach with SkySails.

Brown, A., P. Beiter, D. Heimiller, C. Davidson, P. Denholm, J. Melius, A. Lopez, D. Hettinger,

Bureau of Ocean Energy Management. 2021. "Vineyard Wind 1 Offshore Wind Energy Project Final Environmental Impact Statement Volume 1 (OCS EIS/EIA BOEM 2021-0012)." https://www.boem.gov/sites/default/files/documents/renewable-energy/state-activities/Vineyard-Wind-1-FEIS-Volume-1.pdf.

Calverley, G., 2021. Private communication in stakeholder outreach with Abound.

Cherubini, A., A. Papini, R. Vertechy, M. FontanaS. . 2015. "Airborne Wind Energy Systems: A review of the technologies." *Renewable and Sustainable Energy Reviews* 51, 1461–1476. https://doi.org/10.1016/j.rser.2015.07.053.

Craig, M. T., I. L. Carreño, M. Rossol, B.-M. Hodge, C. Brancucci. 2019. "Effects on power system operations of potential changes in wind and solar generation potential under climate change." *Environ. Res. Lett.* 14, 034014. https://doi.org/10.1088/1748-9326/aaf93b.

Craig, M. T., S. Cohen, J. Macknick, C. Draxl, O. J. Guerra, M. Sengupta, S. E. Haupt, B.-M. Hodge, C. Brancucci. 2018. "A review of the potential impacts of climate change on bulk power system planning and operations in the United States." *Renewable and Sustainable Energy Reviews* 98, 255–267. https://doi.org/10.1016/j.rser.2018.09.022.

Creighton, R., 2021. Private communication in stakeholder outreach with Windlift.

Dalla Longa, T. Kober, J. Badger, P. Volker, C. Hoyer-Klick, I. Hidalgo Gonzalez, H. Medarac, et al. 2018. Wind potentials for EU and neighbouring countries: Input datasets for the JRC-EU-TIMES Model. Publications Office, LU. https://publications.jrc.ec.europa.eu/repository/handle/JRC109698.

De Lellis, M., A. K. Mendonça, R. Saraiva, A. Trofino,Á Lezana. 2016. "Electric power generation in wind farms with pumping kites: An economical analysis." *Renewable Energy* 86, 163–172. https://doi.org/10.1016/j.renene.2015.08.002.

De Lellis, M., R. Reginatto, R. Saraiva, A. Trofino. 2018. "The Betz limit applied to Airborne Wind Energy." *Renewable Energy* 127, 32–40. https://doi.org/10.1016/j.renene.2018.04.034.

De Rose, A., M. Buna, C. Strazza, N. Olivieri, T. Stevens, L. Peeters, D. Tawil-Jamault. 2017. *Technology Readiness Level: Guidance Principles for Renewable Energy Technologies*. European Commission: Petten, The Netherlands. https://op.europa.eu/en/publication-detail/-/publication/d5d8e9c8-e6d3-11e7-9749-01aa75ed71a1.

De Schutter, J., R. Leuthold, T. Bronnenmeyer, R. Paelinck, M. Diehl. 2019a. "Towards a Modular Upscaling Strategy for Utility-Scale Airborne Wind Energy." https://repository.tudelft.nl/islandora/object/uuid%3Ac7970efe-b007-4d26-a00f-14c0cdba0ba1.

De Schutter, J., R. Leuthold, T. Bronnenmeyer, R. Paelinck, M. Diehl. 2019b. "Optimal control of stacked multi-kite systems for utility-scale airborne wind energy." https://doi.org/10.1109/CDC40024.2019.9030026.

Deutsche WindGuard GmbH. 2018. *Capacity Densities of European Offshore Wind Farms*. Bundesamt fur Seeschifffahrt und Hydrographie. https://vasab.org/wp-content/uploads/2018/06/BalticLINes_CapacityDensityStudy_June2018-1.pdf.

Draxl, C., A. Clifton, B.-M. Hodge, J. McCaa. 2015. "The Wind Integration National Dataset (WIND) Toolkit." *Applied Energy* 151, 355–366. https://doi.org/10.1016/j.apenergy.2015.03.121.

Echeverri, P., T. Fricke, G. Homsy, N. Tucker. 2020. *The Energy Kite: Selected Results from the Design, Development, and Testing of Makani's Airborne Wind Turbines*. Makani Technologies, LLC. http://www.energykitesystems.net/AWEC2020Teleconference/PostersAWEC2020/AWEC2020POSTER023DocumentMakaniPowerLLC.pdf.

Eijkelhof, D., S. Rapp, U. Fasel, M. Gaunaa, R. Schmehl. 2020. "Reference Design and Simulation Framework of a Multi-Megawatt Airborne Wind Energy System." *J. Phys.: Conf. Ser.* 1618, 032020. https://doi.org/10.1088/1742-6596/1618/3/032020.

European Commission., 2018. Study on challenges in the commercialisation of airborne wind energy systems. https://op.europa.eu/en/publication-detail/-/publication/a874f843-c137-11e8-9893-01aa75ed71a1/language-en.

Faggiani, P., R. Schmehl. 2018. Design and Economics of a Pumping Kite Wind Park, in: Schmehl, R. (Ed.), *Airborne Wind Energy: Advances in Technology Development and Research*. Springer, Singapore, pp. 391–411. https://doi.org/10.1007/978-981-10-1947-0_16.

Fagiano, L. 2019. "Automation Challenges in Airborne Wind Energy Systems and the Role of Academic Research" Airborne Wind Energy Conference 2019, https://www.youtube.com/watch?v=ZptcYca_JBs.

Firestone, J., D. Bidwell, M. Gardner, L. Knapp. 2018. "Wind in the sails or choppy seas? People-place relations, aesthetics and public support for the United States' first offshore wind project." *Energy Research & Social Science* 40, 232–243. https://doi.org/10.1016/j.erss.2018.02.017.

Fitzgerald, J., B. Bolund. 2012. Technology Readiness for Wave Energy Projects; ESB and Vattenfall classification system, in: *Proceedings of the 4th International Conference on Ocean Energy*, Dublin, Ireland. pp. 17–19. https://www.icoe-conference.com/publication/technology_readiness_for_wave_energy_projects_esb_and_vattenfa ll_classification_system/.

Gambier, A. 2014. *Projekt OnKites: Untersuchung zu den Potentialen von Flugwindenergieanlagen (FWEA); Abschlussbericht.* Technische Informationsbibliothek u. Universitätsbibliothek, Bremerhaven. https://doi.org/10.2314/GBV:81573428X.

Haas, T., J. Meyers. 2017. *Comparison study between wind turbine and power kite wakes. J. Phys.: Conf. Ser.* 854, 012019. https://doi.org/10.1088/1742-6596/854/1/012019.

Håland, A. 2018. *Testing of Kitemill's Airborne Wind Energy System at Lista, Norway. Assessing the Impacts on birds. A pilot study.* https://airbornewindeurope.org/wp-content/uploads/2020/10/NNI-Report-520-2018-Testing-of-Kitemill-AWES-at-Lista_Norway_-Assessment-of-impacts-on-birds_A-pilot-study_-December-2018.pdf.

Harklau, T., 2021. Private communication in stakeholder outreach with Ampyx Power.

Harrison-Atlas, D., A. Lopez, E. Lantz. forthcoming. National land use implications of an expanding wind energy portfolio.

Haupt, S. E., J. Copeland, W. Y. Y. Cheng, Y. Zhang, C. Ammann, P. Sullivan. 2016. "A Method to Assess the Wind and Solar Resource and to Quantify Interannual Variability over the United States under Current and Projected Future Climate." *Journal of Applied Meteorology and Climatology* 55, 345–363. https://doi.org/10.1175/JAMC-D-15-0011.1.

Hayek, N. E. 2020. Airborne Wind Energy: Challenges and Opportunities. Proceedings of IEA Wind, Topical Expert Meeting #102. IEA Wind Task 11. September 23-24, 2020. Online Meeting.

Heilmann, J. 2012. Technical and Economic Potential of Airborne Wind energy M.Sc. Thesis, Utrecht University. http://dspace.library.uu.nl/handle/1874/258716.

Ippolito, M., 2021. Private communication in stakeholder outreach with KiteGen. Isensee, M. 2021. Private communication in stakeholder outreach with KiteKraft.

Jonkman, J., S. Butterfield, W. Musial, G. Scott. 2009. Definition of a 5-MW Reference Wind Turbine for Offshore System Development (Technical Report). Golden, CO: National Renewable Energy Laboratory (NREL). NREL/TP-500-38060. https://www.nrel.gov/docs/fy09osti/38060.pdf.

Junginger, M., and A. Louwen. 2019. *Technological Learning in the Transition to a Low-Carbon Energy System: Conceptual Issues, Empirical Findings, and Use, in Energy Modeling*. Academic Press.

Karnauskas, K. B., J. K. Lundquist, L. Zhang. 2018. "Southward shift of the global wind energy resource under high carbon dioxide emissions." *Nature Geoscience* 11, 38–43. https://doi.org/10.1038/s41561-017-0029-9.

Kaufman-Martin, S., N. Naclerio, P. May, P. Luzzatto-Fegiz. 2021. An entrainment-based model for annular wakes, with applications to airborne wind energy.

Kiteswarms. "Developing the future of airborne wind energy" https://www.youtube.com/watch?v=URyacqV5xI0.

Kruijff, M., 2021. Private communication in stakeholder outreach with Ampyx Power.

Kruijff, M., and R. Ruiterkamp. 2018. "A Roadmap Towards Airborne Wind Energy in the Utility Sector" in: *Airborne Wind Energy; Advances in Technology Development and Research*. pp. 643–662. https://doi.org/10.1007/978-981-10-1947-0_26.

Leuthold, R., J. De Schutter, E. C. Malz, G. Licitra, S. Gros, M. Diehl, M. 2018. "Operational Regions of a Multi-Kite AWE System" in 2018 European Control Conference (ECC). Presented at the 2018 European Control Conference (ECC), pp. 52–57. https://doi.org/10.23919/ECC.2018.8550199.

Leuthold, R., S. Gros, M. Diehl. 2017. "Induction in Optimal Control of Multiple-Kite Airborne Wind Energy Systems." IFAC-PapersOnLine, 20th IFAC World Congress 50, 153–158. https://doi.org/10.1016/j.ifacol.2017.08.026.

Li, Y., E. Kalnay, S. Motesharrei, J. Rivas, F. Kucharski, D. Kirk-Davidoff, E. Bach, N. Zeng. 2018. "Climate model shows large-scale wind and solar farms in the Sahara increase rain and vegetation." *Science* 361, 1019–1022. https://doi.org/10.1126/science.aar5629.

Lopez, A., T. Mai, E. Lantz, D. Harrison-Atlas, T. Williams, G. Maclaurin. 2021. "Land use and turbine technology influences on wind potential in the United States." *Energy* 223, 120044. https://doi.org/10.1016/j.energy.2021.120044.

Losada Carreño, I., M. T. Craig, M. Rossol, M. Ashfaq, F. Batibeniz, S. E. Haupt, C. Draxl, B.-M. Hodge, C. Brancucci. 2020. "Potential impacts of climate change on wind and solar electricity generation in Texas." *Climatic Change* 163, 745–766. https://doi.org/10.1007/s10584-020-02891-3.

Loyd, M. L. 1980. "Crosswind kite power (for large-scale wind power production)." *Journal of Energy* 4, 106–111. https://doi.org/10.2514/3.48021.

Luchsinger, R. H. 2013. Pumping Cycle Kite Power, in: Ahrens, U., M. Diehl, R. Schmehl, eds. *Airborne Wind Energy*. pp. 47-64. https://doi.org/10.1007/978-3-642-39965-7.

Lunney, E., M. Ban, N. Duic, N. A. Foley. 2017. "A state-of-the-art review and feasibility analysis of high-altitude wind power in Northern Ireland." *Renewable and Sustainable Energy Reviews* 68, 899–911. https://doi.org/10.1016/j.rser.2016.08.014.

Maclaurin, G. J., N. Grue, A. Lopez, D. Heimiller. 2019. *The Renewable Energy Potential (reV) Model: A Geospatial Platform for Technical Potential and Supply Curve Modeling* (Technical Report). Golden, CO: National Renewable Energy Laboratory (NREL). NREL/TP-6A20-73067. https://www.nrel.gov/docs/fy19osti/73067.pdf.

Malz, E. 2020. "Airborne Wind Energy - To fly or not to fly?" PhD thesis. Chalmers University of Technology. https://research.chalmers.se/en/publication/518841.

Marsh, R., 2021. Private communication.

Marvel, K., B. Kravitz, K. Caldeira. 2013. Geophysical limits to global wind power. *Nature Climate Change* 3, 118–121. https://doi.org/10.1038/nclimate1683.

Miller, L. M., D. W. Keith. 2018. "Climatic impacts of wind power" Joule 2, 2618–2632. https://doi.org/10.1016/j.joule.2018.09.009.

Miller, L. M., F. Gans, A. Kleidon. 2011. "Jet stream wind power as a renewable energy resource: little power, big impacts." *Earth System Dynamics* 2, 201–212. https://doi.org/10.5194/esd-2-201-2011.

Mulcahy, D., G. Porro. 2016. *Estimating Renewable Energy Economic Potential in the United States. Methodology and Initial Results* (Technical Report). Golden, CO: National Renewable Energy Laboratory (NREL). NREL/TP-6A20-64503. https://www.nrel.gov/docs/fy15osti/64503.pdf.

Musial, W., D. Heimiller, P. Beiter, G. Scott, C. Draxl. 2016. *2016 Offshore Wind Energy Resource Assessment for the United States* (Technical Report). Golden, CO: National Renewable Energy Laboratory (NREL). NREL/TP-5000-66599. https://www.nrel.gov/docs/fy16osti/66599.pdf.

Musial, W., P. Beiter, P. Spitsen, J. Nunemaker, V. Gevorgian, A. Cooperman, R. Hammond, M. Shields. 2020. *2019 Offshore Wind Technology Data Update* (Technical Report). Golden, CO: National Renewable Energy Laboratory (NREL). NREL/TP-5000-77411. https://www.nrel.gov/docs/fy21osti/77411.pdf.

National Renewable Energy Laboratory. 2020. "Annual Technology Baseline." https://atb.nrel.gov/.

National Wind Coordinating Collaborative. 2010. *Wind Turbine Interactions with Birds, Bats, and their Habitats: A Summary of Research Results and Priority Questions*. https://www1.eere.energy.gov/wind/pdfs/birds_and_bats_fact_sheet.pdf.

Optis, M., A. Rybchuk, N. Bodini, M. Rossol, W. Musial. 2020. *2020 Offshore Wind Resource Assessment for the California Pacific Outer Continental Shelf* (Technical Report). Golden, CO: National Renewable Energy Laboratory (NREL). NREL/TP-5000-77642. https://www.nrel.gov/docs/fy21osti/77642.pdf.

Peschel, J. 2019. "Kitepower and the Journey Towards 24/7 Operation." Airborne Wind Energy Conference 201, http://resolver.tudelft.nl/uuid:5bea59d6-be63-46c6-aba5-d60967d7783b.

Peschel, J. 2021. Private communication in stakeholder outreach with KitePower.

Petrick et al. 2021. IEA Wind AWE task proposal. Implementing Agreement for Co-operation in the Research and Development of Wind Energy Systems (IEA Wind TCP). Internal distribution only.

Petrick, K. 2018. "Policies for Airborne Wind Energy - Preparing the grounds for AWE-specific incentive schemes - Scoping Study." Airborne Wind Europe. https://airbornewindeurope.org/policies-for-airborne-wind-energy-scoping-study-now-in-the-eu-open-access-repository/.

Pryor, S. C., R. J. Barthelmie, T. J. Shepherd. 2020. "20% of US electricity from wind will have limited impacts on system efficiency and regional climate." *Scientific Reports* 10, 541. https://doi.org/10.1038/s41598-019-57371-1.

Read, R. 2018. Kite Networks for Harvesting Wind Energy, in: Schmehl, R. (Ed.), *Airborne Wind Energy, Advances in Technology Development and Research*. Springer Singapore, Singapore, pp. 515–537. https://doi.org/10.1007/978-981-10-1947-0_21.

Remington, T. E., Deibert, P.A., Hanser, S. E., Davis, D. M., Robb, L. A., Welty, J. L., 2021. Sagebrush Conservation Strategy–Challenges to Sagebrush Conservation (Open-File Report No. 2020–1125). U.S. Geological Survey, Reston, VA. https://pubs.er.usgs.gov/publication/ofr20201125.

Sanchez, R. 2011. *Technology Readiness Assessment Guide*. U.S. Department of Energy, Washington, DC. https://www.directives.doe.gov/directives-documents/400-series/0413.3-EGuide-04a.

Santos, D., 2021. Private communication in stakeholder outreach with kPower.

Schaefer, D., 2021. Private communication in stakeholder outreach with eWindSolutions.

Schmehl, R. (Ed.). 2018. *Airborne Wind Energy: Advances in Technology Development and Research*. Springer Singapore. https://doi.org/10.1007/978-981-10-1947-0.

Schmehl, R. 2020. "Airborne Wind Energy" PowerWeb webinar lecture https://airbornewindeurope.org/wp-content/uploads/2020/06/20200520-Powerweb.pdf.

Schmehl, R. 2021. Private communication in stakeholder outreach with the Delft University of Technology.

Shields, M., P. Beiter, W. Kleiber. 2021. "Spatial impacts of technological innovations on the levelized cost of energy for offshore wind power plants in the United States." *Sustainable Energy Technologies and Assessments* 45, 101059. https://doi.org/10.1016/j.seta.2021.101059.

Shifman, R. *Airborne Wind Energy: Waiting For Take-Off*, 2019, 19 pages, Bloomberg NEF, Bloomberg Finance L.P.

Stehly, T., P. Beiter, P. Duffy. 2020. *2019 Cost of Wind Energy Review* (Technical Report). Golden, CO: National Renewable Energy Laboratory (NREL). NREL/TP-5000-78471. https://www.nrel.gov/docs/fy21osti/78471.pdf.

The Dutch Offshore Wind Atlas. 2021. The ECN part of TNO, Whiffle, KNMI, and the Ministry of Economic Affairs and Climate Policy. https://www.dutchoffshorewindatlas.nl/.

Tulloch, O., A. Amiri Kazemi, H. Yue, J. Feuchtwang, R. Read. 2019. "Modeling Studies on Tensile Rotary Power Transmission for Airborne Wind Energy Systems." https://repository.tudelft.nl/islandora/object/uuid:fa2a20d6-51d7-4f85-9f64-65b6df1ca3e0?collection=research.

Vautard, R., F. Thais, I. Tobin, F.-M. Bréon, J.-G.-D. de Lavergne, A. Colette, P. Yiou, P. M. Ruti. 2014. "Regional climate model simulations indicate limited climatic impacts by operational and planned European wind farms." *Nature Communications* 5, 3196. https://doi.org/10.1038/ncomms4196.

Vermillion, C., Cobb, M., Fagiano, L., Leuthold, R., Diehl, M., Smith, R., Wood, T., Rapp, S., Schmehl, R., Olinger, D., Demetriou, M., 2021. "Electricity in the air: Insights from two decades of advanced control research and experimental flight testing of airborne wind energy systems," Annual Reviews in Control, https://doi.org/10.1016/j.arcontrol.2021.03.002.

Vimalakanthan, K., M. Caboni, J. G. Schepers, E. Pechenik, P. Williams. 2018. "Aerodynamic analysis of Ampyx's airborne wind energy system." *J. Phys.: Conf. Ser.* 1037, 062008. https://doi.org/10.1088/1742-6596/1037/6/062008.

Vineyard Wind. 2020. *Draft Construction and Operations Plan (Construction and Operations Plan)*. Vineyard Wind LLC. https://www.boem.gov/sites/default/files/documents/renewable-energy/Vineyard%20Wind%20COP%20Volume%20I_Complete.pdf.

Weber, J. 2019. "AirborneMax – Scaling as the Key Issue for Airborne Wind." Airborne Wind Energy Conference 2019. https://repository.tudelft.nl/islandora/object/uuid:ffd965bc-39ed-41cb-8d91-46504f9aef12.

Weber, J., 2012. "WEC Technology Readiness and Performance Matrix–finding the best research technology development trajectory" in: *Proceedings of the 4th International Conference on Ocean Energy*, Dublin, Ireland. https://www.researchgate.net/publication/233810908_WEC_Technology_Readiness_and_Perfor mance_Matrix_-_finding_the_best_research_technology_development_trajectory.

Weber, J., Laird, D., Costello, R., Roberts, J., Bull, D., Babarit, A., Nielsen, K., Bittencourt Ferreira, C., Kennedy, B. 2017. "Cost, Time, and Risk Assessment of Different Wave Energy Converter Technology Development Trajectories," in Proc. Eleventh European Wave and Tidal Energy Conference.

Weber, J., M. Marquis, A. Lemke, C. Draxl, A. Lopez, O. Roberts, A. Cooperman, M. Shields. 2021. *Proceedings of the 2021 Airborne Wind Energy Workshop*. Golden, CO: National Renewable Energy Laboratory (NREL). NREL/TP-5000-80017. nrel.gov/docs/fy21osti/80017.pdf.

Weber, J., R. Costello, J. Ringwood. 2013. "WEC Technology Performance Levels (TPLs)-Metric for Successful Development of Economic WEC Technology." *Proceedings EWTEC 2013*. https://www.researchgate.net/publication/326986433_WEC_Technology_Performance_Levels_TPLs_-_Metric_for_Successful_Development_of_Economic_WEC_Technology.

Wiser, R., M. Bolinger, B. Hoen, D. Millstein, J. Rand, J., G. Barbose, N. Darghouth, W. Gorman, S. Jeong, A. Mills, B. Paulos. 2020. *Wind Energy Technology Data Update: 2020 Edition*. Lawrence Berkeley National Laboratory, Berkley, CA. https://emp.lbl.gov/sites/default/files/2020_wind_energy_technology_data_update.pdf.

Wu, G. C., E. Leslie, O. Sawyerr, D. R. Cameron, E. Brand, B. Cohen, D. Allen, M. Ochoa, A. Olson. 2020. "Low-impact land use pathways to deep decarbonization of electricity." *Environ. Res. Lett.* 15, 074044. https://doi.org/10.1088/1748-9326/ab87d1.

Yip, C. M. A., U. B. Gunturu, G. L. Stenchikov. 2017. "High-altitude wind resources in the Middle East." *Scientific Reports* 7, 9885. https://doi.org/10.1038/s41598-017-10130-6.

Zerweckh, S. 2021. Private communication.

Zillmann, U. 2020. Introduction to Airborne Wind Energy. Airborne Wind Europe. https://airbornewindeurope.org/wp-content/uploads/2020/07/AWEurope_Intro-Airborne-Wind-Energy_2020-03-23.pdf.

Zywietz, D. 2019. What Will it Take for AWE to be Successful in Remote & Mini-grid Applications? Airborne Wind Energy Conference 2019. http://resolver.tudelft.nl/uuid:dbe7de7e-151d-483e-9658-0b4f9f26d26a.

Chapter 2

Wind and Solar Hybrid Power Plants for Energy Resilience*

Caitlyn E. Clark
Aaron Barker
Jennifer King
and James Reilly
National Renewable Energy Laboratory

Notice

This work was authored by the National Renewable Energy Laboratory, operated by Alliance for Sustainable Energy, LLC, for the U.S. Department of Energy (DOE) under Contract No. DE-AC36-08GO28308. Funding provided by the U.S. Department of Energy Office of Energy Efficiency and Renewable Energy Wind Energy Technologies Office. The views expressed herein do not necessarily represent the views of the DOE or the U.S. Government.

This report is available at no cost from the National Renewable Energy Laboratory (NREL) at www.nrel.gov/publications.

U.S. Department of Energy (DOE) reports produced after 1991 and a growing number of pre-1991 documents are available free via www.OSTI.gov.

---

* This is an edited, reformatted and augmented version of Technical Report of the national laboratory of the U.S. Department of Energy, Office of Energy Efficiency & Renewable Energy, Publication No. NREL/TP-5R00-80415, dated January 2022.

In: Wind Energy
Editor: Mehmet Landrey
ISBN: 979-8-89113-528-4

Acknowledgments

We are thankful to all project team members from partnering laboratories on the Microgrids, Infrastructure Resilience, and Advanced Controls Launchpad (MIRACL) project:

- Idaho National Laboratory
- Pacific Northwest National Laboratory
- Sandia National Laboratories.

We also express our sincere gratitude to our industry advisory board members for their valuable insights and recommendations during the March 2020 advisory board meeting: Venkat Banunarayanan (National Rural Electric Cooperative Association), Chris Rose (Renewable Energy Alaska), Rob Wills (Intergrid), Paul Dockrill (Natural Resource Canada), Jeff Pack (POWER Engineers), Arvind Tiwari (GE Global Research), Kristin Swenson (Midcontinent Independent System Operator), Jonathon Monken (PJM), and Scott Fouts (QED Wind Power). In addition, we would like to thank Robert Preus, Dan Olis, Megan Culler, Craig Rieger, Andrew Reiman, Brian Smith, and Paul Veers for their thoughtful review, and Amy Brice and Sheri Anstedt for editorial support. Lastly, we would like to thank Paul Stackhouse and Bradley Macpherson for their assistance in using the NASA POWER tool for our Alaska analysis.

Executive Summary

Wind-solar-storage hybrid power plants represent a significant and growing share of new proposed projects in the United States. Their uptake is supported by increasing renewable energy market share, enhanced technical abilities for dispatch and control, and decreasing costs for wind energy, solar energy, and battery storage. Simultaneously, there is also increased use of generation and storage resources in distributed power systems. The diversification of energy resources through hybridization or spatial distribution provides an opportunity to enhance power system resilience (compared to single-source generation), addressing growing concerns about the reliability of the aging, transforming U.S. electric grid. The question of where to build hybrid plants for resilience value—rather than for bulk power supply—has not been fully explored in previous studies. Therefore, in this study we complete a national complementarity analysis to identify areas in the United States that are particularly suited for wind-solar hybrid power plant development. The authors show the importance of seasonal and diurnal patterns in assessing complementarity and identify that regions in the Great Plains, Midwest, and Southeast are particularly suited for hybrid power plants. We demonstrate the resilience value of hybridization for a reference system based near Memphis, Tennessee, and show optimal sizing of wind, solar, and storage assets given 1.0 and 0.9 critical load factors. Our results indicate that the pairing of wind and solar assets better meets constant load demand and reduces storage requirements compared to using only solar assets. These results will enable future work to integrate complementarity metrics with resilience frameworks. The results also indicate a need for more finely resolved data for local resources, demand, and hazards.

1. Introduction

The electric power industry is rapidly transforming, pushed by aging infrastructure, evolving customer needs, and environmental impact considerations, and pulled by technical advancements, favorable economic conditions, and supportive policy. Utilities seek to modernize grid infrastructure to enhance resilience and operational efficiency and integrate higher levels of distributed energy resources (both generation and storage) (U.S. Department of Energy 2017). Meanwhile, consumer control over energy costs and sources is increasing. Despite this rapid modernization, providing safe, reliable, and affordable electricity to critical infrastructure systems and customers remains a paramount priority (U.S. Department of Energy 2017; Obama 2013). In electric systems, "reliability" means that sufficient, uninterrupted electricity is provided to all customers in nominal operating conditions, or conditions for which that system was designed. For instance, one standard defines electric system reliability as the average likelihood of disconnecting customers to be no more than 1 day in 10 years. Electric systems, however, face threats that exceed nominal operating conditions, such as short-term natural hazards like wildfires, floods, and other extreme weather events, and human-made threats like physical attacks or cyberattacks. The ability of the power system to resist, respond to, and recover from these low-probability, high-impact extreme events defines the resilience of the system.

As grid investments are made, solutions to improve efficiency and grid resilience should be considered, such as diversifying renewable generation sources (by combining wind turbines and solar photovoltaics through hybrid power plants), distributing generation throughout the grid, enhancing grid flexibility by adding storage, and implementing advanced controls.

Diverse, distributed generation and storage is inherently more resilient than a single, centralized generator. If a disruptive event occurs that compromises transmission or distribution lines, limits remote generation capacity, or causes cascading failures, the local availability of distributed power sources enables the system to better handle the disturbance. Factors like market expectations, policy priorities, and regulations (Orrell, Homer, and Tang 2018) make it difficult to assign a value to this enhanced resilience, but valuation is critical for stakeholders to compare distributed hybrid power plants to other grid infrastructure investments in resilience planning. This is especially true for hybrid power plants that combine solar, wind, and storage, given that regulation and market participation of these plants is uncertain (Gorman et al., 2020). Often, large wind- or solar-only power plants are sited based on the strongest available resource, which is not necessarily an optimal method for maximizing hybrid plant resilience benefits. Instead, hybrid power plant siting, especially when considering resilience and efficiency objectives, should consider 1) where the renewable resources are complementary (or noncoincident), 2) where the generated energy can be used locally, and 3) where batteries can provide the most grid benefits (which depends not just on battery placement but also on the ability to participate in capacity markets). To optimally site hybrid power plants and understand their resilience value to the grid, we need to first identify regions where renewable resources are complementary and develop metrics that account for their hybrid and distributed-asset resilience value.

The purpose of this study is to develop a method to identify areas in the United States where wind and solar hybrid plants can provide enhanced resilience value to distributed grid systems via complementary resources. We achieve this through two objectives. First, we aim to determine appropriate resilience metrics that can be used to measure resilience benefits of hybrid and distributed resources at the local or grid level. Second, we aim to identify where those benefits can be best realized from a national resource analysis. To achieve these objectives, we identify regions in the United States where wind and solar resources are complementary and thus are better at providing resilience benefits. Then, we analyze a reference system near Memphis, Tennessee, to investigate how these hybrid plants and their generation would be integrated and dispatched to support increased system resilience. This study is part of a multiyear effort across laboratories aimed at identifying the challenges and opportunities in valuing and integrating distributed wind turbines with solar photovoltaics and energy storage systems for resilience (Reilly 2020).

Section 2 describes relevant complementarity and resilience metrics that account for the resilience value of hybrid or distributed renewable energy assets. The methods used for resource complementarity quantification and reference system analysis are described in Section 3. Section 4 discusses the results of the resource complementarity mapping and describes the resilience value within the selected reference system. Section 5 highlights the main findings of this study and describes directions for future work to realize the resilience benefits of hybrid power plants at a distributed scale.

2. Literature Review

By hybridizing or spatially distributing renewable energy generation assets, the complementarity of the distinct resources can be leveraged to provide energy and grid services more reliably than any of the assets can on their own. Complementarity can be achieved by co-locating generation assets or by creating virtual power plants (coordinating distributed, electrically connected assets). Previous studies have assessed complementarity as a means of overcoming the historical reliability shortcomings of renewable energy generation in transmission systems.

Although similar concepts, reliability and resilience are different in two distinct ways. First, reliability is an outcome (for example, maintaining sufficient electricity to all customers), whereas resilience is a characteristic of the system that enables it to achieve that outcome (for example, the ability of a system to return power to customers after a blackout). Second, reliability refers to an outcome in nominal conditions, or conditions for which the system was designed, and resilience is a characteristic of the system in extreme conditions, or conditions that exceed design expectations. Thus, metrics that measure the reliability of a system can also measure the effectiveness of a resilient system, contingent on the presence of a disruptive event. For instance, loss-of-load probability is a common metric used in reliability standards. When applied in the context of resilience, loss-of-load probability implies a system's ability to resist a disruption and provide reliable electric service.

Previous studies have defined reliability metrics and applied them in a resilience context. Few studies, however, have considered complementarity as a means of improving resilience in distributed systems. Leveraging resource complementarity could provide systems with a greater ability to resist, respond to, or recover from load loss during an extreme event, thus providing resilience value to the system. This is particularly important in smaller systems with fewer distributed generation assets and therefore less redundancy. Metrics are needed that 1) quantify resource complementarity to identify opportunities for hybrid power plant development and 2) can be used in conjunction with resilience frameworks to compare distributed hybrid power plants with other grid resilience investments. In this section, we determine the most appropriate metrics for quantifying the resilience value of distributed and hybrid generation assets to be used in hybrid plant design at the local or distributed grid scale.

2.1. Complementarity Metrics

Previous studies that analyze complementarity often focus on siting and sizing hybrid plants based on hourly or subhourly time series data. These studies frequently measure complementarity via statistical correlation metrics or reliability metrics (Jurasz et al., 2020). Correlation metrics provide insight to resource dependence and are best used to assess hybrid potential from a resource perspective when paired with absolute resource availability. Reliability metrics directly measure how adequate resource complementarity and availability are to meet a load. Although reliability metrics provide a direct way to measure the outcome of a distributed hybrid system's resilience, they require system-specific load data and thus are not suitable for regional or national complementarity analyses.

Pearson's correlation coefficient, r, is the most popular statistical metric used in complementarity analysis. It is used to 1) measure the potential for a hybrid system to efficiently provide generation capacity in renewable energy systems (Li, Agelidis, and Shrivastava 2009; Liu et al., 2013; Miglietta, Huld, and Monforti-Ferrario 2016; Vega- Sanchez et al., 2017; Shaner et al., 2018; Slusarewicz and Cohan 2018), 2) optimize hybrid systems design (Jurasz, Beluco, and Canales 2018a), and 3) optimize dispatch and operations of existing solar and wind generation assets on a grid (Aza-Gnandji et al., 2018; Naeem et al., 2019). Other metrics are used for similar applications, including the Kendall rank correlation coefficient, τ (Xu, Wang, and Liu 2017; Ren et al., 2019), cross-correlation (Li, Agelidis, and Shrivastava 2009; Dos Anjos et al., 2015), and author-defined metrics.

Metrics that relate complementarity to hybrid or distributed system reliability often measure loss of load, power fluctuations, or power smoothing. Loss-of-load probability has been used to assess enhanced system reliability from distributed wind assets (Kahn 1979), complementary solar and wind resources (Jurasz, Beluco, and Canales 2018b), and distributed wind and solar resources (Shaner et al., 2018). Schmidt, Cancella, and Pereira applied this loss-of-load probability metric as a constraint to optimize the size and composition of solar, wind, and hydropower resources to meet demand (Schmidt, Cancella, and Pereira 2016). Electrical output fluctuations have also been used as a reliability metric for complementary wind and solar resources (Gburcik et al., 2006) and distributed photovoltaics plants (Murata, Yamaguchi, and Otani 2009; Marcos et al., 2012). Widén considered output fluctuation reduction from wind and solar energy generation under minor disturbance events, such as clouds passing over solar arrays (Widén 2015). Ramp rate, or how quickly power output changes over time, is also used as a complementarity metric. Ramp rate is used to enable supply and demand balancing, prevent grid effects from rapid loading or discharge, and prevent the draining of system reserves (Tarroja, Mueller, and Samuelsen 2013; Widén 2011; Zhang et al., 2018). The ability of a hybrid or distributed system to reduce fluctuations or ramp rate could be used to measure the added resilience value from grid investments if scaled to consider extreme events.

Power smoothing has also been used in complementarity analyses to measure the added value of assets to system reliability. Power smoothing leverages the variations in generation output across spatially distributed generation assets to provide adequate power during minor disruptive events, such as periods of low wind power production. Previous studies leveraged national or regional analyses to attain maximum smoothing. Small distribution systems may not have the same potential for power smoothing given their geographical extent, but if a system has little load, it could be possible to leverage even small amounts of power smoothing through distribution to achieve greater system resilience. In a resilience context, where low-probability events could cause a proportionally large loss in generation capacity, the ability to ride through events becomes demand-dependent (i.e., what is the level of demand, and how much of it is critical). The concept of power production during critical time windows, or events with lower-than-average capacity factors, is proposed by Berger et al. They consider spatiotemporal complementarity, measuring the time during which renewable energy generation passes below a particular threshold during events of low wind power production (Berger et al., 2020). Sterl et al., extend this metric to the power smoothing value of hybrid power plants. They propose a stability coefficient that measures the added value of using additional solar and wind energy capacity to balance daily power output (Sterl et al., 2018).

2.2. Measuring Resilience

Efforts to measure distributed power system resilience often rely on reliability metrics but apply them through different phases of a disturbance event to measure how well a system resists, responds to, recovers from, or restores after a disturbance. Resilience frameworks are used in some instances to distinguish the phases of the disturbance event. While an extensive body of research exists for resilience framework development and validation, this review focuses on resilience metrics that can be used regardless of the framework employed.

The ability of power systems to supply load (similar to the loss-of-load probability metric) is often used as the metric to quantify resilience in electricity systems. McJunkin and Rieger define metrics for adaptive capacity (the available active and reactive power at a node that could be injected if necessary in response to a disturbance) and insufficiency (the inability of that node to supply active and reactive power when necessary) (McJunkin and Rieger 2019). They also define "resilience threshold," which delineates when a power system transitions between performing satisfactorily and unsatisfactorily given a disturbance and mirrors the threshold concept proposed by Berger et al., (Berger et al., 2020). Cicilio et al., incorporate uncertainty from hazard probability and electrical component performance into their resilience evaluation, using "energy not served" as their metric, while also introducing a resilience framework to compare resilience improvements and derate a system's resilience over time. This methodology accounts for the resilience stages of the event originally defined by Tierney and Bruneau (Tierney and Bruneau 2007) and the resilience threshold under various hardening improvements. Cicilio et al., use "energy not served" as the performance metric to evaluate overall system resilience (Cicilio et al., 2020).

In microgrid system resilience quantification, metrics frequently go beyond load-not-served metrics, such as adaptive capacity and energy not served, to encompass metrics that measure the self-sufficiency of the system, the ability of the system to persist through a hazard or extreme event, and the amount of fuel used to amend system generation to meet load (Nelson et al., 2020). Reiman et al., define metrics for energy use efficiency and ride-through ability to quantify resilience benefits from distributed and hybrid energy resources that normally go unacknowledged in the energy markets (Reiman et al., 2020). Marginal local energy usage efficiency represents the distributed energy both produced and consumed behind the meter, which minimizes the energy required to be purchased from the grid at the full retail electricity rate. This is especially important if the grid is experiencing high electricity rates because of a disturbance event, or if a segment of the grid is islanded. Marginal combined energy usage efficiency indicates the ability of a segment of the grid to contribute energy to the larger grid and potentially serve critical loads. The incremental sustainable ride-through energy ratio describes the ability of a microgrid to serve critical load during an outage, which depends on the combined availability of variable distributed generation and storage. This ratio could be used to optimally size and site distribution-scale hybrid systems for maximum resilience value.

Whereas previous research has advanced our ability to measure and compare the added value of hybrid or distributed energy assets on system resilience, we need to first understand where wind and solar resources are complementary. Previous work focused on U.S. resources has considered 1) site-specific hybrid complementarity or 2) spatial distribution of a single generation technology. A national analysis that highlights hybrid opportunity areas where wind and solar resources are complementary does not exist.

3. Methods

This work aims to identify areas in which there is enhanced opportunity to realize the resilience benefit of hybrid systems by analyzing the complementarity of wind and solar resources across the United States. We demonstrate how this complementarity analysis can be paired with local analysis to evaluate the resilience benefit of distribution-level hybrid power plants through a particular reference system: a microgrid near Memphis, Tennessee.

The national complementarity analysis was completed for the contiguous United States (CONUS), Hawaii, and Alaska for the year 2013. The spatial extent and resolution of the data sampling for wind and solar resources for the analysis are detailed in Figure 1 and Table?

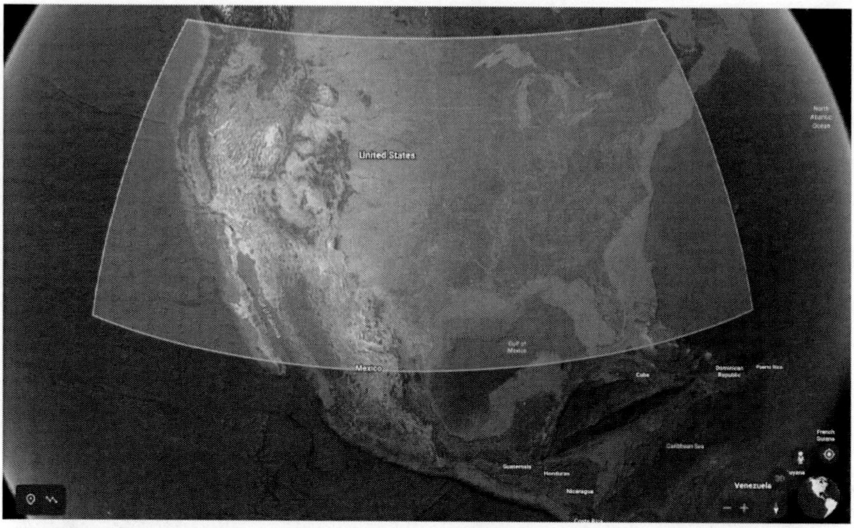

(a) Contiguous United States

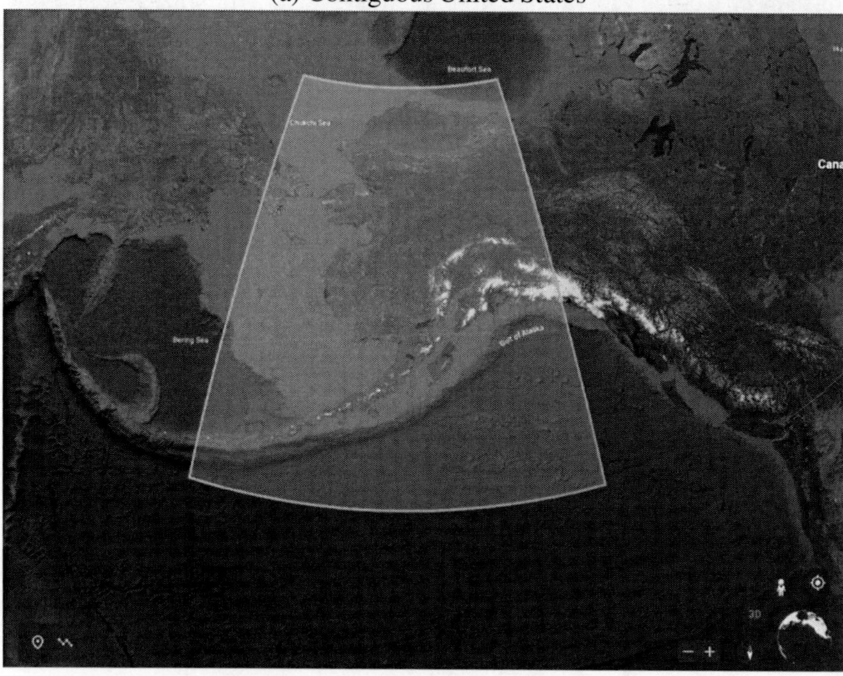
(b) Alaska

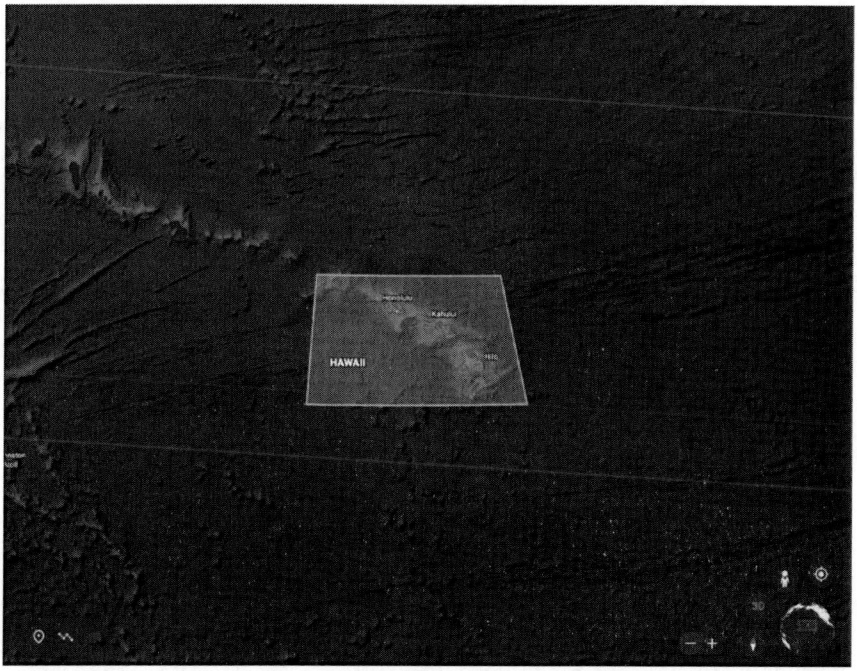

(c) Hawaii

Figure 1. Geographic extent of complementarity analysis. Source: Image ©2021 Google, Map Data ©2021 SIO, NOAA, U.S. Navy, NGA, GEBCO Landsat/Copernicus IBCAO INEGI SK telecom.

Table 1. Geographic extent of complementarity analysis

Area	Latitude (min, max, no. points)	Longitude (min, max, no. points)
CONUS[a]	(23.8, 49.3, 42)	(-129.2, -65.7, 164)
Alaska	(50.0, 72.0, 60)	(-160.2, -154.7, 60)
Hawaii	(22.2, 18.7, 6)	(-175.0, -130.0, 15)

[a] CONUS = Contiguous United States.

For each location within CONUS and Hawaii, the Wind Integration National Dataset Toolkit (WIND Toolkit) (Draxl et al., 2015) was used to collect hourly wind speed and direction data at an assumed hub height of 100 meters (m).

Global horizontal irradiance (GHI) data were used as a representative metric for total solar resource and were collected hourly for each location within CONUS and Hawaii from the National Solar Radiation Database (Sengupta et al., 2018). Because of data limitations at the time of this study, the National Aeronautics and Space Administration (NASA) Prediction of Worldwide Energy Resources (POWER) Project was used to collect wind and solar resource data in Alaska (NASA 2021). NASA POWER uses satellite data and data from the Modern-Era Retrospective analysis for Research and Applications, Version 2 (MERRA-2) as the basis of their platform, so there are discrepancies between the Alaska data and the rest of the data used in this study, but the same metrics were used to measure each resource. Wind speed and direction were measured at 50-m hub heights for the Alaska wind data, and GHI was calculated via Eq. 1:

$$\mathrm{GHI} = \mathrm{DHI} + \mathrm{DNI} * \cos(Z) \tag{1}$$

where DHI is the diffuse horizontal irradiance, DNI is the direct normal irradiance, and Z is the solar zenith angle. We assume ground-reflected radiation is insignificant compared to direct and diffuse radiation.

Using wind speed and GHI, complementarity was measured by calculating the Pearson r coefficient using Eq. 2:

$$r = \frac{\sum(x-\bar{x})(y-\bar{y})}{\sqrt{\sum(x-\bar{x})^2\sum(y-\bar{y})^2}} \tag{2}$$

where x^{-} is the mean of the vector x, y^{-} is the mean of the vector y, and the vectors x and y represent the wind and solar data, respectively. The resulting Pearson r coefficient is on a scale from -1 to 1, with a score of -1 representing perfect negative correlation, a score of 0 representing no correlation, and a score of 1 representing perfect positive correlation. A perfect positive correlation means that the wind and solar resource profiles occur at the same time in a given location, while a perfect negative correlation means that the wind and solar resource profiles are perfectly complementary, occurring inversely at a given location.

Complementarity was measured over four time scales: annual versus monthly, and daily-averaged versus hourly- averaged:

1. The annual, daily-averaged complementarity metric calculates a Pearson r coefficient across the 24 hourly wind and solar resource

values for each day in 2013, and then averages those 365 values for a single, annual value. The annual, daily-averaged complementarity metric does not consider seasonal or diurnal patterns but gives a general indication for overall complementarity of a given location for sizing and siting considerations.

2. The annual, hourly-averaged complementarity metric calculates the complementarity for every hour in 2013, and averages those 8,760 values over the year. The annual, hourly-averaged complementarity metric does not consider seasonality but does consider diurnal patterns.
3. The monthly, daily-averaged complementarity metric calculates a Pearson r coefficient across the 24 hourly wind and solar resource values for each day and averages those values over each month (approximately 30 values), resulting in 12 seasonally dependent complementarity values.
4. The monthly, hourly-averaged complementarity metric calculates the Pearson r coefficient for each hour and averages those values over each month (approximately 720 values). This method results in a metric that considers season and diurnal patterns and is well-suited for assessing the resource ability to serve hourly demand or other grid services.

Although this study focuses on a national complementarity analysis for the United States, complementarity and hybrid power plant ability to meet system demand are sensitive to local conditions (especially at the distribution level). Thus, after we completed a complementarity analysis to identify regions where the resources are especially conducive to hybrid power plant development, we selected a reference system in an opportunity area to demonstrate hybridization of generation assets in a specific, highly complementary system. The chosen system for this analysis is in the southeastern United States, less than 20 miles north of Memphis, Tennessee (Figure 2). It was selected because of the presence of a microgrid and local interest in enhancing grid resilience in the area.

A representative daily profile for wind and solar generation was determined by averaging the resource profile for each day in 2013. We then used REopt® (Cutler et al., 2017), the Renewable Energy Integration and Optimization platform from the National Renewable Energy Laboratory (NREL) to determine the optimal size and mix of technologies for a wind-solar-storage plant. The Hybrid Optimization and Performance Platform

(HOPP) (National Renewable Energy Laboratory 2021b) was used to design and analyze the plant at the component level, giving cost and value estimates for the plant. The hybrid plant sizing was completed for critical load factor values of 1.0 and 0.9, representing minimum load targets of 100% and 90% of the design load, respectively. The technology cost assumptions used for sizing are based on NREL's Annual Technology Baseline 2025 (National Renewable Energy Laboratory 2021a) data for the "moderate" 2025 technology innovation category and are as follows: $1,492 per kilowatt (kW) for wind; $1,095/kW for solar; and $179/kW and $206 per kilowatt-hour (kWh) for battery energy storage rate and storage capacity, respectively.

The wind turbine used in this analysis is a GE 2.5xl, 2.5-megawatt (MW) unit, with parameters that are determined through the NREL System Advisor Model (National Renewable Energy Laboratory 2020) default model generator. The wind turbine exhibits the key performance characteristics displayed in the power-performance curve (Figure 3) and the additional performance characteristics in Table 2.

By default, for utility-scale projects, the System Advisor Model selects the SunPower SPR-E19-310-COM photovoltaics module, a nonbifacial monocrystalline silicon panel with specifications as detailed in Table 3, and module current/voltage curve given in Figure 4.

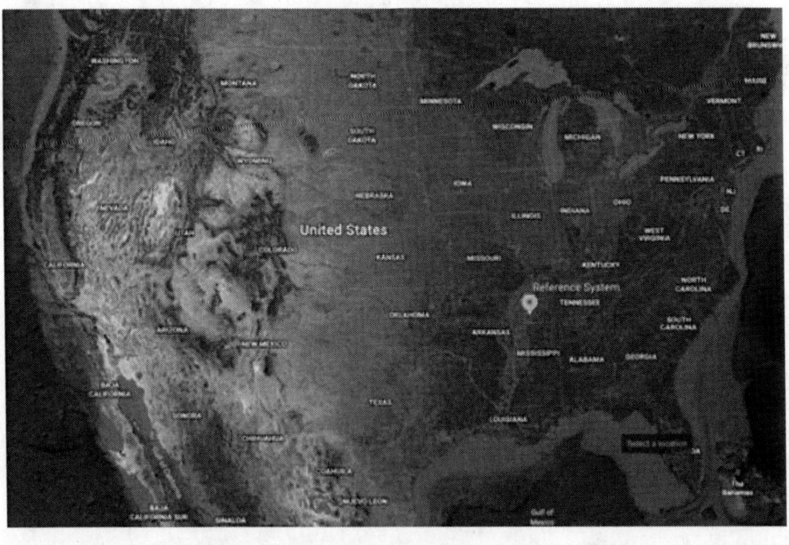

Figure 2. Location of reference system microgrid near Memphis, Tennessee. Source: Image ©2021 Google, Map Data ©2021 SIO, NOAA, U.S. Navy, NGA, GEBCO Landsat/Copernicus IBCAO INEGI SK telecom.

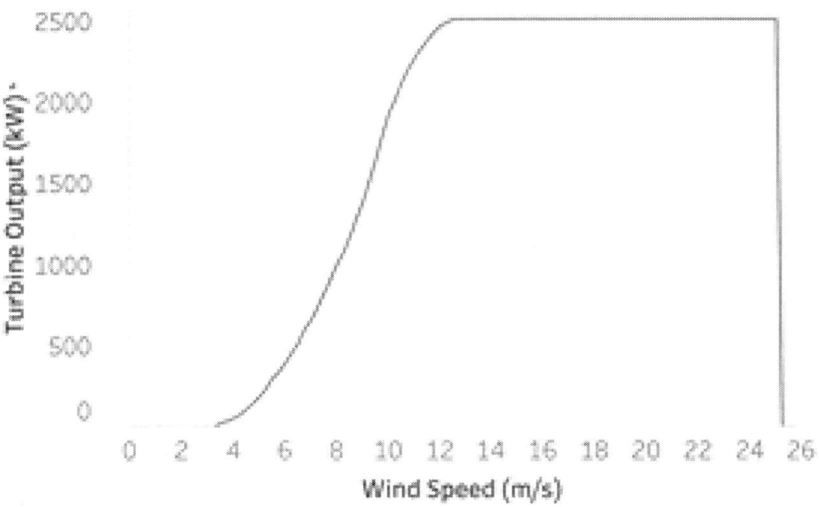

Figure 3. Power curve for the 2.5-megawatt GE 2.5xl wind turbine. Source: System Advisory Model (National Renewable Energy Laboratory 2020).

Table 2. Wind turbine performance characteristics

Parameter	Value
Maximum Rated Power (MW)	2.5
Cut-In Speed (m/s)	4
Cut-Out Speed (m/s)	25
Hub Height (m)	80
Rotor Diameter (m)	100

Table 4. Solar plant efficiency and performance details

Parameter	Value
Solar Albedo (ratio)	0.2
AC Wiring Loss (%)	1
DC Wiring Loss (%)	2
Diode Connection Loss (%)	0.5
Mismatch Loss (%)	2
Nameplate Loss (%)	0
Soiling Loss (%)	5
DC Degradation (%)	0.5
Azimuth (degrees)	180

Parameter	Value
Ground Coverage Ratio (ratio)	0.3
Tilt (degrees)	20
Module Aspect Ratio (ratio)	1.7
Inverter DS Efficiency (%)	96
Inverter PD Efficiency (%)	95

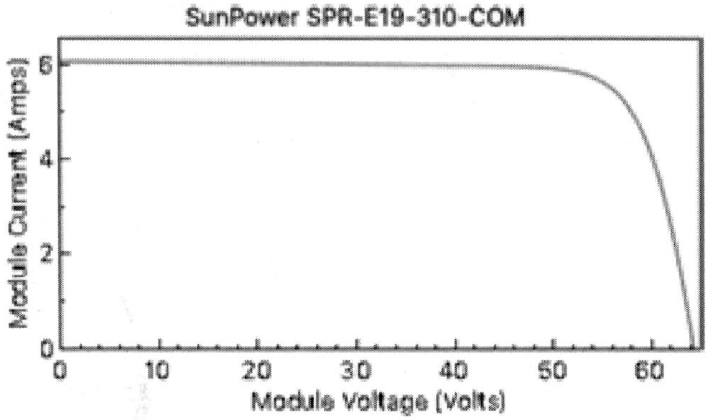

Figure 4. SunPower photovoltaics module power curve.

Table 3. Solar module technical details

Parameter	Value
Nominal Efficiency (%)	19.02
Maximum Power (P_{MP}) (watts)	310.15
Maximum Power Voltage (V_{MP}) (volts)	54.7
Maximum Power Current (A_{MP}) (amps)	5.7
Open Circuit Voltage (V_{OC}) (volts)	64.4

The total system is sized using these modules and then assumes a perfect inverter, which is sized to match the total number of modules and strings required to reach the desired solar install capacity. By default, the System Advisor Model chooses the SMA America SC750CP-US inverter, which is rated at 770 kilowatts (kW) of AC power per inverter, and a DC-to-AC ratio of 1.2. Table 4 describes overall plant performance characteristics and assumptions for system efficiency and losses.

With the assumption that there was no land constraint, the ground coverage ratio of the project was set to 0.3. Accordingly, self-shading was no longer a constraint in the system design discussed earlier. This key assumption helped reduce the total simulation time by avoiding a complex calculation of self-shading losses.

4. Results

In this section, we first discuss the national complementarity analysis based on the methods covered in Section 3 and then follow with a more detailed analysis using our selected real-world reference system located north of Memphis, Tennessee.

4.1. Complementarity Mapping

The figures in the appendices depict the national complementarity analyses. Data are displayed by region as well as by whether they are annual or seasonal, and whether they are averaged daily or hourly. In addition to the complementarity analyses, we have included figures depicting annual average wind speed and GHI (Appendices Appendix B, Appendix D, and Appendix F) to convey the relative abundance of each resource in the given location. Resources that are abundant enough to meet local demand and complementary indicate an opportunity to build hybrid power plants for enhanced resilience.

It is important to note again that the Alaska results are based on different data than those used for CONUS and Hawaii. Therefore, the results from these two data sets should not be directly compared.

4.1.1. Contiguous United States

Annual analyses for CONUS (both daily- and hourly-averaged) are provided in Appendix Appendix A and monthly analyses are provided in Appendix Appendix G. Figures Appendix A.1 and Appendix A.2 show that in areas with complex terrain, such as the Appalachian area in the Northeast and mountainous areas in the West, there is lower complementarity. Areas of higher annual complementarity include the Great Plains, East Texas, the Midwest, and nonmountainous regions of the Northeast. In the daily-averaged analysis (Figure Appendix A.1), there are fewer regions of the United States that are complementary compared to the annual, hourly-averaged analysis (Figure Appendix A.2), indicating a need to use hourly-averaged analyses to fully account for the ability of hybrid resources to meet hourly local demand.

When considering wind and solar resource abundance (Appendix Appendix B), areas in the Southwest, Great Plains, and Great Lakes regions have resources that could meet local demand. There are additional regions in the Southeast and West that have relative resource abundance; however, given the spatial variability of these regions, more detailed local or regional resource and complementarity analyses are needed.

Monthly analyses for CONUS are included in Appendix Appendix G, with the daily-averaged analysis in Figure Appendix G.1 and the hourly-averaged analysis in Figure Appendix G.2. The daily-averaged plots echo the trends of the annual analysis, with regions in the Midwest, Southeast, and nonmountainous Northeast having high resource complementarity, and western regions having more spatially variable complementarity. The temporal variability of complementarity is apparent across all regions, indicating that seasonal changes of wind and solar resources are a major influence on complementarity. The hourly-averaged analyses show trends that are less temporally variable, with most of the nonmountainous regions of the United States showing some complementarity. The West, again, is more spatially variable than the East. In the summer months, the complementarity trends intensify, with correlated areas becoming more correlated, and complementarity areas becoming more complementary. Areas in the Southeast and Midwest as well as those in California's Central Valley show particularly high complementarity during summer months. Mountainous regions, again, are less complementary (for example, regions in the Rocky Mountains, the Appalachian Mountains, the western coastal ranges, the Cascade Mountains, and the Sierra Nevada) and become more so during summer months. This trend intensification (correlated areas becoming more correlated and complementarity areas becoming more complementary) during summer months is thought to be caused by the seasonal changes in diurnal patterns for solar energy, namely, the increase in solar irradiance during summer months attributable to longer days and fewer cloudy days.

4.1.2. Hawaii

In Hawaii, annual analyses (both daily- and hourly-averaged) are provided in Appendix Appendix C and monthly analyses in Appendix Appendix H. For the annual, daily-averaged analysis (Figure Appendix C.1), areas of high complementarity include the center of Kauai, the eastern regions of Oahu, and the west coast of the Island of Hawaii. These regions of complementary resources, however, are not reflected in the hourly-averaged analysis in Figure Appendix C.2, which instead shows few, if any, regions with complementary resources. When considering resource abundance in Appendix Appendix D, annual average wind speed is fairly consistent across the islands (Figure Appendix D.1), whereas the annual average GHI is spatially more variable across the islands (Figure Appendix D.2). Areas of the islands where annual average GHI is low coincide with areas where complementarity is high. Given the spatial extent of the resources, distributed generation assets could take advantage of complementary resources while also leveraging high-resource areas to add power production. Monthly analyses for Hawaii are included in Appendix Appendix H, with the daily-averaged analysis in Figure Appendix H.1 and the hourly-averaged analysis in Figure Appendix H.2. The daily-averaged analysis indicates seasonal variance in complementarity, with more regions that have complementary resources and intensification of that complementarity in fall (September,

October, November) and spring (April, May, June), and fewer regions with complementary resources in summer (July, August) and winter (December, January, February, and March). When observing the hourly-averaged analysis, most of Hawaii is correlated rather than complementary. Overall, Hawaii has abundant wind and solar resources, but they are not complementary on an hourly scale. Thus, Hawaii's resources are more suited for bulk power production and less suited for providing resilience value.

4.1.3. Alaska

In Alaska, annual analyses (both daily- and hourly-averaged) are provided in Appendix Appendix E and monthly analyses are provided in Appendix Appendix I. In the annual, daily-averaged analysis (Figure Appendix E.1), most regions in Alaska have complementary resources. In the annual, hourly-averaged analysis (Figure Appendix E.2), however, the correlation coefficient becomes more positive, with only coastal regions and some areas in the interior and south-central regions showing complementarity. The data indicate that it is especially important to consider diurnal patterns in Alaska when leveraging the complementarity of wind and solar resources for improved resilience value in generation assets.

Monthly analyses for Alaska are included in Appendix Appendix I, with the daily-averaged analysis in Figure Appendix I.1 and the hourly-averaged analysis in Figure Appendix I.2. Whereas complementarity trends in the daily-averaged analysis are less spatially variable than those in some of the western regions of CONUS, they are more temporally variable, with the same regions fluctuating between positive and negative correlation from month to month.

The hourly-averaged analysis shows few regions of complementarity; they mostly occur during winter months when the solar resource is especially low. The spatial and temporal variability indicates that multiple factors are interacting to influence these seasonal patterns.

The influence of diurnal patterns is pronounced in Alaska because of dramatic changes in daylight hours during the year. Seasonal changes in daylight and solar resource are evident in Appendix Appendix I. Figure Appendix I.1 shows the daily-averaged complementarity per month; the data indicate high temporal variability. In the plots in Appendix Appendix I, it is important to note the gray area in North Alaska in December, which indicates a lack of solar data as a result of a lack of sunlight. The changing amount of daylight influences the resource in other months, so that even if a site has complementary resources, it may not have enough solar resource abundance during certain parts of the year to support local demand. The hourly-averaged data, which take into account the diurnal patterns, indicate a less variable complementarity trend. Wind and solar resources are most complementary in specific regions in the winter months and become less complementary during summer months as the solar resource increases significantly (Figure Appendix I.2). This pattern is particularly important when considering energy use. In some regions in Alaska, demand is driven by heat consumption during cold months rather than by air-conditioning consumption during warm months, as is common in other parts of the United States. This demand pattern would result in the peak demand occurring in the months with minimal solar resource. On the contrary, there are also regions in Alaska that experience peak demand during months of high fishery or cruise ship activity, which would result in a better alignment of demand with solar resource availability. In Alaska, more detailed, local analysis that incorporates demand data is required to assess if hybrid power plants can meet local demand.

4.2. Reference System Analysis

In this section, we complete a detailed local analysis based on both resource and demand data for a microgrid north of Memphis, Tennessee. The site is at Naval Support Activity Mid-South and includes a 12-MW direct-current solar array that supplies electricity both to the site and to other Tennessee Valley Authority customers in the area. This site is of particular interest from a resilience perspective because it was built with resilience objectives in mind. The Navy estimates that when solar resource is available, the array could power 1,300 homes in addition to alongside mission-critical operations at the site. Additionally, the solar array is part of a larger installation that has a total direct current capacity of 53 MW.

To demonstrate the potential resilience benefit of hybridization in a distributed-level generation site, we consider a scenario in which Naval Support Activity Mid-South generates its required power via wind and solar assets with storage capacity. Figure 5 assumes a 0.9 critical load factor scenario, wherein the hybrid power plant is optimized to meet 90% of the load with renewable energy generation. The remaining load is met with diesel generation. Figure 6 assumes a 1.0 critical load factor scenario, or a scenario in which 100% of the load is met with the hybrid power plant. We assumed a constant load of 10 MW for the purposes of this study, although in reality, 10 MW is probably closer to the peak load of the site. Table 5 shows the sizing of the hybrid power plant based on a simulation in HOPP.

Table 5. Hybrid optimization and performance platform project sizing results

Hybrid Plant Component Sizing	Critical Load Factor	
	0.9	1.0
Wind Capacity (MW)	21.96	16.89
Solar Capacity (MW)	28.04	33.11
Storage Size (MW)	12.35	18.09
Storage Size (MWh)	821.40	1384.76

When assuming a critical load factor of 0.9, the total solar capacity is sized at approximately 28 MW, the total wind capacity is sized at approximately 22 MW, and the total storage capacity is sized at approximately 12 MW. Figure 5a shows how wind and solar generation complement each other over the course of a representative day at this site, with solar ramping up to peak at midday at just under 10 MW (orange curve), and wind generation following an inverse trend (blue curve). Wind production is highest at night (just over 8 MW) when there is no solar production. Then, as solar production begins to increase in the morning hours, wind production decreases to a midday low. The aggregate generated power between these two sources (pink curve) is smoother than either the solar or wind generation alone, with sustained production at night between 8 and 10 MW, a daytime peak around 12 MW, and an evening minimum at about 17:00, as wind production increases and solar production decreases. Stored power smooths this power generation more, providing 2–4 MW throughout the day (brown curve). Stored power is used mostly at night, beginning in the evening hours during the solar and wind aggregate minimum power production at 17:00. The green curve represents the aggregate power supplied by wind, solar, and storage resources, which is consistently between 12 and 16 MW, with peak supply hours coinciding with peak solar production hours. Because a constant demand is assumed to be 10 MW, both wind and solar must be curtailed throughout the day. Those curtailments, along with the amount of load not met, are shown in Figure 5b. Solar power production, unsurprisingly, is increasingly curtailed during peak production hours. Wind power production is curtailed by 2.5–4 MW per hour. With this plant sizing, there is also a consistent energy shortfall of 250–750 kW throughout the day, as was planned for the 0.9 critical load factor.

When assuming a critical load factor of 1.0, the total solar capacity is sized for approximately 33 MW, the total wind capacity is sized for approximately 17 MW, and the total storage capacity is sized for approximately 18 MW. Figure 6 shows the supplied power from solar and wind generation and from storage assets over the course of a representative day. Because demand is again assumed to be constant at 10 MW, the generation and storage profiles are similar to those of Figure 5, with only a 1 MW difference in estimated constant demand throughout the day. With a critical load factor of 1.0, the optimal hybrid power plant sizing shifts to provide more generated power through solar rather than wind assets, presumably driven by cost. This shift results in solar power production peaking at near 11 MW during the day rather than at less than 10 MW; however, solar curtailment is reduced by nearly 1 MW at peak production. Wind power production also decreases, with nighttime wind production at 7 MW rather than 9 MW, and curtailment reduced to 1–2.75 MW per hour (rather than 2.5–4 MW per hour in the 0.9 critical load factor case). Storage is similar between the two critical load factor cases, with approximately 4 MW per hour supplied during the night and 2 MW per hour supplied during the day.

Both critical load factor cases show the benefit of hybridization in a distribution-level system. Solar power generation alone would not meet the demand in a constant load scenario, as no solar power is generated at night, and the storage capacity required to ensure demand is met is large and expensive, even for the 10 MW load of this microgrid. Hybridizing a microgrid by adding wind power generation enables smoother power output and decreased storage capacity requirements. Further, in a disruptive event like a cloudy day, the system generates enough power to meet demand (shown by the curtailment profiles), especially when wind constitutes a greater share of base generation.

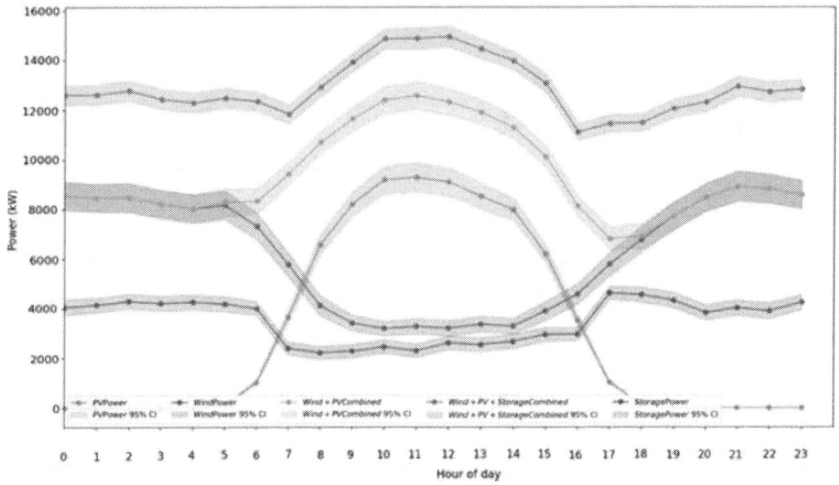

(a) Generation and storage profiles for the 0.9 critical load factor (CI = confidence interval)

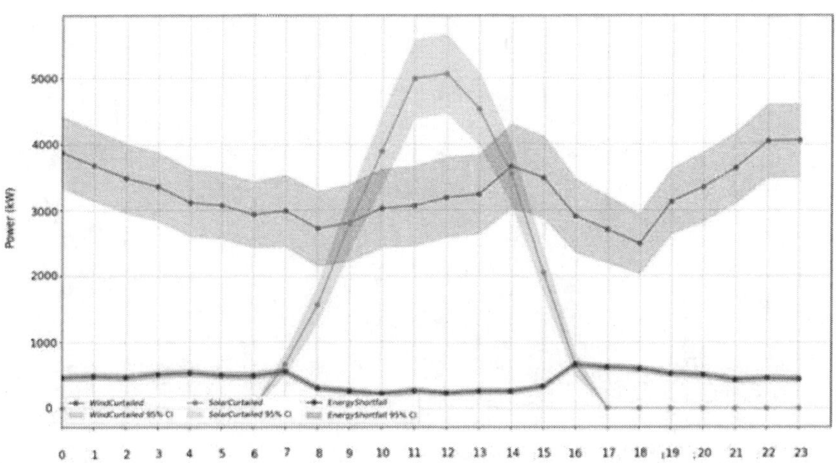

(b) Profiles of wind and solar curtailment and energy shortfall

Figure 5. Hybrid plant performance for Naval Support Activity Mid-South given a 0.9 critical load factor.

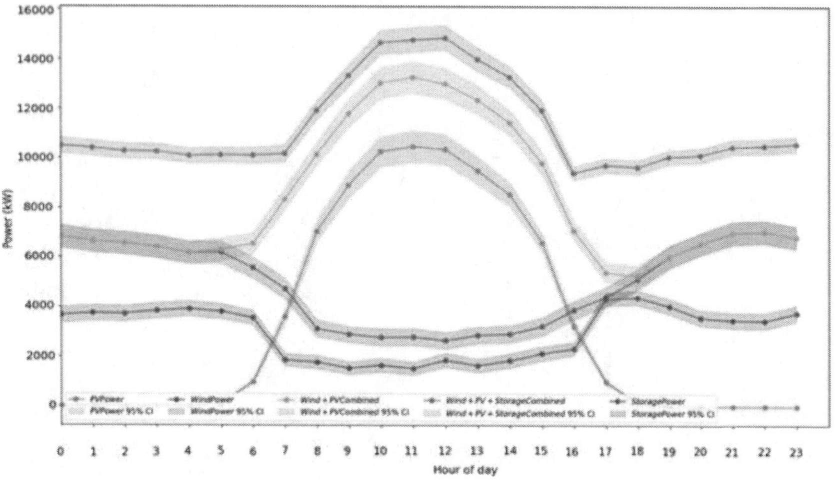

(a) Generation and storage profiles for the 1.0 critical load factor

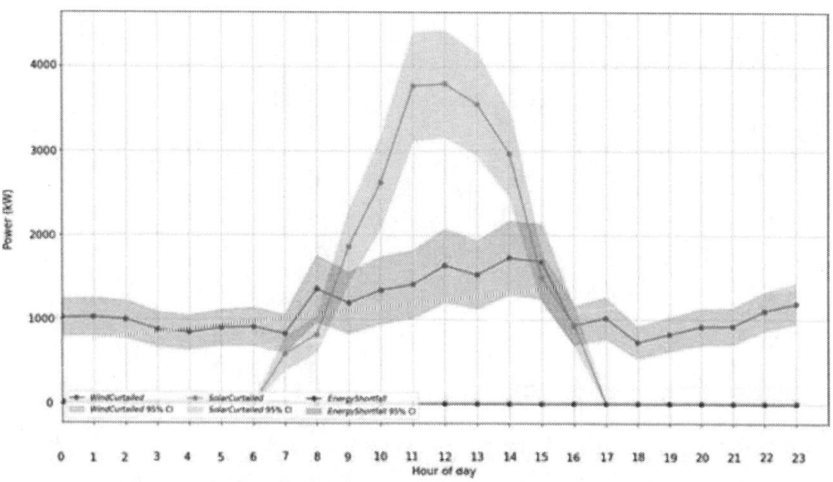

(b) Profiles for wind and solar curtailment and energy shortfall

Figure 6. Hybrid plant performance for Naval Support Activity Mid-South given a 1.0 critical load factor.

Some of the metrics described in Section 2 are relevant to this particular system. First, because we assume this hybrid plant is functionally a microgrid, it is completely autonomous. That is, there is no selling or buying of electricity to or from the larger grid. We also assume that when we optimize the hybrid plant for a critical load factor of 0.9, we meet any energy shortfall with diesel generation. We can therefore quantify diesel generation, which is a direct artifact of the critical factor load setting. Because of this, "load not served" will always be 0 MW. In Figure 5b, the energy shortfall is approximately 250 kW (0.025% of load) during daylight hours, or from 08:00 to 15:00. The energy shortfall increases during nighttime to approximately 500 kW (0.05% of load) between 16:00 and 07:00 the next morning.

Second, because of our assumption of an isolated microgrid, the ability to quantify usage efficiency is somewhat limited. Metrics that take into account the import and export of energy relative to local use are not relevant, but energy use relative to curtailment can provide a proxy to the energy efficiency of the system. This metric enables one to compare the system sizing and efficiency and becomes particularly relevant in systems designed to supply load during long-duration outages or capacity loss, in which the system might be overbuilt to accommodate such an event. Figure 5b depicts the curtailment of both solar and wind assets in the reference system designed for a critical load factor of 0.9. In this system, approximately 72 megawatt-hours (MWh) of wind energy generation and 30 MWh of solar energy generation are curtailed, or approximately 30% of the energy produced is curtailed. In the reference system designed for a critical load factor of 1.0 (Figure 6b), approximately 19 MWh of wind energy generation and 46.5 MWh of solar energy generation is curtailed, resulting in a sum curtailment of 16% of the total energy generation.

Finally, sustainable ride-through is an important statistic that should be quantified for both isolated and connected distributed grid systems. Sustainable ride-through depends not only on detailed local demand data but also on local hazard and outage data. In most studies reviewed, ride-through statistics were offered in reference to a single outage scenario. Although this approach enables the design of distributed renewable energy systems with more resiliency than systems in which an outage event has not been considered, it is limited. Ongoing research at NREL uses a Monte Carlo analysis with varying outage duration and frequency to quantify load not served in transmission systems. This type of approach enables stakeholders to measure system resilience in a more robust way and to incorporate uncertainty (via probabilistic outage scenarios) into their decision-making (e.g., they can answer the question, "How does the probability of outage frequency and duration impact the load not served?"). Although we do not quantify sustainable ride-through in this particular reference system because of a lack of outage data, future work will define probabilistic outage scenarios, quantify the sustainable ride-through required for a system to endure these scenarios, and leverage this information as a basis for resilient hybrid system design.

Conclusions and Future Work

In this study, we aimed to identify regions of the United States that have adequate wind and solar resources, as well as complementarity of those resources, to support hybrid plant development. Daily- and hourly-averaged complementarity analyses show that areas in the Great Plains, Midwest, Southeast, and nonmountainous Northeast are all regions in which there is high complementarity. Regions with complex terrain (Alaska and western regions of CONUS) could also have areas of high complementarity but require local analysis because complementarity is so spatially and temporally variable. Furthermore, areas in the Northeast and Southeast require more local analysis because although complementarity is high, the abundance of one or more resources might be low, limiting the ability to provide resilience value in disruptive events. To exemplify this local analysis, we used the HOPP tool to optimally size a hybrid power plant for a microgrid near Memphis, Tennessee. Results for critical load factors of 0.9 and 1.0 indicate that solar power generation alone would not meet the demand in a constant load scenario, but pairing solar power generation with wind power generation and storage capacity could provide smoother power output and would better meet local demand, with extra capacity available.

Although this work provides a national analysis to identify hybrid power plant opportunities and uses a reference system to demonstrate the benefit that hybrid plants could provide to distribution-level power systems, it is only the first step in understanding how to fully leverage hybrid power plants in U.S. power systems to provide resilience value. To address this aim, future work would require 1) consistent resource data, 2) more detailed local analysis, and 3) consideration of resilience-specific complementarity metrics.

As discussed in Section 4.1, the data used for Alaska and CONUS were not consistent because of the availability of wind and solar resource data. Current efforts to extend WIND Toolkit and the National Solar Radiation Database to include northern climates are underway but were not completed at the time of this study. It is critical to generate consistent data and validate that data via ground measurements to correctly estimate available resources and technically available resources. Arctic climates have many special considerations for the design of renewable energy systems, and understanding the availability of renewable resources is foundational in designing solutions for these regions.

HOPP, which relies on data from the WIND Toolkit and the National Solar Radiation Database, can be amended when new wind and solar resource data become available to take into account the Arctic climate considerations needed to estimate the technical generation ability of renewable energy generation assets. Further, by extending the analysis to include multiple years, interannual variability can be considered, which will ultimately reduce resource uncertainty.

With respect to the need for more detailed local analysis, our national complementarity analysis showed that complementarity is spatially and temporally variable, especially in complex terrain. Although finer spatial resolution might benefit the complementarity analysis in complex terrain, it prompts the question, How fine a resolution is enough?

This study focuses on distribution-level power systems, which are at a local community or county scale. Although this work shows opportunity for hybrid plants in the United States from a resource perspective, specific, local data are needed to understand the ability of a distribution-level hybrid plant to meet the resilience objectives of a local power system given available resources, demand, and hazards. With local demand and hazard data, for instance, the hybrid power plant designed for the reference system in this study could be sized to meet actual demand during both nominal and disruptive conditions instead of an assumed constant load.

These two areas of future work lead to a final future work suggestion. First, by reducing uncertainty in local resources by ensuring consistent, multiyear resource data and analyzing those data at the local level alongside local demand and hazard data, we would be able to estimate power system resilience through reliability-based hybrid metrics and resilience frameworks, as discussed in Section 2. Second, complementarity analysis has thus far used metrics that estimate the hybrid benefit to renewable energy generation reliability. If we considered low-probability events with these metrics—thereby evaluating the resilience benefit rather than reliability—and integrated them with power system resilience frameworks, we would be able to estimate the resilience benefit that distribution-level hybrid power plants could provide a specific community.

Appendix A. Annual Complementarity for 2013, Contiguous United States

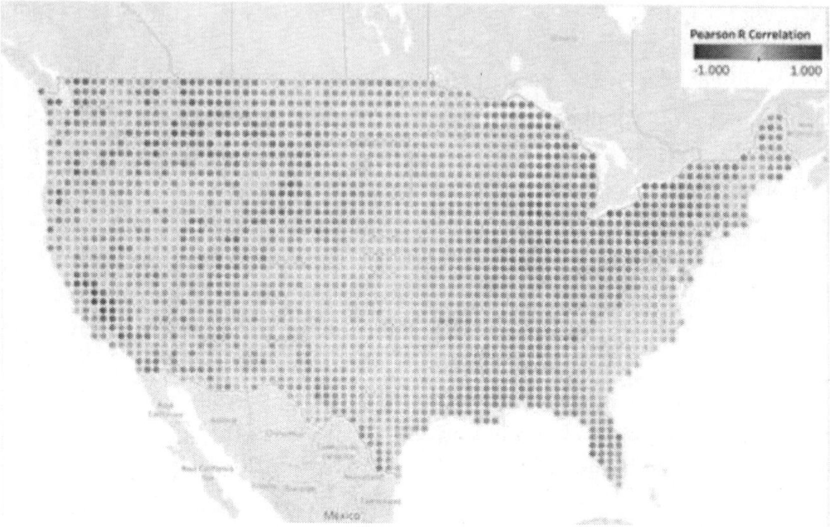

Figure Appendix A.1. Annual, daily-averaged complementarity (represented by the Pearson correlation metric) for 2013 in the contiguous United States.

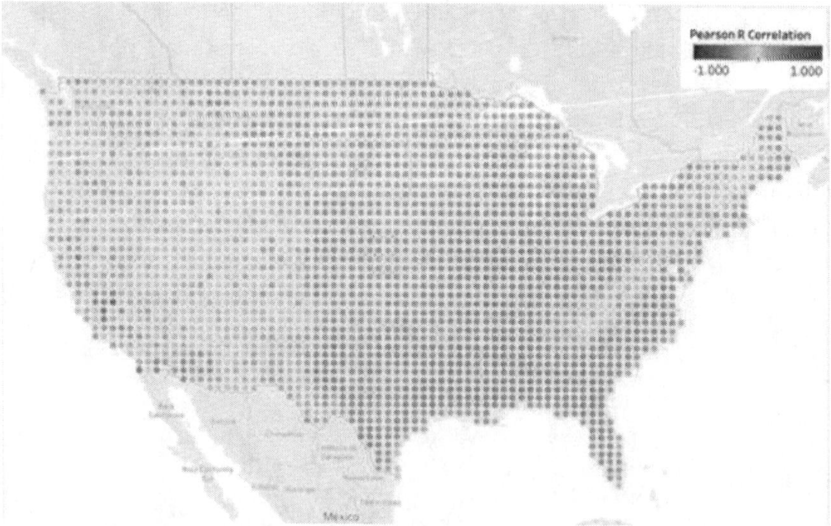

Figure Appendix A.2. Annual, hourly-averaged complementarity (represented by the Pearson correlation metric) for 2013 in the contiguous United States.

Appendix B. Annual Average Wind Speed and Global Horizontal Irradiance for 2013, Contiguous United States

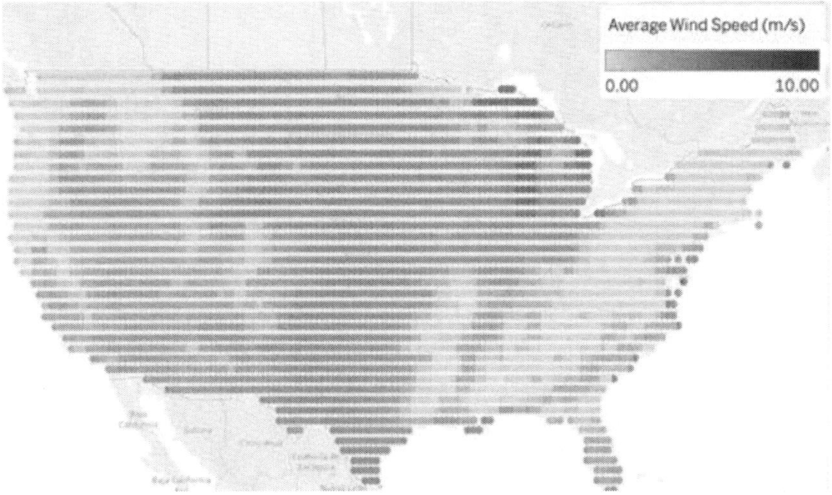

Figure Appendix B.1. Average annual wind speed in meters per second (m/s) for 2013 in the contiguous United States.

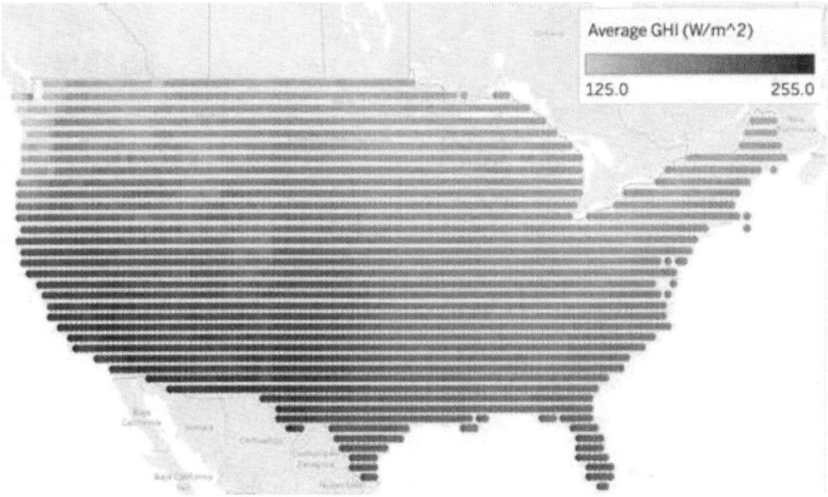

Figure Appendix B.2. Average annual global horizontal irradiance (GHI) in watts per square meter (W/m^2) for 2013 in the contiguous United States.

Appendix C. Annual Complementarity for 2013, Hawaii

Figure Appendix C.1. Annual, daily-averaged complementarity (represented by the Pearson correlation metric) for 2013 in Hawaii.

Figure Appendix C.2. Annual, hourly-averaged complementarity (represented by the Pearson correlation metric) for 2013 in Hawaii.

Appendix D. Annual Average Wind Speed and Global Horizontal Irradiance for 2013, Hawaii

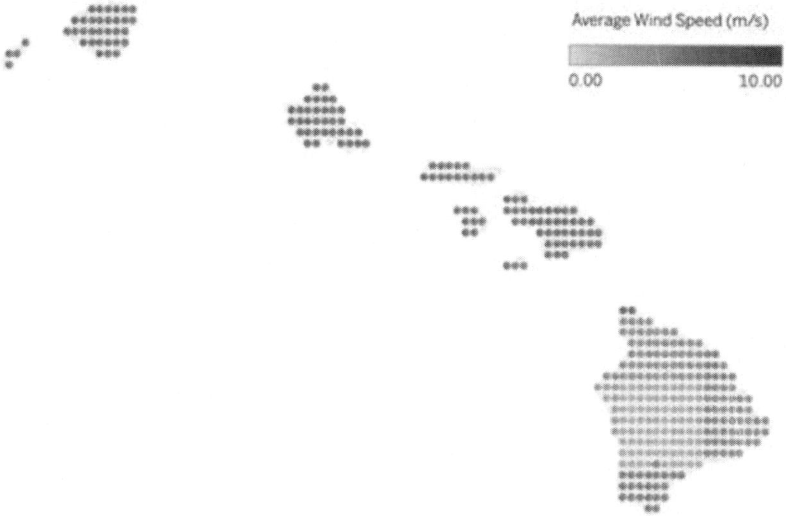

Figure Appendix D.1. Average annual wind speed (m/s) for 2013 in Hawaii.

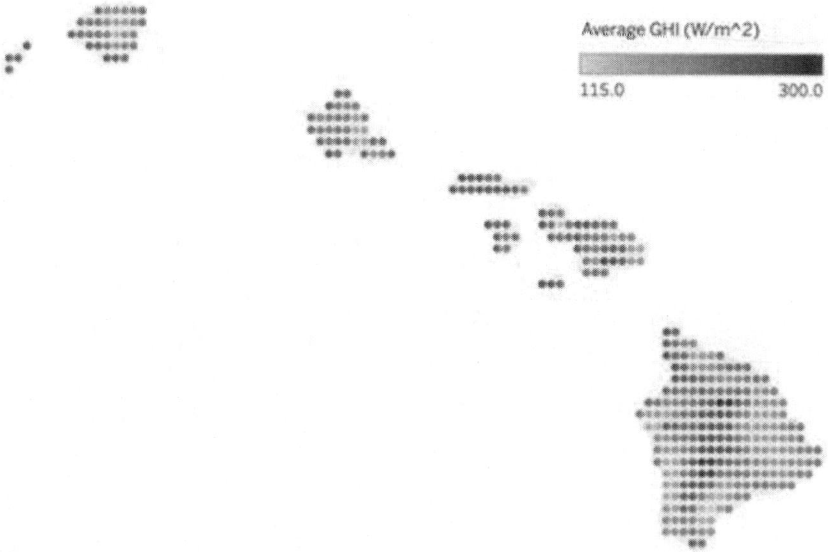

Figure Appendix D.2. Average annual global horizontal irradiance (W/m^2) for 2013 in Hawaii.

Appendix E. Annual Complementarity for 2013, Alaska

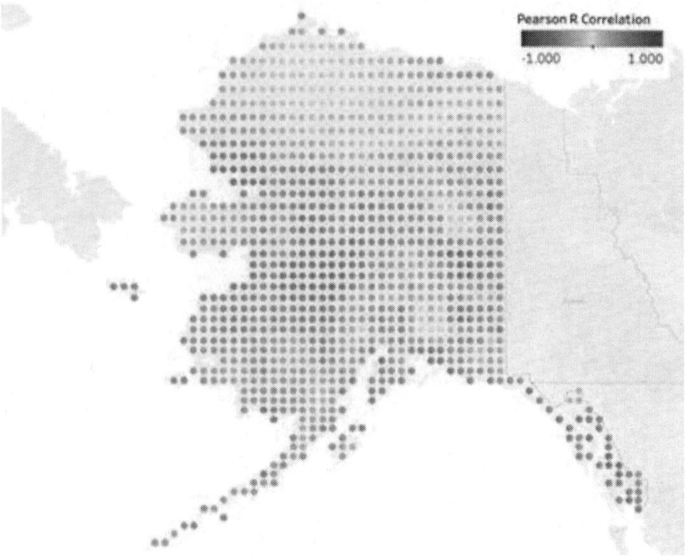

Figure Appendix E.1. Annual, daily-averaged complementarity (represented by the Pearson correlation metric) for 2013 in Alaska.

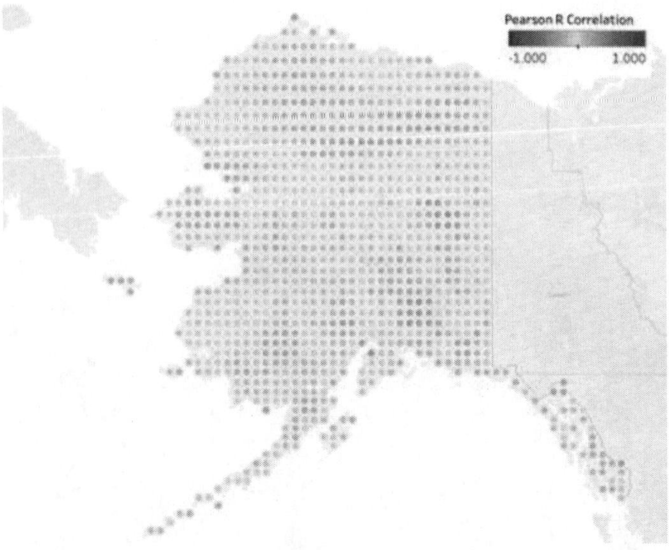

Figure Appendix E.2. Annual, hourly-averaged complementarity (represented by the Pearson correlation metric) for 2013 in Alaska.

Appendix F. Annual Average Wind Speed and Global Horizontal Irradiance for 2013, Alaska

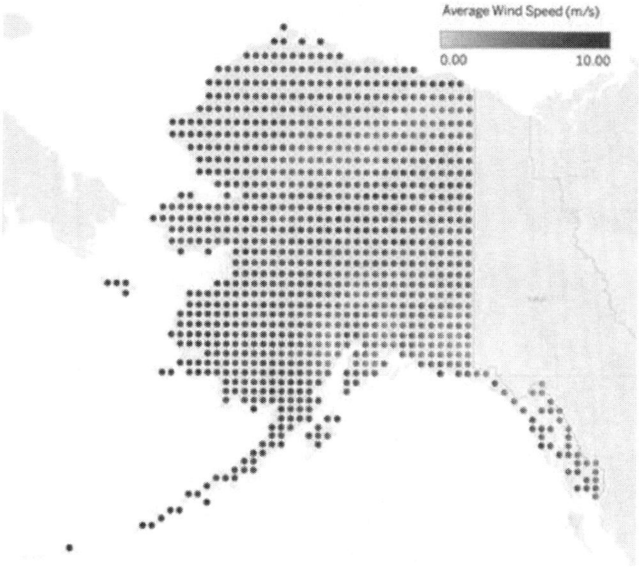

Figure Appendix F.1. Average annual wind speed (m/s) for 2013 in Alaska.

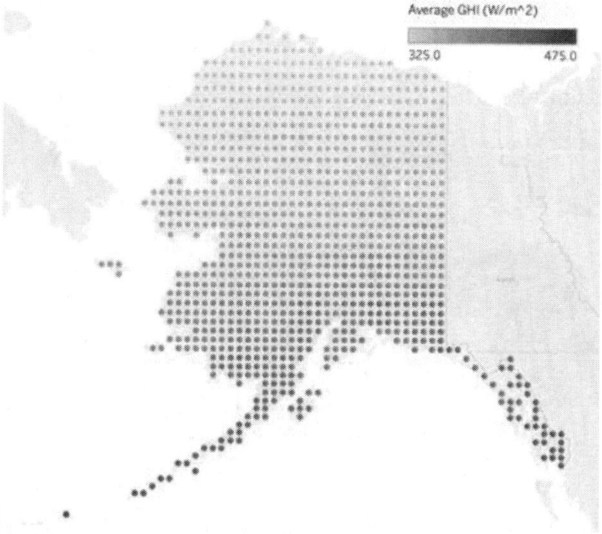

Figure Appendix F.2. Average annual global horizontal irradiance (W/m^2) for 2013 in Alaska.

Appendix G. Complementarity per Month, Contiguous United States

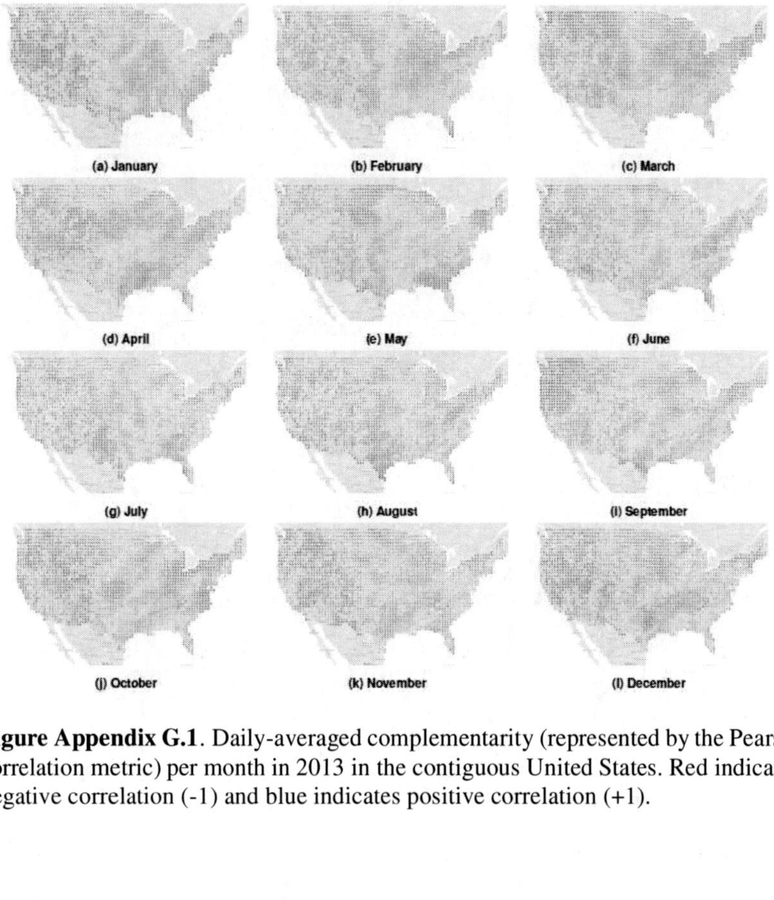

Figure Appendix G.1. Daily-averaged complementarity (represented by the Pearson correlation metric) per month in 2013 in the contiguous United States. Red indicates negative correlation (-1) and blue indicates positive correlation (+1).

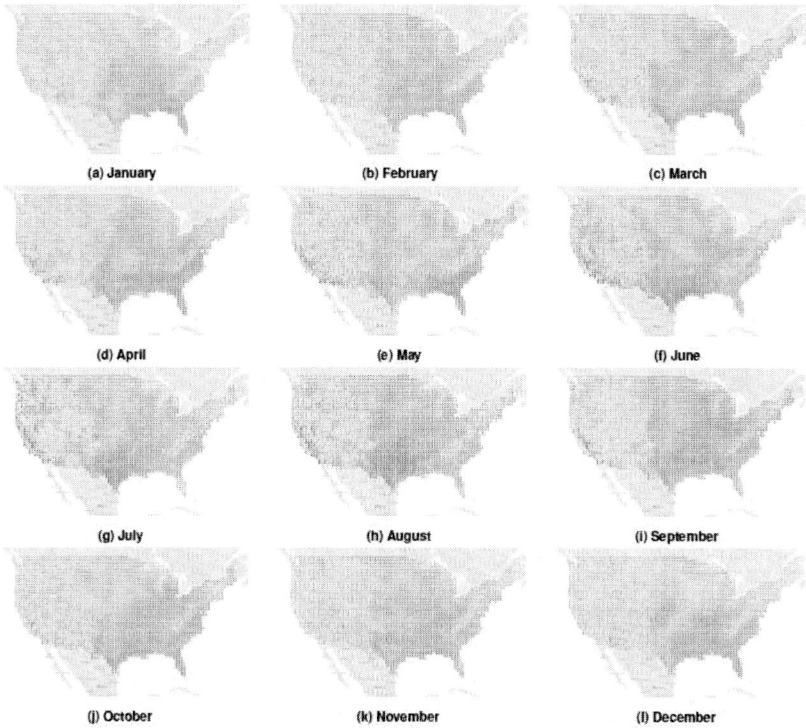

Figure Appendix G.2. Hourly-averaged complementarity (represented by the Pearson correlation metric) per month in 2013 in the contiguous United States. Red indicates negative correlation (-1) and blue indicates positive correlation (+1).

Appendix H. Complementarity per Month, Hawaii

Figure Appendix H.1. Monthly, daily-averaged complementarity (represented by the Pearson correlation metric) for 2013 in Hawaii. Red indicates negative correlation (-1) and blue indicates positive correlation (+1).

Figure Appendix H.2. Monthly, hourly-averaged complementarity (represented by the Pearson correlation metric) for 2013 in Hawaii. Red indicates negative correlation (-1) and blue indicates positive correlation (+1).

Appendix I. Complementarity per Month, Alaska

Figure Appendix I.1. Daily-averaged complementarity (represented by the Pearson correlation metric) per month for 2013 in Alaska. Red indicates negative correlation (-1) and blue indicates positive correlation (+1). The gray area in December signifies the days in which there is 1 hour or less of sunlight, resulting in a null Pearson *r* value.

Figure Appendix I.2. Hourly-averaged complementarity (represented by the Pearson correlation metric) per month for 2013 in Alaska. Red indicates negative correlation (-1) and blue indicates positive correlation (+1). The gray area in December signifies the days in which there is 1 hour or less of sunlight, resulting in a null Pearson *r* value.

References

Aza-Gnandji, M., F. X. Fifatin, A. H. J. Hounnou, F. Dubas, D. Chamagne, C. Espanet, and A. Vianou. 2018. "Complementarity between Solar and Wind Energy Potentials in Benin Republic". Advanced Engineering Forum 28 (June): 128–138. ISSN: 2234-991X. doi:10.4028/www.scientific.net/aef.28.128.

Berger, M., D. Radu, R. Fonteneau, R. Henry, M. Glavic, X. Fettweis, M. Le Du, P. Panciatici, L. Balea, and D. Ernst. 2020. "Critical time windows for renewable resource complementarity assessment". Energy 198. ISSN: 03605442. doi:10.1016/j.energy.2020.117308.

Cicilio, P., L. Swartz, B. Vaagensmith, C. Rieger, J. Gentle, T. McJunkin, and E. Cotilla-Sanchez. 2020. "Electrical grid resilience framework with uncertainty". Electric Power Systems Research 189 (March): 106801. ISSN: 03787796. doi:10.1016/j.epsr.2020.106801. https://doi.org/10.1016/j.epsr.2020.106801.

Cutler, D., D. Olis, E. Elgqvist, X. Li, N. Laws, N. DiOrio, A. Walker, and K. Anderson. 2017. Reopt: A platform for energy system integration and optimization. Tech. rep. September. Golden, CO: National Renewable Energy Laboratory. doi:10.1115/ES2014-6570.

Dos Anjos, P. S., A. S. Alves Da Silva, B. Stošic´, and T. Stošic´. 2015. "Long-term correlations and cross-correlations in wind speed and solar radiation temporal series from Fernando de Noronha Island, Brazil". Physica A: Statistical Mechanics and its Applications 424:90–96. ISSN: 03784371. doi:10.1016/j.physa.2015.01.003. http://dx.doi.org/ 10.1016/j.physa.2015.01.003.

Draxl, C., A. Clifton, B. M. Hodge, and J. McCaa. 2015. "The Wind Integration National Dataset (WIND) Toolkit". Applied Energy 151:355–366. ISSN: 03062619. doi:10.1016/j.apenergy.2015.03.121. http://dx.doi.org/10.1016/j.apenergy.2015.03. 121.

Gburcik, P., V. Gburcik, M. Gavrilov, V. Srdanovic, and S. Mastilovic. 2006. "Complementary regimes of solar and wind energy in Serbia". Geographica Pannonica, no. 10: 22–25. ISSN: 0354-8724. doi:10.5937/geopan0610022g.

Gorman, W., A. Mills, M. Bolinger, R. Wiser, N. G. Singhal, E. Ela, and E. O'Shaughnessy. 2020. "Motivations and options for deploying hybrid generator-plus-battery projects within the bulk power system". Electricity Journal 33 (5): 106739. ISSN: 10406190. doi:10.1016/j.tej.2020.106739. https://doi.org/10.1016/j.tej.2020.106739.

Jurasz, J., F. A. Canales, A. Kies, M. Guezgouz, and A. Beluco. 2020. "A review on the complementarity of renewable energy sources: Concept, metrics, application and future research directions". Solar Energy 195 (April 2019): 703–724. ISSN: 0038092X. doi:10.1016/j.solener.2019.11.087. arXiv: 1904.01667. https://doi.org/10.1016/j. solener.2019.11.087.

Jurasz, J., A. Beluco, and F. A. Canales. 2018a. "The impact of complementarity on power supply reliability of small scale hybrid energy systems". Energy 161:737–743. ISSN: 03605442. doi:10 . 1016/j.energy.2018. 07. 182. https://doi.org/10.1016/j.energy. 2018.07.182.

– . 2018b. "The impact of complementarity on power supply reliability of small scale hybrid energy systems".

Energy 161:737–743. ISSN: 03605442. doi:10.1016/j.energy.2018.07.182. https://doi.org/ 10.1016/j.energy.2018.07.182.

Kahn, E. 1979. "The reliability of distributed wind generators". Electric Power Systems Research 2 (1): 1–14. ISSN: 03787796. doi:10.1016/0378-7796(79)90021-X.

Li, Y., V. G. Agelidis, and Y. Shrivastava. 2009. "Wind-solar resource complementarity and its combined correlation with electricity load demand". 2009 4th IEEE Conference on Industrial Electronics and Applications, ICIEA 2009: 3623–3628. doi:10.1109/ICIEA.2009.5138882.

Liu, Y., L. Xiao, H. Wang, S. Dai, and Z. Qi. 2013. "Analysis on the hourly spatiotemporal complementarities between China's solar and wind energy resources spreading in a wide area". Science China Technological Sciences 56 (3): 683–692. ISSN: 16747321. doi:10.1007/s11431-012-5105-1.

Marcos, J., L. Marroyo, E. Lorenzo, and M. García. 2012. "Smoothing of PV power fluctuations by geographical dispersion". Progress in Photovoltaics: Research and Applications 20 (2): 226–237. ISSN: 10627995. doi:10.1002/ pip.1127. http://doi.wiley.com/10.1002/pip.1127.

McJunkin, T. R., and C. Rieger. 2019. "Resilient control system metrics". Advances in Information Security 75:255– 276. ISSN: 15682633. doi:10.1007/978-3-030-18214-4_12.

Miglietta, M. M., T. Huld, and F. Monforti-Ferrario. 2016. "Local Complementarity of Wind and Solar Energy Resources over Europe: An Assessment Study from a Meteorological Perspective". Journal of Applied Meteorology and Climatology 56 (1): 217–234. ISSN: 1558-8424. doi:10.1175/jamc-d-16-0031.1.

Murata, A., H. Yamaguchi, and K. Otani. 2009. "A method of estimating the output fluctuation of many photovoltaic power generation systems dispersed in a wide area". Electrical Engineering in Japan (English translation of Denki Gakkai Ronbunshi) 166 (4): 9–19. ISSN: 04247760. doi:10.1002/eej.20723.

Naeem, A., N. U. Hassan, C. Yuen, and S. M. Muyeen. 2019. "Maximizing the economic benefits of a grid-tied microgrid using solar-wind complementarity". Energies 12 (3): 1–22. ISSN: 19961073. doi:10.3390/en12030395.

NASA. 2021. The POWER Project. power.larc.nasa.gov/beta.

National Renewable Energy Laboratory. 2020. System Advisor Model. Version 2020.11.29 (SAM 2020.11.29). Golden, CO.

. 2021a. 2021 Annual Technology Baseline. Golden, CO. https://atb.nrel.gov/.

. 2021b. Hybrid Optimization and Performance Platform (HOPP). Version 1.0. https://github.com/nrel/hopp.

Nelson, J., N. G. Johnson, K. Fahy, and T. A. Hansen. 2020. "Statistical development of microgrid resilience during islanding operations". Applied Energy 279 (July): 115724. ISSN: 03062619. doi:10.1016/j.apenergy.2020.115724. https://doi.org/10.1016/ j.apenergy.2020.115724.

Obama, B. 2013. Presidential Policy Directive 21. https://obamawhitehouse.archives.gov/ the-press-office/2013/02/ 12/presidential-policy-directive-critical-infrastructure-security-and-resil.

Orrell, A. C., J. S. Homer, and Y Tang. 2018. "Distributed Generation Valuation and Compensation White Paper". Pacific Northwest National Laboratory, U.S. Department of Energy, no. February. https:// www. districtenergy. org/ HigherLogic/ System/ DownloadDocumentFile. ashx? DocumentFileKey= 0103ebf1 - 2ac9 - 7285 - b49d- e615368725b2{\&}forceDialog=0.

Reilly, J. 2020. MIRACL: Microgrids, Infrastructure Resilience, and Advanced Controls Launchpad. Tech. rep. Deparment of Energy. https://www.nrel.gov/wind/miracl.html {\#}controls.

Reiman, A. P., J. S. Homer, B. Bhattarai, and A. C. Orrell. 2020. "Quantifying technical diversity benefits of wind as a distributed energy resource". 2020 IEEE Power and

Energy Society Innovative Smart Grid Technologies Conference, ISGT 2020. doi:10.1109/ISGT45199.2020.9087665.

Ren, G., J. Wan, J. Liu, and D. Yu. 2019. "Spatial and temporal assessments of complementarity for renewable energy resources in China". Energy 177:262–275. ISSN: 03605442. doi:10.1016/j.energy.2019.04.023.https://doi.org/10.1016/j.energy.2019.04.023.

Schmidt, J., R. Cancella, and A. O. Pereira. 2016. "An optimal mix of solar PV, wind and hydro power for a low- carbon electricity supply in Brazil". Renewable Energy 85 (2016): 137–147. ISSN: 18790682. doi:10. 1016/j.renene.2015.06.010. http://dx.doi.org/10.1016/j.renene.2015.06.010.

Sengupta, M., Y. Xie, A. Lopez, A. Habte, G. Maclaurin, and J. Shelby. 2018. "The National Solar Radiation Data Base (NSRDB)". Renewable and Sustainable Energy Reviews 89 (January 2018): 51–60. ISSN: 18790690. doi:10. 1016/j.rser.2018.03.003. https://doi.org/10.1016/j.rser.2018.03.003.

Shaner, M. R., S. J. Davis, N. S. Lewis, and K. Caldeira. 2018. "Geophysical constraints on the reliability of solar and wind power in the United States". Energy and Environmental Science 11 (4): 914–925. ISSN: 17545706. doi:10.1039/c7ee03029k.

Slusarewicz, J. H., and D. S. Cohan. 2018. "Assessing solar and wind complementarity in Texas". Renewables: Wind, Water, and Solar 5 (1). ISSN: 2198-994X. doi:10.1186/s40807-018-0054-3. https://doi.org/10.1186/s40807-018-0054-3.

Sterl, S., S. Liersch, H. Koch, N. P. Lipzig, and W. Thiery. 2018. "A new approach for assessing synergies of solar and wind power: Implications for West Africa". Environmental Research Letters 13 (9). ISSN: 17489326. doi:10. 1088/1748-9326/aad8f6.

Tarroja, B., F. Mueller, and S. Samuelsen. 2013. "Solar power variability and spatial diversification: implications from an electric grid load balancing perspective". International Journal of Energy Research 37 (9): 1002–1016. ISSN: 0363907X. doi:10.1002/er.2903. arXiv: arXiv:1011.1669v3. http://doi.wiley.com/10.1002/er.2903.

Tierney, K., and M. Bruneau. 2007. "A Key to Disaster Loss Reduction". TR News: 14–18. ISSN: 07386826. http://onlinepubs.trb.org/onlinepubs/trnews/trnews250{_}p14-17.pdf.

U.S. Department of Energy. 2017. "Modern Distribution Grid Report, Volume III: Decision Guide". III:June 28. https://gridarchitecture.pnnl.gov/media/Modern-Distribution-Grid-Volume-III.pdf{\%}0Ahttps://gridarchitecture. pnnl.gov/modern-grid-distribution-project.aspx.

Vega-Sanchez, M. A., P. D. Castaneda-Jimenez, R. Pena-Gallardo, A. Ruiz-Alonso, J. A. Morales-Saldana, and E. R. Palacios-Hernandez. 2017. "Evaluation of complementarity of wind and solar energy resources over Mexico using an image processing approach". 2017 IEEE International Autumn Meeting on Power, Electronics and Computing, ROPEC 2017 2018-Janua (Ropec): 1–5. doi:10.1109/ROPEC.2017.8261637.

Widén, J. 2011. "Correlations between large-scale solar and wind power in a future scenario for Sweden". IEEE Transactions on Sustainable Energy 2 (2): 177–184. ISSN: 19493029. doi:10.1109/TSTE.2010.2101620.

. 2015. "A model of spatially integrated solar irradiance variability based on logarithmic station-pair correlations".

Solar Energy 122:1409–1424. ISSN: 0038092X. doi:10.1016/j.solener.2015.10.043.

Xu, L., Z. Wang, and Y. Liu. 2017. "The spatial and temporal variation features of wind-sun complementarity in China". Energy Conversion and Management 154 (June): 138–148. ISSN: 01968904. doi:10.1016/j.enconman. 2017.10.031. https://doi.org/10.1016/j.enconman.2017.10.031.

Zhang, H., Y. Cao, Y. Zhang, and V. Terzija. 2018. "Quantitative synergy assessment of regional wind-solar energy resources based on MERRA reanalysis data". Applied Energy 216 (February): 172–182. ISSN: 03062619. doi:10. 1016/j.apenergy.2018. 02.094. https://doi.org/10.1016/j.apenergy.2018.02.094.

Chapter 3

Opportunities for Hybrid Wind and Solar PV Plants in India*

Marty Schwarz
Amy Rose
Patrick Brown
Ilya Chernyakhovskiy
and David Palchak

The share of variable renewable energy (VRE) on India's grid has grown significantly in recent years, and the government has ambitious plans to accelerate its growth. The country surpassed 100 GW of renewable capacity in 2021 and plans to reach 500 GW by 2030 (Gupta 2021; IndBiz 2021). Wind and solar PV are expected to play a major role in achieving this goal (Chernyakhovskiy et al., 2021; Central Electricity Authority 2020). One strategy to increase wind and solar photovoltaic (PV) deployment is through the co-location of wind and solar PV plants to form a single hybrid power plant. By building wind and solar PV in the same location, hybrid plants have the potential to reduce transmission infrastructure costs and variability in the output power profile, compared to a stand-alone plant with a single technology. This is because the times at which the two technologies generate electricity are complementary in India; in general, wind speeds tend to increase during and after sunset hours. However, questions remain regarding the potential savings from hybridizing wind and solar PV plants in India and

---

* This is an edited, reformatted and augmented version of National Renewable Energy Laboratory Publication No. NREL/TP-6A40-81545, dated February 2022.

In: Wind Energy
Editor: Mehmet Landrey
ISBN: 979-8-89113-528-4

the locations for these opportunities. This resource analysis aims to address these questions and take a first step toward quantifying the opportunities for hybrid wind and solar PV plants in India.

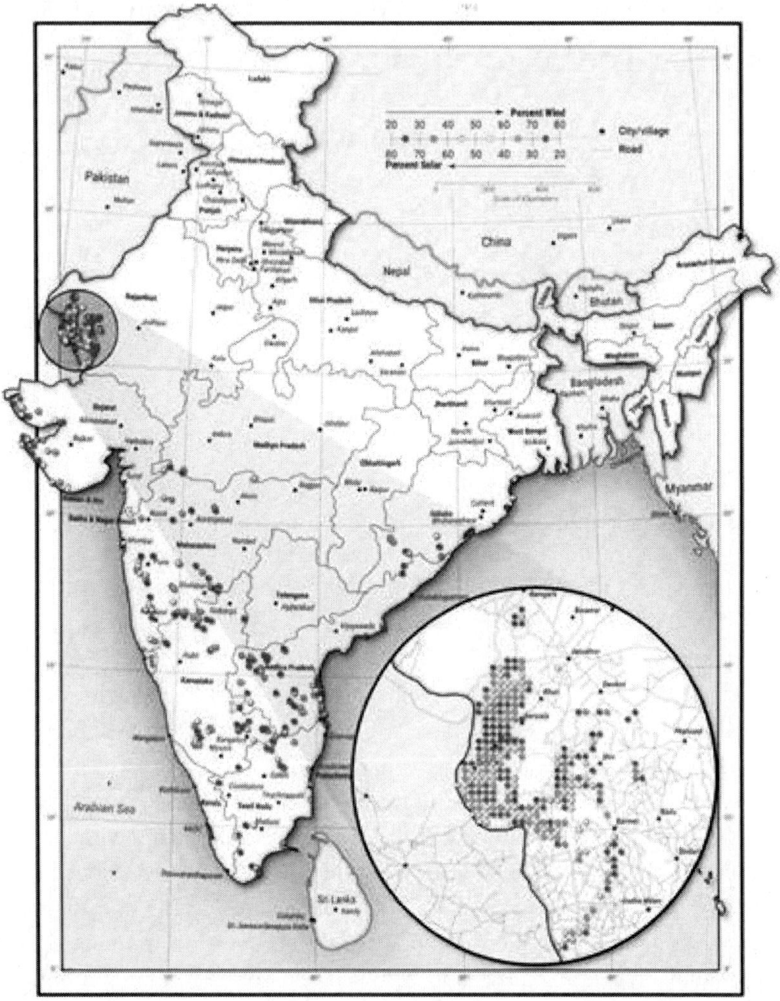

Figure 1. Estimated locations for hybrid wind and solar PV plants. Red dots indicate a higher proportion of solar PV, and blue dots indicate a higher proportion of wind. Only sites with a solar PV percentage between 20% and 80% are shown, to illustrate the criterion from MNRE's 2018 national hybrids policy. Map by Billy Roberts, NREL.

To identify potential locations for hybrid plants, we compared the levelized cost of energy (LCOE) of hybrid and stand-alone plants at all locations in India where VRE development could occur. As detailed in Textbox 1, our calculation of LCOE includes the capital cost, fixed operational and maintenance cost, and cost of interconnection. Because hybrid plants share a single interconnection, transmission costs savings may lead to a lower LCOE. For a given location, if the LCOE of a combined wind and solar PV plant is lower than the LCOE of either stand-alone technology, then a hybrid plant may provide some benefit. We find that the best locations for hybrid plants exhibit both (1) a high interconnection cost and (2) a wind capacity factor between roughly 34% and 38%. As shown in Figure 1, suitable locations for hybrid plants are most prevalent in western Rajasthan, western Gujarat, and a scattering of sites across southern India. The map shows the locations that, after optimizing for the mix of solar PV and wind at each site, theoretically meet the criterion from India's Ministry of New and Renewable Energy (MNRE) national 2018 policy for hybrid plants (MNRE 2018). The policy requires that the nameplate capacity of one technology is at least 25% the capacity of the other to qualify as a hybrid. In total, roughly 49 GW of wind capacity and 32 GW of solar capacity could achieve a lower LCOE in a hybrid configuration than in a stand-alone configuration at the same location. The results in Figure 1 also show the relative fractions of wind to solar PV that are optimal in each location. Red grid cells indicate sites where the lowest LCOE would be achieved with a solardominant hybrid plant, while blue cells represent locations favorable for wind-dominant hybrids. White cells show areas where an even proportion of wind and solar PV is optimal. Of these hybrid suitable sites, 63% are solar-dominant.

Textbox 1. Minimum LCOE calculation

A relatively simple method for quantifying the energy cost of a particular VRE plant is to calculate the total cost of producing a unit of energy. This metric, known as LCOE, is shown in Equation 1.[a]

$$\text{Lifetime LCOE}\left(\frac{INR}{MWh}\right)=\frac{\text{capital expenditure}+\text{interconnection cost}+\sum_{t=0}^{30\text{ years}}\text{fixed operational \& maintenance cost}_t*\text{discount rate}_t}{\text{average annual capacity factor}*8760\text{ hours}*\sum_{t=0}^{30\text{ years}}\text{discount rate}_t}$$

We used the National Renewable Energy Laboratory's (NREL's) Renewable Energy Potential (reV) tool with high resolution solar irradiance and wind speed data from the National Solar Radiation Database and Wind Integration National Dataset Toolkit to estimate capacity factor (CF) profiles for 30 km x 30 km grid cells across India and to calculate the LCOE. The geospatial electrical network data in reV was used to calculate the cost, in lakh/MW, of interconnecting new capacity in each location given the availability of nearby transmission.[b,c] We used the following assumptions of technology costs in 2030, the year for the 450-GW renewable energy target: capital costs of 554 and 334 lakh/MW, and fixed operational and maintenance costs of 23 and 10 lakh/MW for wind and solar PV, respectively. The cost and CF assumptions were used in Equation 1 to calculate the LCOE of stand-alone wind and solar PV for each grid cell.

We used a brute force optimization across a range of hybrid plant configurations—from 100% solar to 100% wind—to identify the type of plant with the lowest LCOE for each grid cell. The CF profile for each hypothetical hybrid plant, shown in the denominator in Equation 1, was equal to the profiles of the individual resources multiplied by their respective ratios with the interconnection capacity. These ratios of generator-to-interconnection capacity were the decision variables in the optimization, and we varied them from 0 to 2. In any hour, if the power output from a hybrid configuration exceeded the interconnection capacity, we limited the output to the interconnection capacity. The configuration with the lowest LCOE was assumed to be the optimal ratio of wind and solar PV capacity in each location.

[a] Our determination of where VRE development could occur is based on a set of land exclusions, including urban, rural, and protected areas, croplands, forest, grassland, bare lands, wetlands, water bodies, and airports. For more details, see Chernyakhovskiy et al., 2021 (Table A-15).

[b] These interconnection costs are common among NREL's other long-term planning studies in India, such as Chernyakhovskiy et al., 2021 and Rose et al., 2021.

[c] One Iakh is 100,000 rupees.

The cost-savings benefit of hybrid plants compared to standalone plants is relatively small, but this simple method does not capture the value of the electricity generated by looking at energy prices, nor does it quantify the potential of hybrids to provide other value streams such as firm capacity and reserves. Further, because the work does not compare solar PV and wind hybrids to alternative generation technologies or storage systems, it cannot be considered a holistic costbenefit analysis.

Hybrid Site Optimization: Two Examples

In cases where the wind and solar resource quality are comparable or hourly resource profiles are complementary (anti-correlated), combining the two technologies can result in more annual energy, relative to investment and operating cost, compared to developing either stand-alone resource. Figure 2 illustrates the diurnal average power output for two hypothetical hybrid plant sites with an assumed interconnection capacity of 200 MW. In the first example, on the left, the wind and solar PV nameplate capacities are similar. Even with a total installed capacity of over 300 MW, the complementary wind and solar generation profiles only exceed the 200-MW interconnection capacity during ~1% of hours, resulting in less than 1% clipped energy. Energy clipping, also known as curtailment, happens when the output from the VRE power plant exceeds the interconnection capacity, and the excess cannot be injected into the grid.

Because the LCOE optimization chooses the ideal generator- to-interconnection ratio, less than 1% of the annual energy is curtailed at all sites across India.

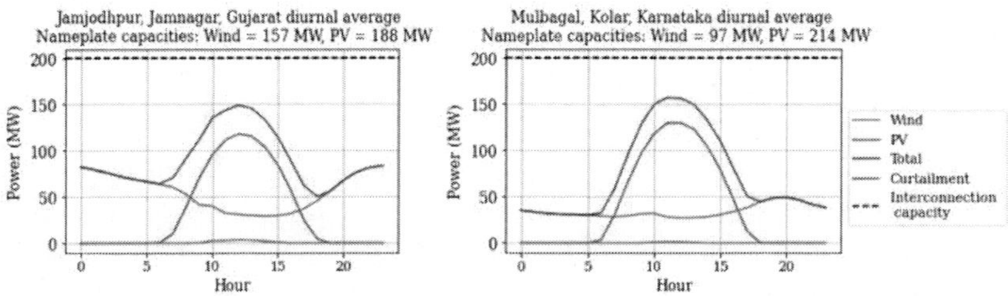

Figure 2. Diurnal average power output curves for two hypothetical LCOE-optimized hybrid plants.

The second example, on the right, illustrates a site where a solar-dominant location would benefit from adding a small amount of wind. The additional wind reduces the LCOE of the plant by 1% compared to building a stand-alone solar PV plant. This results in energy clipping during 5% of hours, but the curtailed energy is still less than 1% of the annual total.[1]

Impact of Interconnection Costs

The economic benefit of hybrid wind and solar PV plants comes largely from avoiding new substation and spur line investments in areas where the two technologies can maximize the use of the same network infrastructure.

Therefore, the estimated cost of interconnection can have a large impact on the potential locations for hybrid plants.[2]

To better understand this impact, we calculated the suitable locations for hybrid plants under a variety of interconnection cost assumptions. Figure 3 shows the change in results across India when the assumed interconnection costs at each site are doubled. The number of suitable hybrid sites almost triples, and the total nationwide hybrid potential increases from 49 to 87 GW wind and from 32 to 99 GW solar PV.

Interconnection costs can also impact the optimal mix of wind and solar PV at hybrid locations. Figure 4 shows the optimal hybrid plant configuration for a hypothetical site in western Rajasthan across a range of interconnection costs. When the interconnection cost is zero, an all-wind or all-solar PV plant always has a lower LCOE than a hybrid plant. This is because we only quantify cost savings from sharing an interconnection and ignore the other potential cost savingsfrom values streams like firm capacity and reserves.

As the interconnection cost increases, it becomes less expensive to build a hybrid plant, because the energy generation profiles are complementary and more electricity can be generated without exceeding the interconnection limit. At this particular site, this economic threshold occurs above an interconnection cost of about 75 lakh/MW.

The default assumption for this site, marked by the green dotted line, is 84 lakh/MW. At this interconnection cost, our optimization found a hypothetical hybrid plant with a capacity split of 74% solar PV and 24% wind to minimize lifetime LCOE.

---

[1] Dynamic line ratings have been proposed as a method to decrease curtailment by temporarily allowing more power to flow from the hybrid plant to the grid (IRENA 2020), but that analysis is beyond the scope of this study.

[2] Interconnection costs do not capture contractual fees and tariffs.

As the hypothetical interconnection cost increases above 125 lakh/MW, the lowest LCOE plant configuration becomes one with a more even split between wind and solar PV (between 50% and 60% PV percentage).

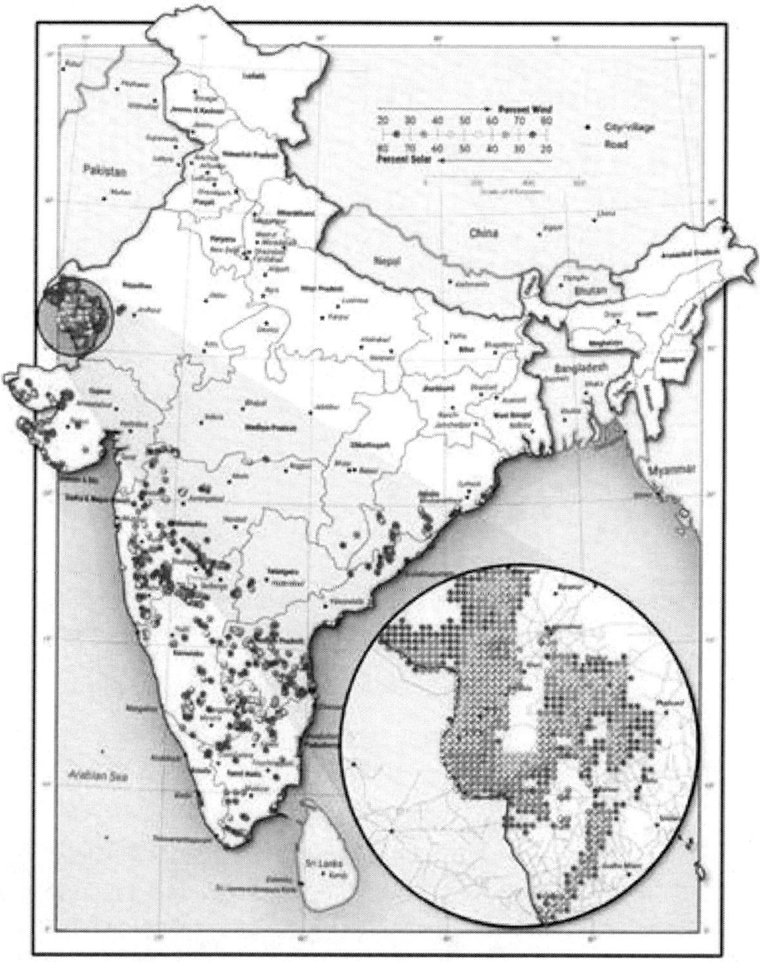

Map by Billy Roberts, NREL.

Figure 3. Estimated locations for LCOE-optimized sites, with the interconnection cost doubled from the original assumption.

Increased interconnection costs also affect the design of stand-alone plants. With high interconnection costs, it becomes efficient to use the spur line at its full capacity as often as possible. This means it may be more cost-effective to oversize the generation capacity, even if it results in some energy clipping. In general, though, a hybrid plant with complementary wind and solar PV generation profiles will result in less clipping for the same level of overbuilt generation capacity compared to a standalone wind or solar PV plant.

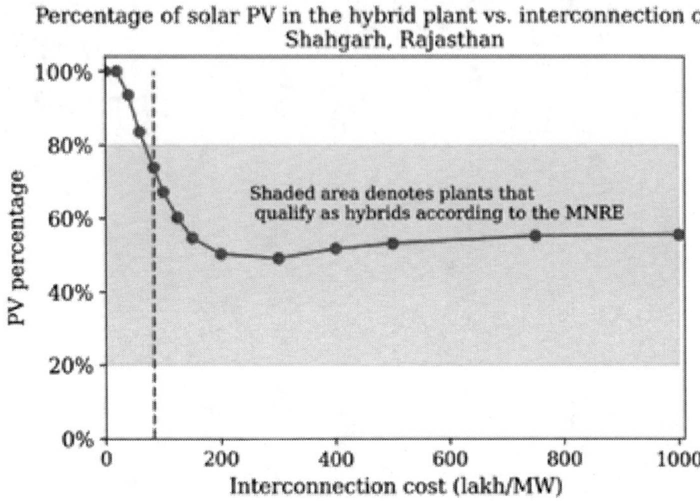

Figure 4. Impact of interconnection cost on the optimal wind/ solar PV ratio for a hypothetical hybrid site in Rajasthan. The default interconnection cost assumption for this site in our analysis, marked by the green dotted line, is 84 lakh/MW. The shaded area denotes plants that qualify as hybrids according to the MNRE.

Impact of Wind Resource

The second major factor that impacts whether a site may be financially suitable for hybrids in India is the strength of the wind resource. Figure 5 plots the impact of the wind capacity factor (CF) on the technology mix of each hypothetical hybrid site. The CF on the x-axis represents the average strength of the wind resource at each site. The colors of the dots indicate the technology mix.

Below an average wind CF of around 32%, stand-alone solar PV plants are the lowest LCOE option for most sites. This is because the projected capital cost of solar PV in 2030 is lower than that of wind. If the wind resource is not significantly stronger than the solar PV resource at a given site, a stand-alone solar PV site will always result in the plant configuration with the lowest LCOE.

However, as the wind CF increases above 32%, it is cost- effective to add wind capacity to stand-alone solar PV plants, resulting in solar PV-dominant hybrid plants. Most (82%) of the theoretical plants that qualify as hybrids according to the MNRE's 2018 policy occur between a stand-alone wind CF of 34% and 38%. Above a wind CF of 39%, which represents 14% of all sites, mostly stand-alone wind plants achieve the lowest LCOE.

Future Work

This analysis is an initial step toward quantifying the opportunities for wind-solar PV hybrid plants in India's future energy mix. Future work can expand both the scope and scale of this analysis. For example, this analysis only considered greenfield projects and did not consider opportunities to hybridize any of the existing 44 GW of wind and 40 GW of solar PV plants already commissioned across the country.

Existing stand-alone wind plants may be able to fit solar PV panels in the gaps surrounding wind turbines, presenting an opportunity to save money on land purchasing costs. However, wind turbines generally cannot fit into the footprint of existing solar PV plants.

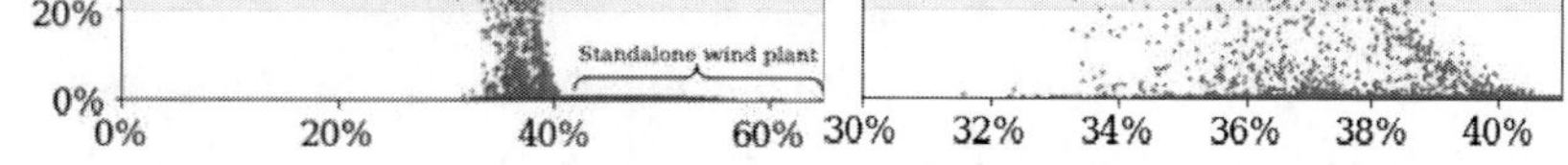

Figure 5. Impact of optimal wind capacity factor on the PV percentage in the LCOE-optimized plant, for the full range across all sites (left) and zoomed in on hybrid sites only (right). The colors represent the same PV percentage bins from Figures 1 and 3. The shaded red area denotes plants that qualify as hybrids, according to the MNRE.

Further, this study does not consider hybrid plants that utilize energy storage systems. Storage can improve the capacity credit and reliability of the hybrid plant as well as allow for more custom plant configurations to maximize the value of a hybrid plant. Storage can also potentially unlock additional ancillary service value streams, to which stand-alone VRE plants usually cannot contribute. Future analysis could run a similar LCOE optimization but could include existing locations for wind and solar PV plants as well as energy storage technologies.

This work only evaluates a few key aspects of hybrid plants: namely, the interconnection capital cost and the ideal wind- to-solar PV ratio. This simplified approach allows simultaneous evaluation of all sites across India, but more detail would be needed to design individual plants. Developers can utilize tools such as NREL's open-source Hybrid Optimization and Performance Platform to optimize the individual components of their hybrid plants, as well as run a detailed financial analysis (Dykes et al. 2020).[3]

Hybrid power plants could offer other benefits beyond lower LCOE. Future work can investigate the impact of hybridization on a plant's contribution to resource adequacy, ramping needs, and other ancillary services. These services will become increasingly important as VRE contribution in the electricity generation mix increases. Quantifying these grid benefits would require evaluation of time-synchronous plant output and system-wide electricity load, for example, in a production cost model.

As India looks increasingly to VRE to meet its electricity needs, wind and solar PV hybrids have the potential to provide lower energy costs compared to stand-alone technologies in specific locations. Technical assessments such as these can inform policymakers, regulators, electric utilities, and project developers as they seek to identify the role for hybrid plants in India's electricity mix.

Learn More

Find out about NREL's power system planning activities in India: Supporting India's States with Renewable Energy Integration.

Please contact SouthAsiaSupport@nrel.gov with any questions.

[3] https://www.nrel.gov/wind/hybrid-energy-systems-research.html.

References

Central Electricity Authority. 2020. Report on Optimal Generation Capacity Mix for 2029-30. Delhi, India: Central Electricity Authority. https://cea.nic.in/old/reports/ others/ planning/irp/Optimal_mix_report_2029-30_FINAL.pdf.

Chernyakhovskiy, Ilya, Mohit Joshi, David Palchak, and Amy Rose. 2021. Energy Storage in South Asia: Understanding the Role of Grid-Connected Energy Storage in South Asia's Power Sector Transformation. Golden, CO, USA: National Renewable Energy Laboratory. NREL/TP-5C00-79915. https://www.nrel.gov/docs/ fy21osti/79915.pdf.

Dykes, Katherine, Jennifer King, Nicholas DiOrio, Ryan King, Vahan Gevorgian, Dave Corbus, Nate Blair, Kate Anderson, Greg Stark, Graig Turchi, and Patrick Moriarty. 2020. Opportunities for Research and Development of Hybrid Power Plants. Golden, CO, USA: National Renewable Energy Laboratory. NREL/TP-5000-75026. https:/www.nrel.gov/docs/fy20osti/75026.pdf.

Gupta, Uma. 2021. "India hits 100 GW renewables milestone." PV Magazine International. https://www.pv-magazine.com/2021/08/13/india-hits-100-gw-renewables-milestone/.

IRENA. 2020. Innovation landscape brief: Dynamic line rating. Abu Dhabi, UAE: International Renewable Energy Agency. https://www.irena.org/-/media/ Files/IRENA/Agency/Publication/2020/Jul/IRENA_Dynamic_line_rating_2020. pdf?la=en&hash=A8129CE4C516895E7749FD495C32C8B818112D7C.

MNRE. 2018. National Wind-Solar Hybrid Policy. https://mnre.gov.in/img/ documents/uploads/2775b59919174bb7aeb00bb1d5cd269c.pdf.

Rose, Amy, Ilya Chernyakhovskiy, David Palchak, Marty Schwarz, Sam Koebrich, and Mohit Joshi. 2021. Pathways for Tamil Nadu's Electric Power Sector: 2020-2030. Golden, CO: National Renewable Energy Laboratory. NREL/TP-6A20-78266. https:// www.nrel.gov/docs/fy21osti/78266.pdf.

IndBiz: Economic Diplomacy Divison. 2021. "India Accelerates towards Mission 500GW by 2030." Ministry of External Affairs - Government of India. https://indbiz. gov.in/india-accelerates-towards-mission-500gw-by-2030/.

Chapter 4

Achieving American Leadership in the Wind Supply Chain*

U.S. Department of Energy

Summary

Land-based and offshore wind are expected to be a cornerstone for achieving U.S. clean electricity generation objectives, including 100% clean electricity by 2035 and an 30 gigawatts (GW) of offshore wind by 2030. Meeting these goals will require significant expansion of domestic supply chains and installation of unprecedented amounts of wind capacity.

To meet this challenge, researchers compiled a report examining the U.S. land-based and offshore wind energy supply chain, including raw materials, processed materials, components, sub-components, and recycling (see Figure 1).

Key Findings and Opportunities

The report identifies several key U.S. wind supply chain challenges and opportunities to help advance grid decarbonization and create jobs and economic opportunity for U.S. workers:

---

* This is an edited, reformatted and augmented version of the US Department of Energy, Office of Energy Efficiency and Renewable Energy, Publication No. DOE/OP-0013-1, dated February 24, 2022.

In: Wind Energy
Editor: Mehmet Landrey
ISBN: 979-8-89113-528-4

- Accelerate offshore wind supply chain development. The domestic supply chain for offshore wind towers, blades, nacelles, and substructures is in its infancy. Several manufacturers have announced the intent to begin production at U.S. facilities, but growth needs to accelerate to meet the 30 GW by 2030 goal with primarily domestic supply.

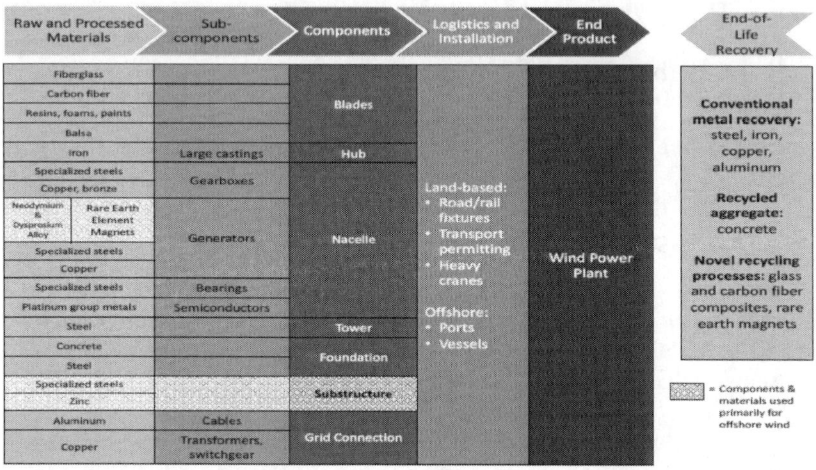

Figure 1. Land-based and offshore wind supply chain overview.

- *Maintain and expand the land-based wind supply chain.* Domestic content in land-based wind turbine blades has declined in recent years. Keeping blade and other component production facilities in the United States while scaling up the land-based wind-component supply chain will protect and expand American jobs and support the Administration's clean energy goals.
- *Develop offshore wind port and vessel infrastructure*. Offshore wind will require specialized ports and vessels. Port development could spur clustering opportunities, such as the establishment of new manufacturing facilities to supply offshore wind components. There are only a handful of vessels globally that can install the large offshore wind turbines that will be used in U.S. projects, and only one under construction that will meet the requirements of the Jones Act. Although the U.S. offshore wind industry can leverage international vessels to accelerate near-term deployment, global demand will be high, and building ports and vessels in the United States would create

new jobs and opportunities for U.S. maritime industries and help developers of U.S. offshore wind projects avoid bottlenecks and delays.

- *Improve supply chain competitiveness through R&D.* Expansion in wind technology R&D could increase the competitiveness of the U.S. supply chain for offshore and land-based wind and allow U.S. suppliers to scale up. Topics could include blade manufacturing automation, additive manufacturing of large castings and forgings, and modularization and onsite manufacturing of large components such as blades and towers. Such research would also result in domestic intellectual property development, accelerate clean energy deployment, and fundamentally alter U.S. competitiveness. In areas such as large castings and forgings, R&D would have benefits beyond wind—for example, both hydropower and nuclear energy require large castings that cannot currently be produced in the United States.
- *Expand training programs to support the wind workforce.* Finding and training workers with a wide range of specialized skills for building and maintaining wind energy facilities are critical challenges. Wind-related job totals in the United States reached 116,800 full-time wind workers in 2020. Meeting the unprecedented expansion required to meet the Administration's clean energy goals will require training and development for a range of professions including design, installation, operations and maintenance, component and subcomponent manufacturing and assembly, and materials production.
- *Improve land-based wind logistics.* Current blades, towers, and nacelles are becoming too large to efficiently transport over existing road and rail networks, and moving components is further challenged by varied state and local transport permit requirements. Making components more modular, along with policy and regulatory solutions, could reduce the challenges presented by overland transport of these components.

Policy Next Steps

The following policies would help to meaningfully address the challenges and opportunities identified above. A complete discussion of policy opportunities can be found in the report America's Strategy to Secure the Supply Chain for a Robust Clean Energy Transition: U.S. Department of Energy Response to Executive Order 14017: America's Supply Chains.

- Enact legislation to provide tax incentives to support domestic clean energy manufacturing and deployment, including incentives for clean energy production, building new manufacturing facilities, and for the ongoing operation of those facilities.
- Prioritize financing of offshore wind ports and vessels through Department of Transportation Maritime Administration programs and the DOE Loan Programs Office.
- Fund transportation improvements and standardize transport permitting requirements for large wind components that need to cross jurisdictional boundaries, in partnership with the Department of Transportation and state and local governments.
- Engage government and the private sector to expand research, development, and demonstration of technologies to enhance U.S. wind supply chain competitiveness and reduce logistical requirements.

Download the full document and the corresponding other documents that are part of the DOE response to the supply chain executive order at: www.energy.gov/policy/supplychains.

Chapter 5

Materials Used in U.S. Wind Energy Technologies: Quantities and Availability for Two Future Scenarios*

Annika Eberle[1]
Aubryn Cooperman[1]
Julien Walzberg[1]
Dylan Hettinger[1]
Richard F. Tusing[1]
Derek Berry[1]
Daniel Inman[1]
Senu Sirnivas[1]
Melinda Marquis[1]
Brandon Ennis[2]
Evan Sproul[2]
Ryan Clarke[2]
Joshua Paquette[2]
Thomas Hendrickson[3]
William Morrow[3]
Sujit Das[4]

* This is an edited, reformatted and augmented version of National Renewable Energy Laboratory, operated by Alliance for Sustainable Energy, LLC, for the U.S. Department of Energy (DOE), Technical Report, NREL/TP-6A20-81483, dated August 2023.

In: Wind Energy
Editor: Mehmet Landrey
ISBN: 979-8-89113-528-4

Matthew Korey[4]
Parans Paranthaman[4]
Robert Norris[4]
Lillie Ghobrial[4]
Sridhar Seetharaman[5]
and Yuri Korobeinikov[5]
[1]National Renewable Energy Laboratory
[2]Sandia National Laboratories
[3]Lawrence Berkeley National Laboratory
[4]Oak Ridge National Laboratory
[5]Arizona State University

Acknowledgments

The authors are grateful for the support of the following individuals in preparing this report:

- Sheri Anstedt, National Renewable Energy Laboratory (NREL)
- Joshua Bauer, NREL
- Mike Derby, U.S. Department of Energy's Energy Efficiency and Renewable Energy Wind Energy Technologies Office (DOE EERE WETO)
- Aletta Dsouza, Arizona State University
- Patrick Gilman, DOE EERE WETO
- Sherif Khalifa, NREL
- Eric Lantz, NREL
- Nicole Leon, NREL
- Abigayle Moser, NREL
- Owen Roberts, NREL
- Matt Shields, NREL
- Brian Smith, NREL
- Liam Watts, NREL
- Jarett Zuboy, NREL

List of Acronyms

ATB	Annual Technology Baseline
DOE	U.S. Department of Energy
GW	gigawatt
kg	kilogram
m	meter
MW	megawatt
NREL	National Renewable Energy Laboratory
PVC	polyvinyl chloride
REMPD	Renewable Energy Materials Properties Database
t	metric tons (tonnes)
USGS	United States Geological Survey

Executive Summary

Wind energy is one of the fastest-growing sources of renewable energy. The Biden administration's decarbonization goals (i.e., carbon-pollution-free electricity by 2035 and net-zero economy by 2050) will require at least a threefold increase in the U.S. wind energy deployment rate from current average levels of 10 gigawatts/year (United States Department of State and the United States Executive Office of the President 2021). Increased deployment of wind energy technologies will influence the demand for raw and processed materials that are required to manufacture and operate wind power plants and could therefore impact national resource use and physical materials availability, including critical materials.

Prior research has performed cross-technology assessments of critical material requirements for renewable energy technologies under low-carbon and clean energy futures, explored material requirements and supply chain constraints for specific types of clean energy technologies, and evaluated how deploying these technologies might influence the demand for certain critical materials (e.g., critical minerals). However, no studies have yet developed detailed estimates for material needs associated with U.S. land-based and offshore wind deployment under plausible high-deployment scenarios that would be needed to achieve decarbonization goals.

In this report, the authors explore how material needs for wind energy might change under two U.S. wind deployment scenarios: Current Policies and High Deployment. The Current Policies scenario represents a business-as-usual level of wind energy deployment, and the High Deployment scenario includes high levels of wind energy deployment consistent with achieving the goal of 100% clean electricity by 2035 and net-zero emissions economywide by 2050. We use the Renewable Energy Materials Properties Database (REMPD) to project the amount and types of materials that will be needed for wind energy deployment in the United States under each scenario from 2020 through 2050. We then analyze potential U.S. vulnerabilities linked to physical materials availability and provide some initial recommendations about new technologies that could mitigate resource constraints for wind energy technologies.

We find that the projected annual U.S. demand for materials to construct wind power plants from 2020 through 2050 is anticipated to be less than 2% of global production in 2020 for most materials. Key exceptions include balsa, carbon fiber, glass fiber, nickel, and the rare-earth elements dysprosium and neodymium (Figure ES-1). Our results show that demand for balsa and carbon fiber for U.S. wind energy could reach or exceed current levels of global production in the High Deployment scenario. In addition, there will likely be continued demand for carbon fiber from other countries and sectors that are not considered in this study. Thus, increased domestic or foreign production of carbon fiber will likely be required to achieve U.S. decarbonization goals. Annual production of balsa depends on the amount of land that can be devoted to growing balsa in suitable climate regions. If demand for balsa in wind energy applications begins to exceed production, other materials such as polymer foams may substituted. Demand for glass fiber, nickel, and rare earth elements for U.S. wind energy in the High Deployment scenario peaks from 2038 to 2044 and approaches, respectively, 88%, 35%, and 50% of 2020 global production of these materials. Although these amounts are within current production levels, if production does not increase there may be more competition for access to these materials because of accelerating wind energy deployment worldwide.

We also consider the scale of demand for materials to support U.S. wind energy deployment in relation to domestic production. The amount of nickel used in U.S. deployment of wind energy technologies is already larger than the amount produced domestically. Several critical minerals used for wind energy including gallium, natural graphite, tin, and some elements used in steel alloys (e.g., chromium, manganese, niobium, and titanium) are not mined in the United States. Multiple strategies can be applied to secure supply or limit demand for these materials, including diversifying import sources to minimize supply chain risk, increasing reuse and recycling, modifying wind turbine designs to reduce material demand, substituting alternative materials where possible, and developing domestic sources if they exist. These strategies are also relevant to materials for which demand from wind energy after 2030 is projected to represent a significant share (greater than 20%) of current domestic production, including cobalt, praseodymium, copper, and aluminum.

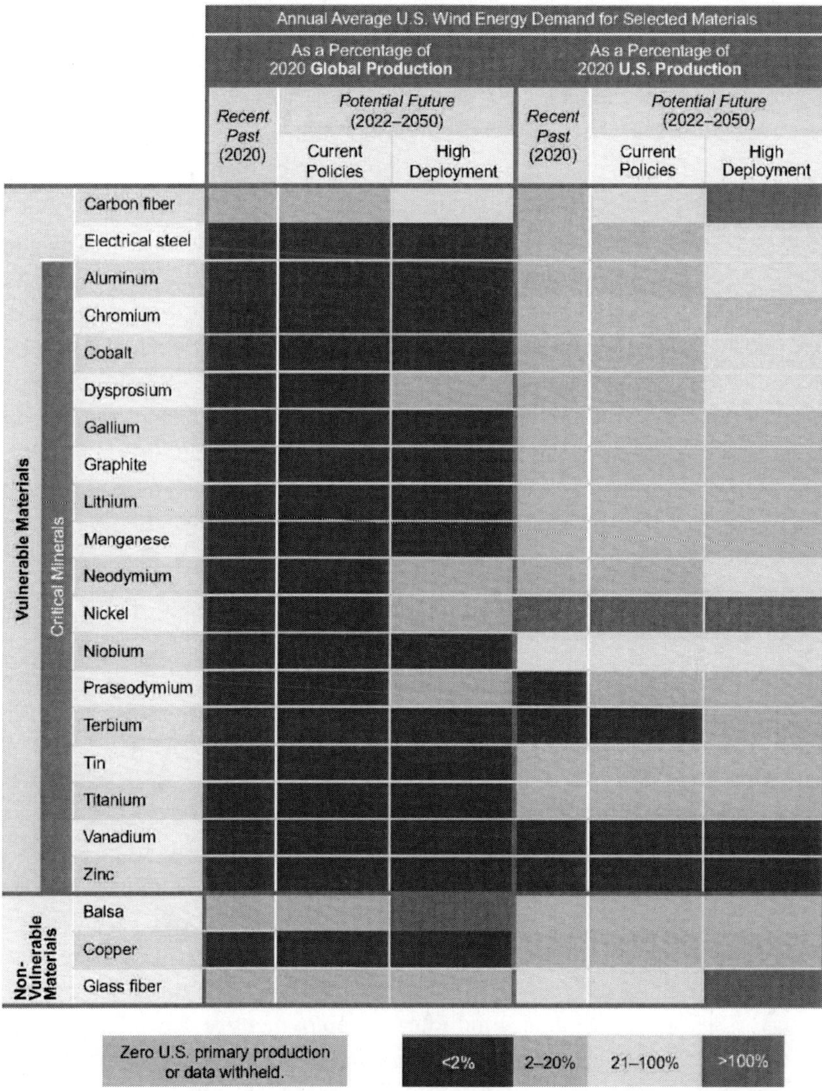

Figure ES-1. Annual U.S. wind energy demand for selected materials as compared to global and U.S. production in 2020.

Although the material demand projections presented in this study provide some insight into which materials may pose challenges for wind energy development, there is scope for additional work to better understand likely constraints and possible solutions. The current work assumes relatively limited changes in wind power plant design and related material requirements. Future work could explore potential impacts of technology innovations on material requirements and incorporate feedback between supply chain constraints and wind plant design to identify technology evolution pathways that avoid material supply bottlenecks.

This study considers only current production and known reserves of wind energy materials. Therefore, a more complete investigation of future material supply chains could identify additional supply risks (for example, due to changes in demand from other countries and industries, or declining mine production) and opportunities for new or expanded production sources. Further research could extend the REMPD to include additional supply chain risk metrics (e.g., likelihood of foreign supply disruption, dependency of U.S. manufacturers on foreign suppliers, and ability of U.S. manufacturers to withstand a supply disruption) and perform a more detailed analysis of supply risks. Other avenues for future work include expanding the REMPD by adding information on wind energy externalities (e.g., emissions from manufacturing and transportation), incorporating other renewable energy technologies (e.g., geothermal plants, marine and hydrokinetic plants, hydrogen electrolyzers, or battery energy storage systems) into the database, and performing a cross-technology analysis of material requirements.

1. Background

Renewable energy deployment is increasing globally. In the United States, generation from utility-scale wind and solar increased by 11% and 20%, respectively, from 2020 to 2021 (Lawrence Livermore National Laboratory 2021). This growth is expected to continue and may need to increase to achieve domestic objectives for clean energy.[1] For example, the United States' long-term strategy to reach net-zero greenhouse gas emissions by 2050 identifies a need for annual wind energy deployment of 25 to 30 gigawatts (GW) per year, which is approximately three times recent annual average

[1] Refer to Appendix A for more details about the Biden administration's clean energy goals, including the Hydrogen Shot.

deployment levels (United States Department of State and the United States Executive Office of the President 2021).

Increased deployment of clean energy technologies will affect the demand for raw and processed materials that are required to manufacture and operate these technologies, thus impacting resource use and physical materials availability. Certain materials that play an important role in the economy and are at risk of supply disruption have been designated as critical (or vulnerable) materials in various regions, including the United States (National Research Council 2008; Achzet and Helbig 2013; Graedel and Reck 2016; Hofmann et al. 2018; Schrijvers et al. 2020).[2] The availability of these critical materials may be further limited as global resource use increases as a result of population growth and economic development (United Nations Environment Program [UNEP] 2016).

Prior research has identified critical material requirements for a broad set of clean energy technologies and examined how increased deployment of these technologies might influence the demand for certain critical materials (e.g., critical minerals) (American Physical Society Panel on Public Affairs and Materials Research Society 2011; Atwater et al. 2011; Bauer et al. 2010; Fraunhofer Institute for Systems and Innovation Research ISI et al. 2013; International Energy Agency [IEA] 2021; Junne et al. 2020; World Bank Group 2017). Several prior studies focused on international or European demand for energy materials (Fraunhofer Institute for Systems and Innovation Research ISI et al. 2013; IEA 2021; Junne et al. 2020; World Bank Group 2017). Two others provided a broad assessment of availability and risks without detailed quantification of the projected material needs for U.S. energy technologies (American Physical Society Panel on Public Affairs and Materials Research Society 2011; Atwater et al. 2011). And, at least one study, the U.S. Department of Energy's (DOE's) *Critical Minerals and Materials Strategy* (Bauer et al. 2010), performed a more detailed analysis of the supply and demand of critical materials in the context of U.S. clean energy technologies and estimated future demand for four specific components: permanent magnets in wind turbines and electric vehicles, advanced batteries in electric vehicles, thin-film semiconductors in solar photovoltaic power systems, and phosphors in high-efficiency lighting systems.

In addition to these cross-technology assessments of critical material requirements, prior work also analyzed material requirements and supply chain constraints for individual types of clean energy technologies (Ardani et

---

[2] See Appendix B for a brief history of critical materials research.

al. 2021; Baars et al. 2021; Cao et al. 2019; Carrara et al. 2020; Dunn et al. 2021; Nassar et al. 2016; DOE 2022; Wilburn 2011; J. Yang et al. 2020). For example, global and regional estimates of the materials needed to fulfill current and future wind energy deployment have been provided in several studies (Cao et al. 2019; Carrara et al. 2020; IEA 2021; Nassar et al. 2016; Wilburn 2011; J. Yang et al. 2020). These studies vary in their estimates of material types and quantities required per megawatt (MW) of installed capacity; some of this variation stems from differences in material requirements between wind turbines of different sizes or configurations. For example, the generator type determines whether the wind turbine requires rare-earth permanent magnets, and the level of rare-earth element demand linked to wind energy deployment has been estimated in various studies (Alves Dias et al. 2020; Fishman and Graedel 2019; Fraunhofer Institute for Systems and Innovation Research ISI et al. 2013; Habib et al. 2014; Hoenderdaal et al. 2013; Li et al. 2020; Ren et al. 2021). In addition, comparative studies have quantified material impacts from alternative wind turbine configurations and wind blades (Cooperman et al. 2021; Carrara et al. 2020; Guezuraga et al. 2012; Ozoemena et al. 2018; Schreiber et al. 2019). At least one study has also evaluated critical material requirements for solar photovoltaic systems in the context of potential high-deployment scenarios in the United States (Ardani et al. 2021).

However, no studies have yet estimated detailed material requirements of U.S. land-based and offshore wind energy technologies under plausible high-deployment scenarios, such as the carbon-pollution-free power sector by 2035 and net-zero economy by 2050 that would be required to achieve the Biden administration's federal sustainability goals (The White House 2021). Because many wind energy materials are sourced and processed globally, U.S. energy security and economic health are vulnerable to disruptions in the supply of these materials outside our borders. Thus, it is important to develop a more detailed understanding of how U.S. demand for wind energy materials might change under various deployment scenarios.

1.1. Motivation and Goals

Building on the Energy Act of 2020's guidance to establish a physical materials property database for wind energy (Consolidated Appropriations Act 2021), this report provides a detailed analysis of wind material requirements at the scales required to achieve the Biden administration's objectives. Here, we explore how material requirements for wind energy might change under two scenarios, analyze potential U.S. vulnerabilities linked to physical materials availability, and provide some initial recommendations about new technologies that could mitigate resource constraints for wind energy technologies. We use a newly developed tool for estimating material requirements associated with renewable energy technologies—the Renewable Energy Materials Properties Database (REMPD) (NREL 2023)—to project the amount and types of materials that will be needed for wind energy deployment in the United States from 2020 through 2050. The analysis performed here incorporates variability in wind turbine designs and technological improvements and includes a review of the geographical origin of wind turbine materials and availability. These results could help inform better planning to mitigate potential material supply risks for wind energy deployment.

The goals of this report are to: (1) improve understanding of the constraints and vulnerabilities that exist for physical materials availability and manufacturing supply chains under two wind energy deployment scenarios, and (2) identify how new technologies could mitigate resource constraints. We describe our methodology in Section 2, summarize results for material requirements to 2050 in Section 3, provide an overview of opportunities for innovation to alter material requirements for future wind power plants in Section 4, and summarize our conclusions in Section 5. We perform our analysis at a high level for all materials used in wind energy technologies and examine a subset of vulnerable materials (defined in Section 1.2) in greater detail.

1.2. Definition of Vulnerable Materials

Within this report, the term "vulnerable materials" encompasses all materials used in wind power plants that are at risk of supply chain disruption, including critical minerals. The Energy Act of 2020 defines a critical mineral as:

> "Any mineral, element, substance, or material designated as critical under subsection (c) except fuel minerals; water, ice, or snow; common varieties of sand, gravel, stone, pumice, cinders, and clay."

The Energy Act of 2020 (2020) also defines how the Secretary of the Interior, acting through the Director of the United States Geological Survey

(USGS), should establish a list of critical minerals, to be revised at least every 3 years. These minerals should meet the following criteria:

- They are essential to the economic or national security of the United States
- Their supply chain is vulnerable to disruption (e.g., due to military conflict, foreign political risks, or sudden demand growth)
- They serve an essential function in the manufacturing of a product (e.g., energy technology, defense, electronics); the absence of which would have significant consequences for the economic or national security of the United States.

A more detailed description of the assessment methodology for material criticality used by the USGS is provided by Nassar and Fortier (2021). In its most recently published list, the USGS identified 50 critical minerals (USGS 2022).

Table 1 lists the vulnerable materials considered here along with the reason for their designation and role in wind energy technologies. The list includes 2 vulnerable materials and 17 critical minerals (the 33 other minerals on the 2022 USGS list play little to no role in wind energy generation facilities[3]). The two vulnerable materials that are not on the USGS list of critical minerals are carbon fiber and electrical steel. Carbon fiber is used to provide structural strength for wind turbine blades, and electrical steel is used in power generators in the nacelle and in transformers. The constituents of carbon fiber and electrical steel are largely not critical by themselves (e.g., electrical steel comprises iron, silicon, carbon, and some aluminum). However, the global capacity for manufacturing these materials is limited. For example, the manufacturing process for electrical steel is metallurgically specialized, and the technical details of the process are highly guarded in the industry. The United States currently relies on Canada and Mexico to augment its domestic production of electrical steel. As a result, carbon fiber and electrical steel are at high risk for supply chain disruption and are thus included in our list of vulnerable materials.

---

[3] The 33 minerals included in the 2022 United States Geological Survey list of critical minerals that play a minor or no role in wind energy technologies are antimony, arsenic, barite, beryllium, bismuth, cerium, cesium, erbium, europium, fluorspar, gadolinium, germanium, hafnium, holmium, indium, iridium, lanthanum, lutetium, magnesium, palladium, platinum, rhodium, rubidium, ruthenium, samarium, scandium, tantalum, tellurium, thulium, tungsten, ytterbium, yttrium, zirconium.

Table 1. Vulnerable materials and their role in wind energy technologies

Type of Material(s)	Reason for Vulnerable Material Designation	Primary Role in Wind Energy Technologies
Carbon fiber	High risk of supply chain disruption	Structural elements in wind turbine blades
Electrical steel	High risk of supply chain disruption	Power generators, transformers
Aluminum	2022 USGS critical mineral	Power cables, nacelle/tower internal equipment
Chromium, cobalt, manganese, nickel, niobium, titanium, vanadium	2022 USGS critical mineral	Steel alloying elements
Graphite, lithium, nickel	2022 USGS critical mineral	Batteries
Dysprosium, neodymium, praseodymium, terbium	2022 USGS critical mineral	Rare-earth permanent magnets
Gallium	2022 USGS critical mineral	Wide bandgap semiconductors for power electronics
Tin	2022 USGS critical mineral	Bronze
Zinc	2022 USGS critical mineral	Anticorrosion coatings (galvanization)

2. Methodology to Assess Material Quantities

In this report, we use the REMPD (NREL 2023) to assess the types and quantities of materials required to construct wind energy technologies. We also compare material demands for wind energy requirements to available materials and provide some initial insights about how new technologies could potentially mitigate resource constraints.

2.1. Overview of the REMPD

The REMPD is a relational database developed using the open-source database server PostgreSQL (PostgreSQL Global Development 2022). A publicly available version of the database can be found at: https://apps.openei.org/REMPD/. A summary of the REMPD including capabilities, definitions, and metrics used in the database are provided in Cooperman et al. (2023). For ease of reference, we provide a high-level overview of the database here and reproduce two figures from Cooperman et al. (2023) (refer to that report for more details). We also provide more information about the REMPD's scenario analysis capabilities, including the system boundary for scenario analysis.

2.1.1. Data Taxonomy

The REMPD uses a six-tiered approach to collect and organize data on the material requirements and properties associated with renewable energy technologies. The database currently includes data for both wind and solar energy technologies. In this report, we describe the database taxonomy in the context of wind energy technologies as that is the focus of our analysis. However, the database does provide additional data on solar technologies (refer to Cooperman et al. [2023] for more details).

The top tier of the REMPD data taxonomy comprises all components and materials required to construct all facilities in the selected category (e.g., all wind power plants in the United States). The next level captures finished components, such as the wind turbine, substation, and electrical cables. Each component is associated with relevant subassemblies and subcomponents (e.g., the pitch drive in a wind turbine is a subcomponent of the hub subassembly). The next tier includes the finished materials, or primary processed materials, such as steel, that are required to manufacture the component, subassembly, and/or subcomponent. The lowest tier provides the raw materials, which also include some secondary processed materials (e.g., glass) that are required to manufacture the finished materials. This taxonomy allows the database to capture all material requirements for energy technologies and break down the material requirements by component, which allows users to explore where materials are used within each technology and help identify opportunities for reducing material requirements.

In some cases, due to data constraints and the desire for the database to focus primarily on material quantities, the REMPD does not have all data at the subassembly or subcomponent levels. However, in all cases, the REMPD does include data for the finished and raw materials associated with each finished component. For example, substation data are only broken down by material type and are not subdivided at the subassembly or subcomponent levels; wind turbine data are subdivided into multiple subassemblies and subcomponents.

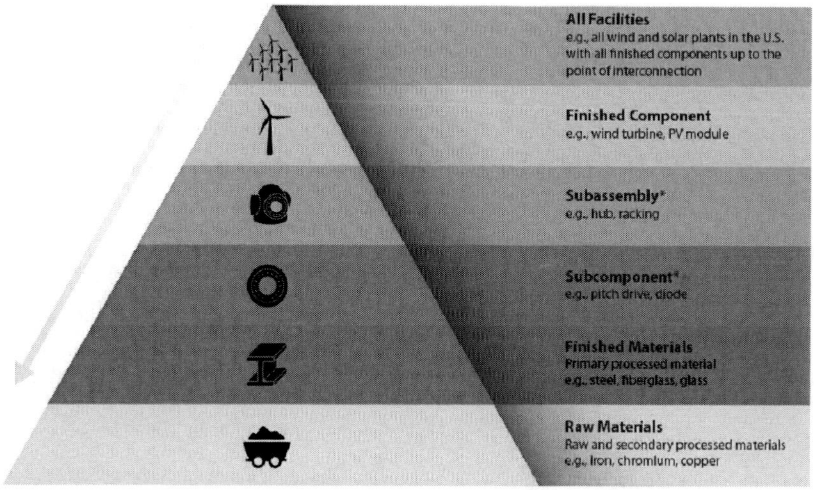

Figure 1. Taxonomy used to organize data in the REMPD. Asterisks note that not all components in the database have data at the subassembly and subcomponent levels. These two tiers are populated based on available data (i.e., whether the materials needed for each component can be disaggregated to the subassembly and/or subcomponent levels, or if they are instead reported at a higher level, such as the finished component level). These data could be added to the database in the future. *Illustration by Nicole Leon, National Renewable Energy Laboratory (NREL).*

2.1.2. Wind Energy System Components

Figure 2 illustrates the wind energy system components that are included in the REMPD and used in the analysis performed here. The five types of components are:

1. Wind turbines, which comprise four subassemblies: the hub, blades, nacelle, and tower
2. Foundation (for land-based wind systems) or substructure (for offshore wind systems)
3. Array and export cables
4. Site access roads (for land-based systems)
5. Substations.

We include all material requirements for the wind turbine (i.e., the nacelle/drivetrain, generator, tower, blades, and hub) and all balance-of-system materials needed to the point of interconnection (i.e., land-based foundation or offshore substructure, electrical cables, substations, and other site parts [e.g., roads for land-based wind power plants]). We exclude material requirements for capital equipment associated with transporting and installing the components as well as materials needed for operating, maintaining, and decommissioning the wind plant (e.g., we do not include the materials needed to construct cranes or other construction equipment and we do not include the fuel required to transport the materials to and from facilities throughout the wind plant life cycle).

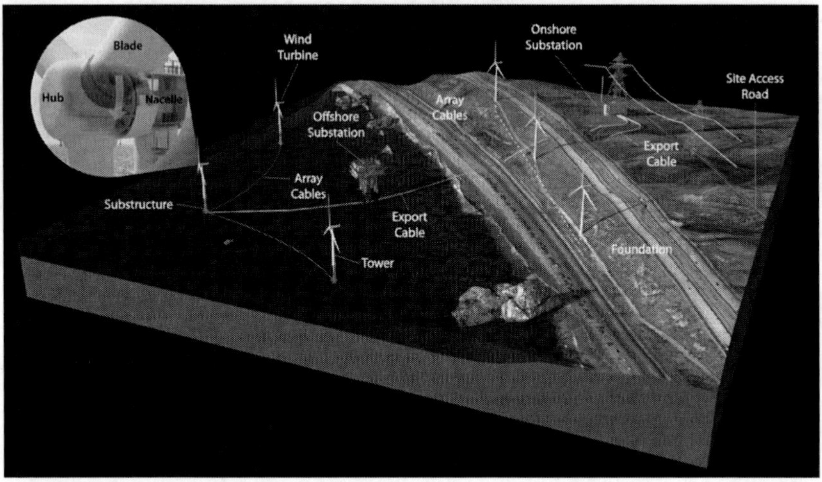

Figure 2. System components included in our analysis of wind energy material requirements.
Illustration by Joshua Bauer, NREL.

2.1.3. System Boundary for Scenario Analysis

Materials for wind energy technologies come from diverse, global supply chains. These supply chains will continue to evolve in the future as materials selection and availability change. The system boundary used in the REMPD allows the database to capture two types of materials—raw and finished materials—at two levels: the foreground system and background system (Figure 3). Finished materials include primary processed materials (e.g., steel), which are required to manufacture the component, subassembly, or subcomponent. Raw materials include materials that are required to manufacture the finished materials (e.g., critical minerals and some secondary processed materials [e.g., glass]). The foreground system captures processes and materials that are under direct control or decisive influence of renewable energy manufacturers and developers, such as the quantity and type of rare-earth elements used in permanent magnets. Inputs to the foreground system include both raw and processed materials because manufacturers may exert control over materials selection at both levels. Foreground material quantities come from original equipment manufacturers and published literature. The background system includes the upstream processes that are required to extract and process raw materials and manufacture processed materials used in the foreground system). The REMPD system boundary excludes materials used for processes after manufacturing, such as transportation, installation, operations and maintenance, or decommissioning.

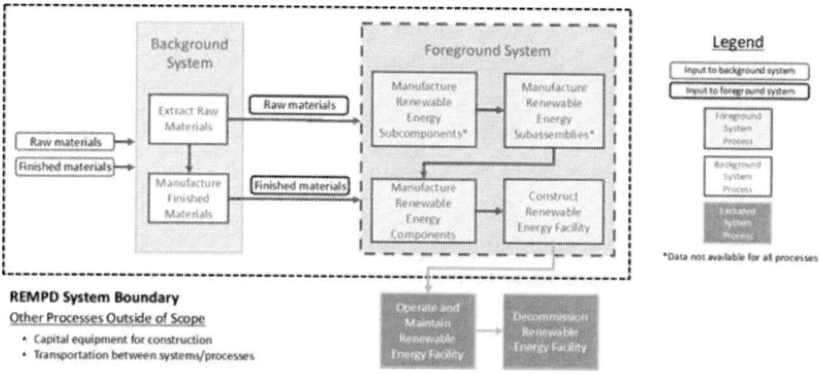

Figure 3. System boundary used in the REMPD.
Illustration by Annika Eberle, NREL.

To identify and characterize the material quantities in the background system, we employ tools from life cycle assessment. Life cycle assessment allows researchers to model the inputs (e.g., materials and fossil-fuel consumption) and outputs (air pollutant emissions, discharges to water) of a product or process's life cycle from resource extraction through manufacturing, use, and disposal. In the REMPD, we use life cycle inventories generated for life cycle assessment to estimate the background material for the foreground system.[4] As an example, the foreground material steel is used in wind system towers, land-based foundations, and offshore wind substructures. Using existing life cycle inventories, the specific background material requirements for steel are characterized, which include iron, manganese, nickel, titanium, and chromium, among others. This approach positions future work on the REMPD to calculate other types of resource requirements and

[4] Refer to Appendix C for a list of sources used for the life cycle inventory data.

impacts (e.g., global warming potential, emissions to water, energy consumption) using the background material data available in the REMPD.

2.1.4. Scenario Analysis Capabilities

We leverage the REMPD's scenario analysis capabilities (illustrated in Figure 4) to evaluate material quantities and assess and compare material properties (e.g., availability). Performing scenario analysis using the REMPD involves defining a scenario (a combination of three inputs: a capacity projection, a plant configuration, and a technology configuration) and specifying scaling relationships that allow a user to vary material quantities based on plant configurations and capacity projections. The REMPD then connects that scenario definition to the required renewable energy components, subassemblies, and subcomponents and identifies the required foreground and background materials associated with them. The REMPD uses this information, along with the scaling relationships, to evaluate material quantities. It also joins the required materials with their associated material properties. For any given scenario, a researcher can calculate the amount of all materials needed to construct a single wind turbine, wind plant, or all wind power plants in the United States. Likewise, once a vulnerable material is identified, all scenarios relying on the material can be identified. The relationships defined in the REMPD allow researchers to discover vulnerabilities in the supply chain by connecting materials to countries of origin and national availability, among other characteristics.

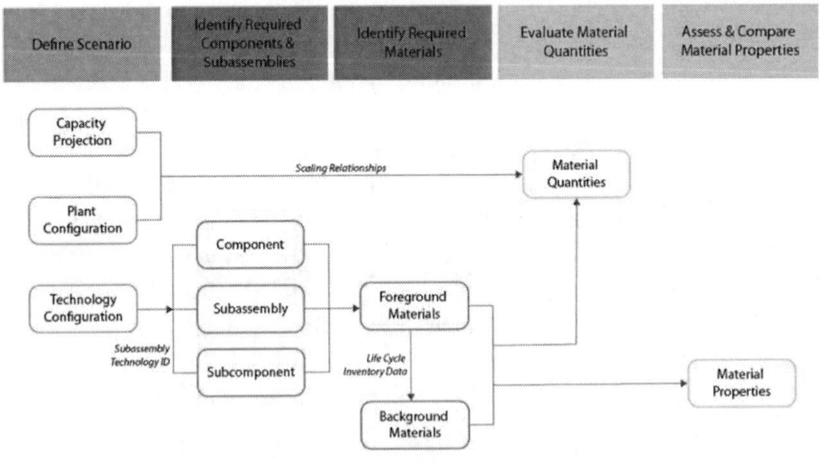

Figure 4. Simplified flow of analysis performed using the REMPD. *Illustration by Nicole Leon, NREL.*

The technology configuration in a scenario definition is used to identify the required components, subassemblies, and subcomponents. These entities are associated with known foreground material requirements, which are linked to background materials via life cycle inventory data. The capacity projection and plant configuration in the scenario definition is combined with the foreground and background material requirements to compute the total material quantities for all facilities. The foreground and background materials are also linked to material properties (such as countries of origin).

2.2. Scenario Definitions

In this report, we explore how the material requirements for wind energy technologies might change under two future wind deployment scenarios: Current Policies and High Deployment. Using the REMPD, we define our analysis scenarios using a combination of three factors:

1. *Capacity projection,* which defines the annual amount of capacity (in MW) that is anticipated each year over the period of interest.
2. *Plant configuration,* which describes the quantitative properties (e.g., the wind turbine rating, wind plant capacity, rotor diameter, and hub height) associated with each type of facility (e.g., offshore versus land-based wind), which can vary over time.
3. *Technology configuration,* which identifies the market share for each type of technology that is used within each facility and allows for the exploration of technology innovations (e.g., superconducting direct-drive generators).

Table 2 summarizes how we define each of these three factors for our two analysis scenarios.[5] The Current Policies scenario represents a medium level of wind energy deployment, consistent with median estimates of technology costs and electric-sector policies as of September 2022 (including the Inflation Reduction Act [IRA]), with limited changes in plant configurations (e.g., wind turbine size) and no significant technology innovations beyond conventional technology. The High Deployment scenario includes high levels of wind energy deployment that are aligned with the Biden administration's decarbonization goals and incorporates significant changes in wind plant

[5] Refer to Section 3.1.1 through 3.1.3 for more details about each factor.

configurations (e.g., large-scale increases in turbine size), but with limited materials-related innovations applied.

Table 2. Scenario definitions for material quantities and availability analysis

	Current Policies Scenario	High Deployment Scenario
Generic description	Limited changes to plant configurations and medium levels of deployment, with no significant materials-related technology innovations.	Significant technology innovations enable large-scale increases in wind turbine size and high levels of deployment, with limited materials-related technology innovations.
Capacity projection[a]	Mid-Case scenario from National Renewable Energy Laboratory's (NREL's) "Standard Scenarios" (Gagnon et al. 2022), which represents a medium level of wind energy deployment, as required to satisfy electricity demand. The scenario assumes no new decarbonization policies and no deployment of nascent technologies.	All Options scenario from Denholm et al. (2022), which achieves 100% clean electricity by 2035 and puts the United States on a path to net-zero emissions economywide by 2050.
Plant configuration[c]	Linear interpolation of wind turbine and plant characteristics from the 2022 Annual Technology Baseline[d] (ATB) Base scenario (year 2020) to the Conservative scenario (year 2030) and linear extrapolation of 2020–2030 scaling trends through 2050 (up to a maximum hub height of 200 meters (m), rotor diameter of 331 m, and turbine rating of 25 MW for offshore wind and a maximum hub height of 140 m for land-based wind).	Linear interpolation of wind turbine and plant characteristics from the 2022 ATB[d] Base scenario (year 2020) to the Advanced scenario (year 2030) and linear extrapolation of 2020–2030 scaling trends through 2050 (up to a maximum hub height of 200 m, rotor diameter of 331 m, and turbine rating of 25 MW for offshore wind and a maximum hub height of 140 m, rotor diameter of 210 m, and turbine rating of 8 MW for land-based wind).
Technology configuration[e]	Low materials-related technology innovation (Low Innovation technology configuration), which represents current technology (e.g., thermoset blades).	Moderate materials-related technology innovation (Moderate Innovation technology configuration), including segmented blades and carbon-fiber spar caps for land-based systems, advanced steel towers (spiral welding) for 25% of land-based systems, and hybrid tower systems for 25% of land-based systems.

[a] Refer to Section 3.1.1 and Figure 5 for more details. Note that Congress passed the Inflation Reduction Act (IRA) after this analysis was performed. As a result, the wind capacity projections used here do not reflect any increased deployment of wind power that may result from the implementation of the IRA.

[c] Refer to Section 3.1.2, Table 3, and Table 4 for more details.

[d] The ATB provides a consistent set of cost and performance data for energy analysis (refer to National Renewable Energy Laboratory [2022] for details).

[e] Refer to Section 3.1.3 and Table 5 for more details.

To capture changes in material intensity that might occur as wind plants and turbines increase in size, the REMPD multiplies the fractional contribution of materials by type (e.g., % concrete, % steel, % carbon fiber) by a scaling relationship related to the plant configuration (e.g., the plant size, number of turbines, rotor diameter). Appendix D provides more details about the scaling relationships used in this analysis.

Using our two analysis scenarios, we can assess what bottlenecks might arise in wind-energy-related materials under current technology assumptions with limited material evolution. Future work could leverage the REMPD to explore how technology innovations could further reduce material usage through high-performing materials, alternate materials (with large supply chains), and/or alternate design approaches (e.g., 100% bio-derived blade).

2.2.1. Capacity Projections

To estimate the future demand for wind energy materials, we consider two different deployment trajectories. The Current Policies scenario uses the Mid-Case deployment projection with no nascent technologies from National Renewable Energy Laboratory's (NREL's) 2022 Standard Scenarios (Gagnon et al. 2022). The High Deployment scenario uses the All Options deployment projection that achieves 100% clean electricity by 2035 and puts the United States on a path to achieve net-zero emissions economywide by 2050 (Denholm et al. 2022). The deployment projections used here were generated using the Regional Energy Deployment System (ReEDS) capacity expansion model (Ho et al. 2021) using the following assumptions[6]:

- *Electricity demand.* The Current Policies scenario assumes moderate growth (close to 1% per year) in electricity demand in response to provisions in the IRA that incentivize electrification of, for example, vehicles and heating. The High Deployment scenario assumes accelerated demand for electrification, with demand growing at a rate of 3.4% per year.
- *Policy.* The Current Policies scenario assumes no change to policies affecting the electric sector beyond September 2022. It incorporates estimated cost impacts due to tax credits that were implemented in the IRA. The High Deployment scenario is designed to meet the Biden administration's goal of 100% clean electricity by 2035. By combining carbon capture and the electrification of sectors that

[6] Refer to Gagnon et al. (2022) and Denholm et al. (2022) for more details.

currently rely on fossil fuels, this scenario enables net-zero emissions by 2050.

- *Retirements.* Deployment projections account for power plant retirements when determining the electricity demand in each year. Wind power plants are assumed to have a service life of 30 years.

The wind capacity projections under each scenario are shown in Figure 5. These projections include utility-scale land-based and offshore wind power plants. Land-based wind capacity represents the majority of the projected installations: 85% of the cumulative capacity in 2050 in the Current Policies scenario, and 91% in the High Deployment scenario. Average annual capacity additions for all utility-scale wind energy between 2022 and 2036 are approximately 21 GW/year in the Current Policies scenario and 68 GW/year in the High Deployment scenario.

These additions represent significant growth for annual United States wind energy deployment, which averaged below 10 GW per year from 2016 to 2020 (Wiser et al. 2021). In the Current Policies scenario, annual wind energy installations decrease to approximately 10 GW per year from 2038 through 2050. In contrast, annual wind energy installations in the High Deployment scenario increase to an average of 84 GW per year from 2038 through 2050.

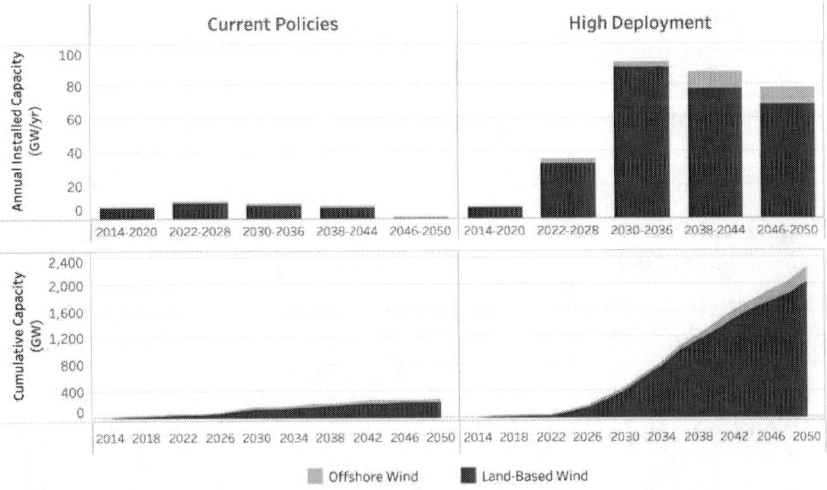

Figure 5. Wind-energy-generating capacity projections for the Current Policies and High Deployment scenarios.

2.2.2. Plant Configuration

As described in Section 2.1, the REMPD considers all wind system components that are used in typical utility-scale land-based and offshore wind power plants. These system boundaries remain consistent across the Current Policies and High Deployment scenarios. The basic wind plant system architecture includes three blades joined to a hub that is connected to the nacelle, which contains the drivetrain, generator, power electronics, and auxiliary equipment, and is mounted on a tower.[7] The hub, blades, nacelle, and tower comprise the wind turbine, which is supported by a foundation (on land) or substructure (offshore). Offshore substructures may be rigidly fixed to the seafloor (e.g., monopile or lattice structures), or floating structures may be held in place by mooring lines and anchors. The balance-of-plant components include electrical cables connecting each wind turbine to one or more substations, and site roads (on land).

Table 3. Plant configuration for the current policies scenario explored here (derived using NREL's ATB [NREL 2022]*)

	Land-Based Wind			Offshore Wind		
	2020	2030	2050	2020	2030	2050
Plant capacity	202 MW	200 MW	205 MW	1,000 MW	1,008 MW	1,000 MW
Turbine rating	2.8 MW	4 MW	6.4 MW	8 MW	12 MW	20 MW
No. of turbines	72	50	32	125	84	50
Hub height	90 m	110 m	140 m	102 m	136 m	200 m
Rotor diameter	125 m	150 m	200 m	159 m	214 m	324 m
Specific power	228 watts (W)/square meter (m^2)	226 W/m^2	204 W/m^2	403 W/m^2	334 W/m^2	243 W/m^2
Plant lifetime	30 years	30 years	30 years	30 years	30 years	30 years

* Values were calculated using a linear interpolation of wind turbine and plant characteristics from the Base scenario (year 2020) to the Conservative scenario (year 2030) using ATB 2022 data (NREL 2022) and a linear extrapolation of 2020-2030 trends through 2050 (up to a maximum hub height of 200 m, rotor diameter of 331 m, and turbine rating of 25 MW for offshore wind and a maximum hub height of 140 m for land-based wind).

Wind turbine sizes have grown significantly over the past several decades, and this growth is expected to continue (Musial et al. 2021; Wiser et al. 2021). Material requirements for wind energy systems depend on several factors, including the rated capacity of turbines and power plants, the size of individual components (notably the rotor diameter and tower height), and the selection

[7] Refer to Figure 2 for a representation of the components included in this analysis.

of materials or technology type for subcomponents where multiple alternatives have gained market share. The plant configuration in the REMPD allows users to define the size and height of the turbines and other parameters associated with a wind plant configuration. The technology configuration (discussed in the next section) allows users to define other factors that affect material requirements (e.g., the type of generator that is used) and are prescribed within each analysis scenario.

Table 4. Plant configuration for the high deployment scenario explored here (derived using NREL's ATB [NREL 2022]*)

	Land-Based Wind			Offshore Wind		
	2020	2030	2050	2020	2030	2050
Plant capacity	202 MW	203 MW	200 MW	1,000 MW	1,008 MW	1,000 MW
Turbine rating	2.8 MW	7 MW	8 MW	8 MW	18 MW	25 MW
No. of turbines	72	29	25	125	56	40
Hub height	90 m	135 m	140 m	102 m	161 m	200 m
Rotor diameter	125 m	200 m	210 m	159 m	263 m	331 m
Specific power	228 W/m^2	223 W/m^2	230 W/m^2	403 W/m^2	331 W/m^2	291 W/m^2
Plant lifetime	30 years	30 years	30 years	30 years	30 years	30 years

* Values were calculated using a linear interpolation of wind turbine and plant characteristics from the Base scenario (2020) to the Advanced scenario (2030) using ATB 2022 data (NREL 2022) and a linear extrapolation of 2020-2030 trends through 2050 (up to a maximum hub height of 200 m, rotor diameter of 331 m, and turbine rating of 25 MW for offshore wind and a maximum hub height of 140 m, rotor diameter of 210 m, and turbine rating of 8 MW for land-based wind).

In our analysis, wind plants are assumed to have wind turbines with dimensions and characteristics as specified in Table 3 and Table 4 (NREL 2022). We develop these plant configurations based on NREL's 2022 Annual Technology Baseline (ATB) (NREL 2022). To define the configurations over time, we linearly interpolate between the configuration defined by the 2022 ATB Base scenario, which represents current wind energy technology, and two future scenarios: 2022 ATB Conservative and 2022 ATB Advanced. The results of these interpolations are summarized in Table 3 for our Current Policies scenario and Table 4 for our High Deployment scenario. As shown in Table 3 and Table 4, land-based wind technology in 2020 involves a 202-MW wind plant comprising seventy-two 2.8-MW wind turbines. Offshore wind technology in 2020 involves a 1-GW plant with one hundred and twenty-five 8-MW wind turbines. The 2020 technology configurations are the same across both our analysis scenarios. However, from 2020 to 2050, the plant configurations for the Current Policies and High Deployment scenarios differ. From 2020 to 2030, turbine and plant characteristics are linearly interpolated from the ATB 2022's Base scenario data in 2020 to the ATB 2022's Conservative data in 2030 for our Current Policies scenario (or from the ATB's 2022 Base scenario in 2020 to the ATB's 2022 Advanced scenario data in 2030 for our High Deployment scenario). From 2030 to 2050, we assume a linear extrapolation of 2020-2030 trends through 2050 (up to a maximum hub height of 200 meters (m), rotor diameter of 331 m, and turbine rating of 25 MW and a maximum hub height of 140 m, rotor diameter of 210 m, and turbine rating of 8 MW for land-based wind).

2.2.3. Technology Configuration

Table 5. Technology configurations for material quantities and availability analysis

	Low Innovation (Used in Our Current Policies Scenario)	Moderate Innovation (Used in Our High Deployment Scenario)
Land-Based Wind		
Foundation	Concrete spread foot	Concrete spread foot
Tower	Transportable tubular steel can	50% transportable tubular steel can 25% spiral-welded steel 25% hybrid steel and concrete
Generator	100% high-speed geared	100% high-speed geared
Blades	Fiberglass/thermoset shell Fiberglass/thermoset spar cap 50% balsa core 50% foam core	Segmented blade tip Fiberglass/thermoset shell Carbon fiber/thermoset spar cap 50% balsa core 50% foam core
Offshore Wind		
Substructure	Steel monopile	80% steel monopile 15% steel jacket 2% concrete gravity base 3% steel semisubmersible
Tower	Tubular steel can	Tubular steel can
Generator	100% permanent-magnet synchronous generator	100% permanent-magnet synchronous generator
Blades	Fiberglass/thermoset shell Carbon fiber/thermoset spar cap 50% balsa core 50% foam core	Fiberglass/thermoset shell Carbon fiber/thermoset spar cap 50% balsa core 50% foam core

We use the technology configuration in the REMPD to incorporate technology changes over time. These technology configurations are developed by assigning market shares for alternative materials and subcomponent technologies based on available data, projections, and expert opinion. There are a variety of technology innovations that could modify the material requirements for wind energy technologies.[8] In this analysis, we use two technology configurations: 1) Low Innovation, which is based on current technology, and 2) Moderate Innovation, which includes three materials-related technology innovations. Both configurations are summarized in Table 5.[9]

2.2.3.1. Low Innovation

As shown in Table 5, under the Low Innovation technology configuration, we assume that land-based and offshore wind plants rely on current technology with no materials-related technology innovations over time. For example, land-based wind towers and foundations use conventional designs: a spread-foot foundation made from concrete and steel rebar, and a transportable tubular steel tower. The blade material quantities assume that fiberglass is used throughout the blade, with a conventional thermoset resin. The blade core material is assumed to be 50% balsa wood and 50% foam. When assessing material properties, we assume polyvinyl chloride (PVC) foam is used in the core, although other polymers such as polyethylene terephthalate may be substituted. Materials used in the nacelle vary depending on the type of generator. Worldwide, nearly 30% of wind turbines use permanent-magnet synchronous generators, whereas 70% use high-speed geared generators, and around 2% use electrically excited synchronous generators (European Commission et al. 2020b). In the United States, the share of permanent-magnet synchronous generators is lower at approximately 2%, with most wind turbines using high-speed geared generators. In the Low Innovation technology configuration, we assume 100% of land-based systems use high-speed geared generators.

For offshore wind technology, the Low Innovation technology configuration assumes fixed-bottom steel monopile foundations with tubular steel towers. Given the limited number of offshore wind turbines currently in U.S. waters, the generator type is drawn from data provided by original equipment manufacturers, which assume offshore turbines will be constructed largely using permanent-magnet synchronous generators. Fiberglass is used for the blade shell, with carbon fiber used for the spar caps. Like land-based wind turbines, 50% of blades are assumed to use balsa wood while the remainder use PVC foam cores.

---

[8] Refer to Table 10 for a list of potential innovations that might impact each wind energy component.

[9] Refer to Section 2.2.3.1 and 2.2.3.2, respectively, for more details about each configuration.

2.2.3.2. Moderate Innovation

The Moderate Innovation technology configuration includes spiral-welded towers, hybrid towers, and segmented blades with carbon-fiber spar caps. Spiral-welded towers enable on-site manufacturing of larger (taller) towers that would be difficult to transport using traditional tubular steel can towers. Hybrid towers are constructed using a hybrid of concrete and steel to help enable taller towers for land-based systems. In the Moderate Innovation technology configuration, we assume 25% of land-based systems will use hybrid towers. We also assume that 25% of land-based towers will be spiral welded. The remaining 50% of land-based towers and all offshore towers are assumed to be assembled from conventional transportable tubular steel "can" segments. In the Moderate Innovation technology configuration, we assume 100% of future land-based systems will use segmented blades. Segmented blades incorporate carbon spar caps and segmentation to enable the transport of longer blades for land-based systems (refer to Appendix D.7 for more details). No other technology innovations (e.g., changes in generator types or other technology changes) are included in the Moderate Innovation technology configuration.

3. Projected Material Needs for U.S. Wind Energy Systems

There are more than 200 unique foreground system materials and more than 1,700 unique background system material flows in the proprietary version of the REMPD.[10] To improve the interpretability of this analysis, we group these materials into seven major categories:

1. Concrete
2. Road aggregate (crushed rock, stones, and gravel)
3. Steel (including electrical steel)
4. Composites and polymers (including carbon fiber in carbon-fiber-reinforced polymers)
5. Cast iron
6. Other metals and alloys (including 16 critical minerals)
7. Other materials (including graphite).

---

[10] Refer to Section 2.1.3 and Figure 3 for a definition of the foreground and background systems.

We use these categories to discuss the results of our analysis and perform a more detailed analysis on 19 vulnerable materials (defined in Table 1). The vulnerable materials that we explore fall into different material categories (e.g., electrical steel is included with other types of steel in the "Steel" category and most critical minerals are included in the "Other metals and alloys" category).

In this section, we summarize material intensities of current and potential future wind energy technologies, discuss the projected material needs for U.S. wind energy systems over time, compare projected needs for all materials to the amount of material that is currently produced, and compare vulnerable material needs to projected availability.

3.1. Material Intensities of Current and Potential Future Wind Energy Technologies

Figure 6 illustrates the total material intensities of current and potential future[11] wind energy technologies and how the material requirements break down by material category. For example, current land-based wind power plants require about 1,200 metric tonnes (t) of material per megawatt, comprised (by mass) of approximately 53% road aggregate, 34% concrete, 9% steel, 2% composites and polymers, 1% cast iron, 1% other metals and alloys, and less than 1% other materials. Future land-based wind plants may contain a larger proportion of concrete due to bigger foundations required for larger and taller turbines and more concrete in hybrid towers.

These changes could shift the material breakdown of the future land-based wind plants by mass to 46% concrete, 39% road aggregate, 10% steel, 3% composites and polymers, 1% cast iron, 1% other metals and alloys, and the remainder other materials. Offshore wind plants currently require about 300 t of material per megawatt, comprised (by mass) of 87% steel, 5% other metals and alloys, 4% composites and polymers, 3% cast iron, and 1% other materials. Shifts in material requirements for land-based wind in the High Deployment scenario are due to the technology configuration changes and moderate materials-related technology that are assumed in this scenario; the High Deployment scenario does not predict major changes to the material requirements for future offshore wind plants. Concrete is used in the High

[11] Potential future technology is represented using results from the High Deployment scenario in the year 2050 (see Table 2 for more details about the High Deployment scenario definition).

Deployment scenario for some offshore wind substructures; it represents a large fraction of the material use in those facilities but only 3% of the average material intensity because concrete gravity-base substructures make up only 2% of predicted installations.

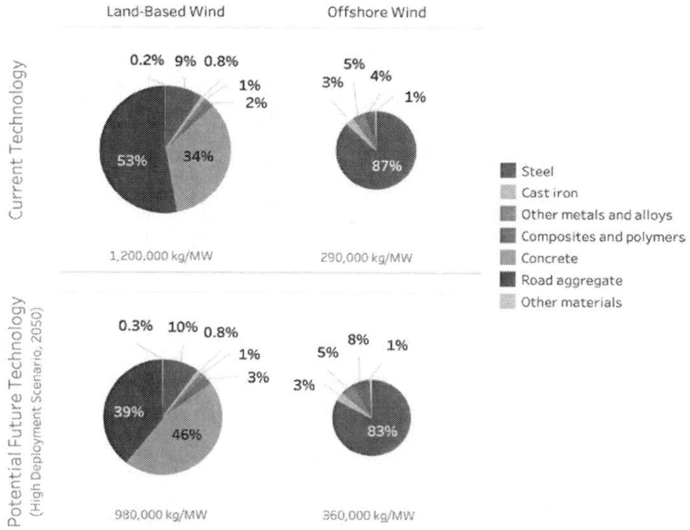

Figure 6. Material intensities of current and potential future wind energy technologies.

Table 6 and Table 7, respectively, provide more details about the average material intensities of current and future wind energy technologies shown in Figure 6. The tables break down the material intensities by facility, component, subassembly, and material type. For example, a current land-based wind turbine tower requires approximately 66 t of steel, 2.7 t of other metals and alloys, 0.1 t of composites and polymers, and 0.2 t of other materials per megawatt of installed capacity (Table 6). Current offshore towers require about 31 t of steel, 1.4 t of other metals and alloys, 0.1 t of composites and polymers, and 0.2 t of other materials per megawatt of installed capacity. The material intensity of future offshore towers in the High Deployment scenario increases slightly but the relative contribution of each material remains similar (Table 7). However, future land-based towers could require 15 t of concrete in addition to 52 t of steel, 22 t of other metals and alloys, 0.1 t of composites and polymers, and 0.2 t of other materials per megawatt of installed capacity due to the adoption of hybrid steel-concrete towers.

Table 6. Average material intensity of current wind energy technologies (2020)

Average Material Intensity (metric tonnes [t]/megawatt [MW]) of Current Wind Energy Technologies by Material Category									
	Component	Subassembly	Steel	Cast iron	Other metals and alloys	Composites and polymers	Concrete	Road aggregate	Other materials
Land-Based Wind	Array and export cables	Total*			3.6	4.4			
	Foundation	Total*	20.8		1.0	<0.01	398.1		<0.01
	Roads	Total*						613.0	
	Substation	Total*	1.0	<0.01	0.2	0.1	0.6		0.2
	Turbine	Blades	0.3		<0.01	15.8			1.1
		Hub	4.8	2.5	0.2	0.1			<0.01
		Nacelle	17.0	6.3	3.8	1.0			0.3
		Tower	66.0		2.7	0.1			0.2
Offshore Wind	Array and export cables	Array cable	0.6		2.6	0.7			0.2
		Export cable	<0.01		0.5	0.5			0.1
		Onshore cable			<0.01	<0.01			
	Substructure	Monopile	144.5		4.9				
		Transition piece	51.5		1.8				
	Substation	Substation equipment	0.3		0.1	<0.01			0.2
		Support structure	7.0		0.2				
	Turbine	Blades	0.3		<0.01	10.4			0.9
		Hub	3.0	0.9	0.2	<0.01			<0.01
		Nacelle	11.5	7.8	2.6	0.7			0.4
		Tower	30.9		1.4	0.1			0.2

* Only total component data are available for these components; these data are not broken down by subassembly.

Table 7. Average material intensity of potential future wind energy technologies (as defined based on technology projections and expert input used in the high deployment scenario, 2050)

Material Intensity (t/MW) of Potential Future Wind Energy Technologies by Material Category									
	Component	Subassembly							
			Steel	Cast iron	Other metals and alloys	Composites and polymers	Concrete	Road aggregate	Other materials
Land-Based Wind	Array and export cables	Total*			3.6	4.4			
	Foundation	Total*	22.8		1.1	<0.01	434.4		<0.01
	Roads	Total*						383.9	
	Substation	Total*	1.0	<0.01	0.2	0.1	0.6		0.2
	Turbine	Blades	0.8		0.1	22.9			1.7
		Hub	4.8	2.5	0.2	0.1			<0.01
		Nacelle	16.7	6.2	3.8	1.0			0.3
		Tower	52.8		2.2	0.1	15.2		0.2
Offshore Wind	Array and export cables	Array cable	0.6		2.6	0.7			0.2
		Export cable	<0.01		0.5	0.5			0.1
		Onshore cable			<0.01	<0.01			
	Substructure	Pile/jacket/floater	218.1		7.4				
		Gravity base			<0.01		11.5	2.3	<0.01
	Substation	Substation equipment	0.3		0.1				0.2
		Support structure	7.0		0.2				
	Turbine	Blades	0.5		<0.01	19.2			1.7
		Hub	4.9	1.5	0.3	<0.01			<0.01
		Nacelle	11.4	7.7	2.6	0.7			0.4
		Tower	50.9		2.4	0.2			0.4

* Only total component data are available for these components; these data are not broken down by subassembly and instead provide a total value equal to the material requirements for all subassemblies associated with the component.

Array and export cables for land-based systems mostly comprise composites and polymers, along with other metals and alloys (mostly aluminum). Offshore array and export cables also contain lead and galvanized steel. Foundations in land-based wind plants comprise mostly concrete with a small amount of steel reinforcement, and substructures in offshore systems are built almost entirely from steel. Roads are only used in land-based systems and are comprised of aggregate. Offshore substations use steel as the primary structural material; electrical equipment within the substation uses more steel (including electrical steel), along with other metals and alloys (e.g., copper) and a small fraction of composites and polymers and other materials. Land-based substations use these materials along with concrete and cast iron.

3.1.1. Variation in Material Intensities for Vulnerable Materials

Figure 7 shows how the material intensities for vulnerable materials could differ between current and potential future wind energy technologies (see Tables E-1 and E-2 for the underlying data used to develop this figure).

For current and potential future technology in both land-based and offshore wind systems, the vulnerable materials with the highest average material intensities are aluminum, carbon fiber, chromium, electrical steel, manganese, and nickel. Aluminum is mostly used in the cables, tower, and nacelle. Carbon fiber is used solely in the wind turbine blades. Chromium is primarily used in the nacelle and foundation in land-based wind plants and in the nacelle and hub in offshore wind plants. Electrical steel is used mostly in the nacelle and substation. Manganese and nickel are primarily used in the tower and foundation (land-based) or substructure (offshore).

As shown in Figure 7, the average material intensities for vulnerable materials used in land-based wind plants do not change much for current versus potential future technology, except for carbon fiber. The material intensity for carbon fiber is higher for potential future technology because blade lengths are expected to increase more quickly than rated power, and the High Deployment scenario assumes the use of segmented blades, which are expected to require more carbon fiber. For offshore wind, longer blades and lower specific power also contribute to an increase in the average material intensity for carbon fiber. Material intensities of chromium, manganese, and nickel all increase for offshore wind because these vulnerable materials are used in steel manufacturing and more steel is required per megawatt to build larger turbines with taller towers and substructures in the High Deployment scenario.

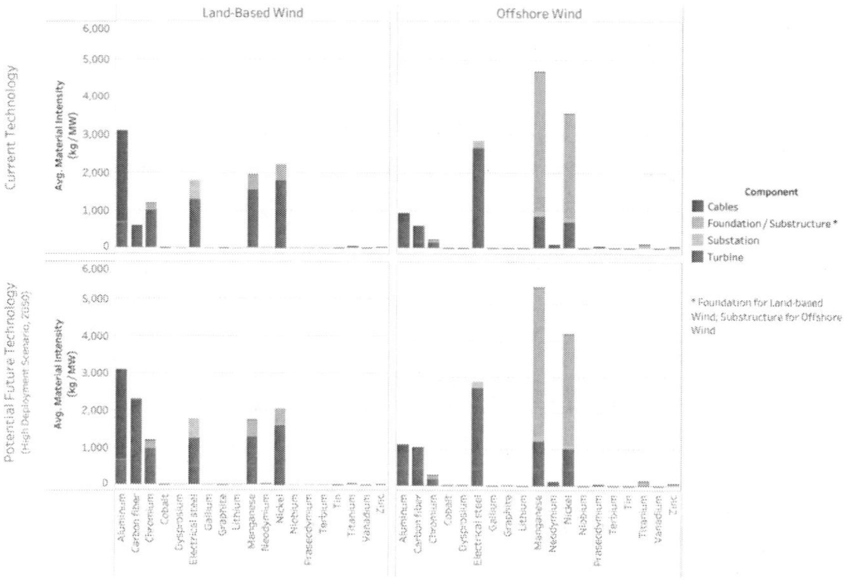

Figure 7. Average material intensities for vulnerable materials as determined using the Current Policies and High Deployment scenarios.

3.2. Projected U.S. Wind Energy Demand for Materials

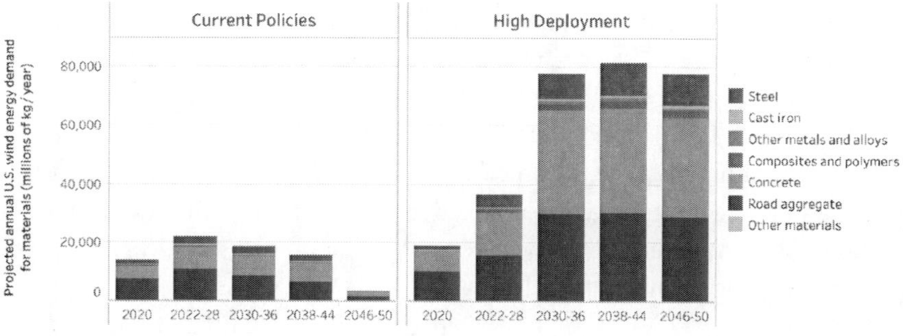

Figure 8. Average quantity per year of all materials required for land-based and offshore wind technologies as determined using the Current Policies and High Deployment scenarios.

Figure 8 illustrates the projected U.S. wind energy demand for our seven categories of wind energy materials as estimated using the assumptions outlined in the Current Policies and High Deployment scenarios (see Table 2 for more details about how we defined these scenarios).

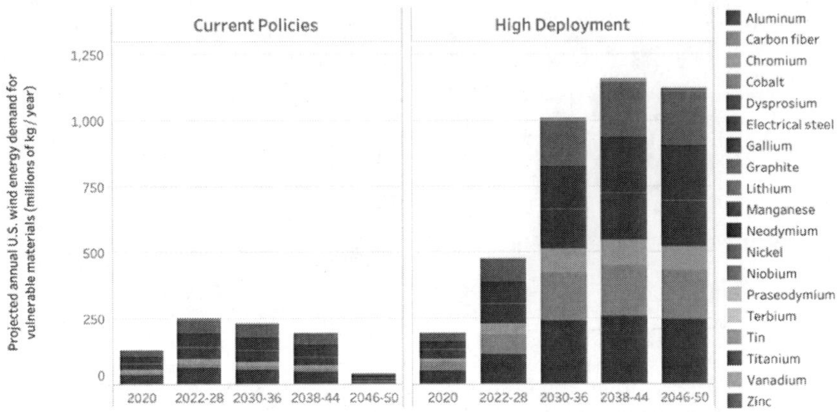

Figure 9. Average quantity per year of vulnerable materials required for land-based and offshore wind technologies as determined using the Current Policies and High Deployment scenarios.

Figure 9 shows the projected U.S. wind energy demand for a subset of these materials, specifically vulnerable materials. In both Figure 8 and Figure 9, average annual material requirements generally follow capacity projection trends (see Figure 5). Material requirements are primarily driven by land-based wind capacity additions, which are more than seven times higher than offshore wind. Land-based wind plants are also more material-intensive than offshore facilities because they require large quantities of concrete to construct foundations and aggregate for roads.[12] As a result, the total material requirements for land-based wind energy technologies are 25 to 35 times greater than for offshore wind.

3.2.1. Projected Changes in Material Intensity over Time

The trends in material quantities shown in Figure 8 are further explained by examining the change in material intensity of each technology over time (Figure 10). As shown in Figure 10, land-based wind plants require a total of 1,000-1,200 t of material per megawatt compared to 300-350 t of material per

[12] See Section 3.1, including Figure 6, Table 6, and Table 7 for more details on the differences in material intensities between land-based and offshore wind energy technologies.

megawatt for offshore wind.[13] Land-based wind energy projects show the most change in material intensity over time. This variation is primarily driven by road aggregate requirements, which decrease over time because fewer miles of roads are needed per unit of capacity to access larger wind turbines. In addition, the amount of concrete required per megawatt for land-based foundations generally increases as the rotor size and hub height increase. There is also a slight increase in the material intensity of offshore wind over time in both scenarios due to greater material requirements for taller towers.

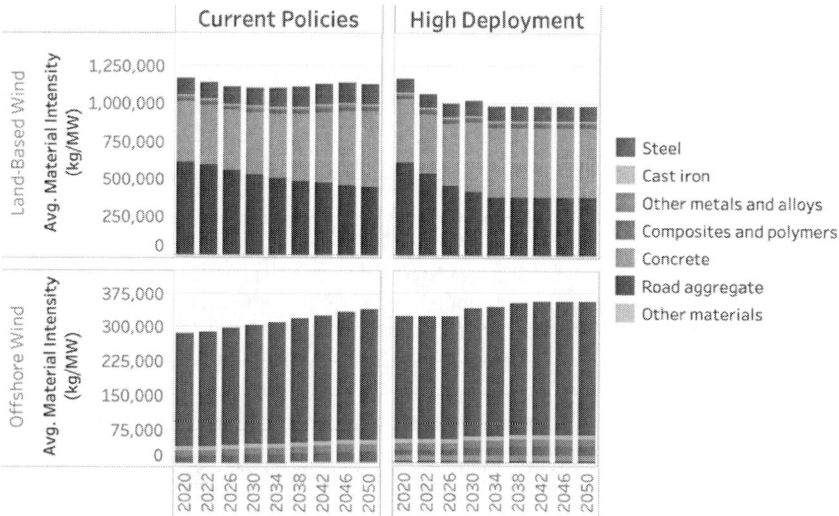

Figure 10. Annual material intensity for land-based and offshore wind in the Current Policies and High Deployment scenarios.

3.3. Projected U.S. Wind Energy Demand for Materials Compared to Current Production

To put the projected material needs for U.S. wind energy technologies into context, we compared the U.S. wind energy demand for nonvulnerable materials in each analysis scenario to global and U.S. production of each material in 2020 (Figure 11).

[13] See Figure 6, Table 6, and Table 7 for more details about the differences in material requirements and the breakdown of material quantities by material type, facility, component, and subassembly.

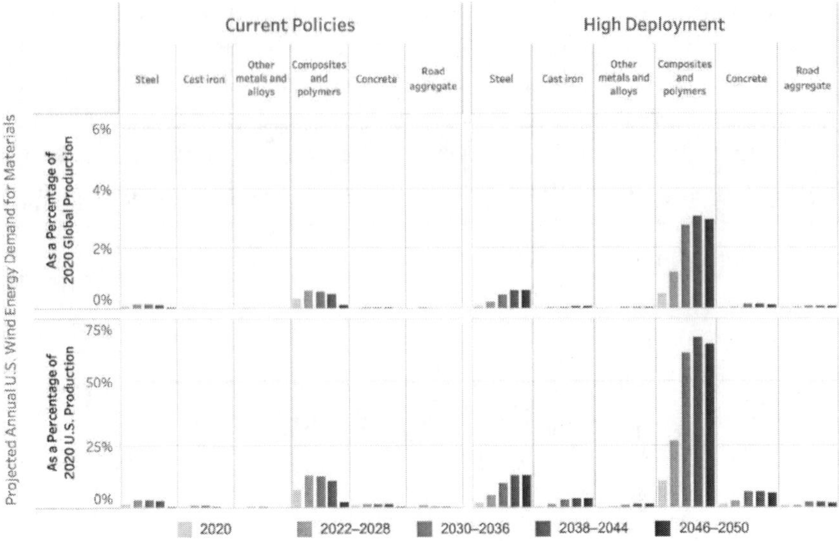

Figure 11. Projected annual U.S. wind energy demand for nonvulnerable materials, as estimated in the Current Policies and High Deployment scenarios as a percentage of 2020 production.
Results are presented for global (top row) and United States (bottom row) production.

Future availability of these materials depends on many factors, including the mineral resources, cost of extraction, and global level of demand from various industries. Current production levels provide an initial data point from which to estimate production in future years.

In both scenarios, the quantity of nonvulnerable materials needed to satisfy U.S. wind energy demand comprise less than 4% of the 2020 global production (top row of Figure 11). The quantities of cast iron, other nonvulnerable metals and alloys, and road aggregate needed to satisfy U.S. wind energy demand through 2050 also comprise less than 5% of the total annual U.S. production of these materials in both scenarios (assuming future average U.S. production levels remain the same as in 2020; bottom row of Figure 11).

However, the projected annual U.S. wind energy demand for composites and polymers, steel, and concrete could exceed 5% of 2020 U.S. production of these materials (bottom row of Figure 11). Under the Current Policies scenario, demand for steel and concrete remains below 5% of current U.S. production, whereas demand for composites and polymers from 2022 through 2036 represents approximately 12% of the total amount of these materials produced in the United States in 2020. In the High Deployment scenario, demand for composites and polymers, steel, and other metals and alloys is even higher. From 2038 to 2044, U.S. wind energy demand for composites and polymers, steel, and concrete could consume 67%, 13%, and 6% of the amount of U.S. production of these materials in 2020, respectively. Thus, it is important to consider how to mitigate potential supply risks for these nonvulnerable materials—especially composites and polymers—to enable high levels of wind energy deployment.

Demand for specific materials (e.g., vulnerable materials such as carbon fiber, nickel, and dysprosium) may comprise an even higher portion of U.S. production. For vulnerable materials, we performed a more detailed analysis and examined how the material needs for wind energy technologies in each scenario compare to U.S. production, global production, U.S. reserves, and global reserves (Table 8, Figure 12, Figure 13, and Figure 14). Table 8 provides the production and reserve values for each vulnerable material and additional information about the significant uses of these vulnerable materials along with a breakdown of material quantities by country of origin. Figures 12-14 illustrate how the material requirements compare to production and reserves. The distribution of countries supplying raw materials to the United States varies in many instances from global production levels, reflecting trade ties between individual companies or nations. Comparisons against global production provide a high-level overview of how the magnitudes of total global production compare to material needs for wind energy technologies.

As shown in Figure 12 and Figure 13, the projected annual U.S. wind energy demand for vulnerable materials is anticipated to require less than 10% of global 2020 production for all materials in the Current Policies scenario. However, in the High Deployment scenario, U.S. wind energy demand for carbon fiber could reach 101% of 2020 global production (Figure 12) and demand for neodymium could reach 12% of 2020 global production from 2038 to 2044 (Figure 13); U.S. wind energy demand for all other materials is below 10% of global production in the High Deployment scenario.

From a domestic perspective, U.S. wind energy demand for nickel, electrical steel, and carbon fiber could require up to 317%, 18%, and 25% of 2020 levels of U.S. production of these materials between 2022 and 2028 in the Current Policies scenario. In the High Deployment scenario, demand for wind energy materials is highest after 2030. From 2030 to 2045, U.S. wind energy demand for nickel, electrical steel, and carbon fiber could approach, respectively, 1,200%, 75%, and 400% of 2020 levels of U.S. production in the High Deployment scenario.

Table 8. Uses, sources, production, reserves, and projected wind material needs for vulnerable materials

Material	Other Significant Uses[a]	Country of Origin		Total Global Production (millions of kilograms [kg] in 2020)	Total Global Reserves (millions of kg)	Ranges of Projected Material Needs for U.S. Wind Energy Technologies (millions of kg)	
		Global Production	*U.S. Import Sources (2016–2019)*			*Current Technology (Current Policies Scenario, 2020)*	*Potential Future Technology (High Deployment Scenario, 2050)*
Carbon fiber	Transportation (aerospace, automotive, marine), consumer goods (pressure vessels, sports equipment)	United States (28%) Japan (13%) China (13%) Turkey (12%) Hungary (5%) Taiwan (5%) Others (24%)	Data not available	192	N/A	7–28	240–260
Electrical steel	Machinery and appliances (transformers, motors, inductors)	South Korea (14%) China (14%) Japan (12%) Germany (11%) Russia (10%) Others (39%)	Japan (21%) Korea (21%) France (13%) Austria (11%) China (6%) Others (28%)	20,000	N/A	18–63	190–570
Critical Minerals							
Aluminum	Transportation (aviation and automotive), consumer goods, packaging, construction, electrical, machinery and appliances	China (57%) Russia (6%) India (5%) Canada (5%) Others (27%)	Canada (50%) United Arab Emirates (10%) Russia (9%) China (5%) Others (26%)	65,200	32,000,000	34–49	300–420

Material	Other Significant Uses[a]	Country of Origin		Total Global Production (millions of kilograms [kg] in 2020)	Total Global Reserves (millions of kg)	Ranges of Projected Material Needs for U.S. Wind Energy Technologies (millions of kg)	
		Global Production	*U.S. Import Sources (2016–2019)*			*Current Technology (Current Policies Scenario, 2020)*	*Potential Future Technology (High Deployment Scenario, 2050)*
Chromium	Steel (stainless and heat-resisting steel), other steel alloys	South Africa (36%) Turkey (22%) Kazakhstan (19%) India (7%) Finland (6%) Others (10%)	South Africa (39%) Kazakhstan (8%) Mexico (6%) Russia (6%) Others (41%)	37,000	570,000	14–47	110–390
Cobalt	Alloys (superalloys, other alloys), chemicals, steel	Congo (73%) Russia (5%) Others (22%)	Norway (20%) Canada (14%) Japan (13%) Finland (10%) Others (43%)	165	7,600	0.03–0.07	0.3–0.6
Dysprosium[b]	Magnets, ceramics and glass, battery alloys, catalysts	China (58%) United States (16%) Burma (13%) Australia (9%) Others (4%)	China (80%) Estonia (5%) Japan (4%) Malaysia (4%) Others (7%)	2.4	44	0.02–0.09	0.3–0.8
Gallium	Electronics (integrated circuits, optoelectronic devices)	China (97%) Others (3%)	China (55%) United Kingdom (11%) Germany (10%) Others (24%)	0.33	100	0.0006–0.002	0.005–0.01

Table 8. (Continued)

Material	Other Significant Uses[a]	Country of Origin		Total Global Production (millions of kilograms [kg] in 2020)	Total Global Reserves (millions of kg)	Ranges of Projected Material Needs for U.S. Wind Energy Technologies (millions of kg)	
		Global Production	*U.S. Import Sources (2016–2019)*			*Current Technology (Current Policies Scenario, 2020)*	*Potential Future Technology (High Deployment Scenario, 2050)*
Graphite (natural)	Metal products (bearings, brake lining, lubricants), rubber	China (79%) Brazil (7%) Others (14%)	China (33%) Mexico (23%) Canada (17%) India (9%) Others (18%)	970	320,000	0.04–0.2	0.3–1.1
Lithium	Batteries, ceramics and glass, lubricating greases	Australia (48%) Chile (26%) China (16%) Argentina (7%) Others (3%)	Argentina (55%) Chile (36%) China (5%) Others (4%)	83	22,000	0.008–0.04	0.06–0.22
Manganese	Steel	South Africa (34%) Australia (18%) Gabon (18%) China (7%) Others (23%)	Gabon (20%) South Africa (19%) Australia (15%) Georgia (10%) Others (36%)	19,000	1,500,000	22–35	220–410
Neodymium[b]	Magnets, ceramics and glass, battery alloys, catalysts	China (58%) United States (16%) Burma (13%) Australia (9%) Others (4%)	China (80%) Estonia (5%) Japan (4%) Malaysia (4%) Others (7%)	40.8	1,200	0.5–1.9	6–18
Nickel	Steel (stainless and heat-resisting steel),	Indonesia (31%) Philippines (13%) Russia (11%)	Canada (42%) Norway (10%) Finland (9%)	2,500	95,000	26–56	240–550

Material	Other Significant Uses[a]	Country of Origin		Total Global Production (millions of kilograms [kg] in 2020)	Total Global Reserves (millions of kg)	Ranges of Projected Material Needs for U.S. Wind Energy Technologies (millions of kg)	
		Global Production	*U.S. Import Sources (2016–2019)*			*Current Technology (Current Policies Scenario, 2020)*	*Potential Future Technology (High Deployment Scenario, 2050)*
	superalloys, batteries	New Caledonia (8%) Australia (7%) Canada (7%) China (5%) Others (18%)	Russia (8%) Other (31%)				
Niobium	Steel, superalloys	Brazil (90%) Canada (10%)	Brazil (66%) Canada (22%) Others (12%)	65	18,000	0.004–0.005	0.03–0.06
Praseo-dymium[b]	Magnets, ceramics and glass, battery alloys, catalysts	China (58%) United States (16%) Burma (13%) Australia (9%) Others (4%)	China (80%) Estonia (5%) Japan (4%) Malaysia (4%) Others (7%)	14.4	370	0.006–0.01	0.8–1.5
Terbium[b]	Magnets, ceramics and glass, battery alloys, catalysts	China (58%) United States (16%) Burma (13%) Australia (9%) Others (4%)	China (80%) Estonia (5%) Japan (4%) Malaysia (4%) Others (7%)	0.5	10	<0.0001	0.006–0.012

Table 8. (Continued)

Material	Other Significant Uses[a]	Country of Origin		Total Global Production (millions of kilograms [kg] in 2020)	Total Global Reserves (millions of kg)	Ranges of Projected Material Needs for U.S. Wind Energy Technologies (millions of kg)	
		Global Production	*U.S. Import Sources (2016–2019)*			*Current Technology (Current Policies Scenario, 2020)*	*Potential Future Technology (High Deployment Scenario, 2050)*
Tin	Alloys, coatings (tinplate), chemicals, metal products (solder)	China (32%) Indonesia (20%) Burma (11%) Peru (8%) Congo (7%) Bolivia (6%) Brazil (6%) Others (10%)	Indonesia (24%) Malaysia (21%) Peru (20%) Bolivia (17%) Other (18%) Scrap: Canada (99%)	260	4,900	0.002–0.004	0.02–0.05
Titanium	Steel, superalloys	China (53%) Japan (21%) Russia (13%) Kazakhstan (7%) Others (6%)	Japan (90%) Kazakhstan (7%) Others (3%)	230	750,000	0.6–0.9	6–10
Vanadium	Steel, other alloys, catalysts	China (67%) Russia (19%) South Africa (8%) Brazil (6%)	Canada (26%) China (14%) Brazil (10%) South Africa (9%) Others (41%)	105	24,000	<0.0001	<0.0001

Material	Other Significant Uses[a]	Country of Origin		Total Global Production (millions of kilograms [kg] in 2020)	Total Global Reserves (millions of kg)	Ranges of Projected Material Needs for U.S. Wind Energy Technologies (millions of kg)	
		Global Production	*U.S. Import Sources (2016–2019)*			*Current Technology (Current Policies Scenario, 2020)*	*Potential Future Technology (High Deployment Scenario, 2050)*
Zinc	Coatings (galvanization), rubber, chemicals, paint, agriculture	China (34%) Australia (11%) Mexico (5%) Peru (11%) United States (6%) India (6%) Others (27%)	Peru (98%) Others (2%)	12,000	250,000	0.4–1.3	3–11

[a] Other significant uses than wind energy technologies. Data are primarily drawn from the USGS *Metals and minerals: U.S. Geological Survey Minerals Yearbooks* (most recent available, 2018-2022) and are supplemented with data from the National Ready Mixed Concrete Association (https://www.nrmca.org/), the UN Comtrade Database (https://comtrade.un.org/data/), BloombergNEF (2020), and Carrara et al. (2020).

[b] The source and other significant uses information reported for dysprosium, neodymium, praseodymium, and terbium correspond to data for all rare-earth compounds and metals (they are not specific to each of the individual elements) because these data are not available at the level of individual elements.

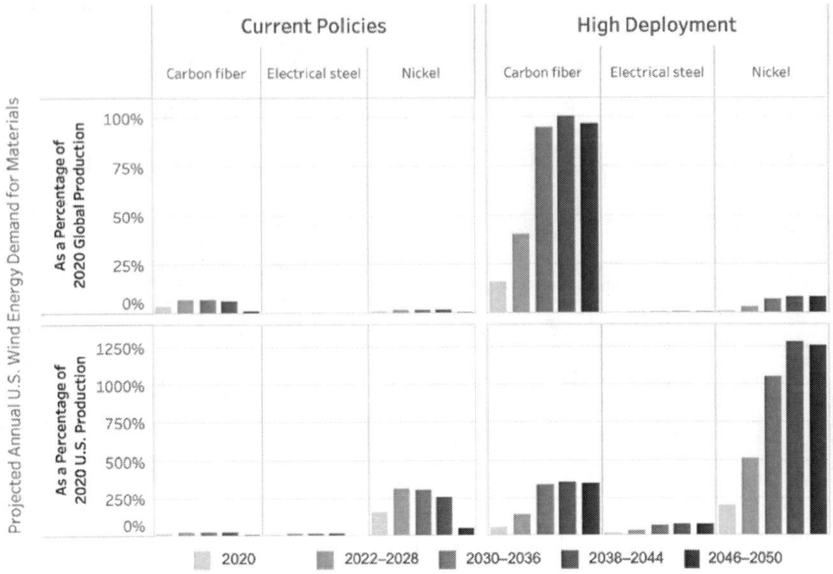

Figure 12. Projected U.S. wind energy demand for carbon fiber, electrical steel, and nickel, as estimated in the Current Policies and High Deployment scenarios as a percentage of 2020 production.
Results are presented for global (top row) and United States (bottom row) production.

Figure 13 illustrates the projected annual U.S. wind energy demand as a percentage of production in 2020 for a subset of vulnerable materials excluding carbon fiber, electrical steel, and nickel. Demand for these materials is compared with global production in the top portion of Figure 13 and U.S. production in the bottom portion of the figure.[14] Here, we can see that U.S. wind energy demand for this subset of vulnerable materials comprises less than 12% of 2020 global production in both the Current Policies and High Deployment scenarios. However, in the High Deployment scenario, demand for neodymium and dysprosium reaches 69% and 55%, respectively, of 2020 U.S. production from 2038 to 2044; demand for praseodymium, cobalt, and aluminum also rises to 20%-35% of 2020 domestic production levels.

Overall, these results indicate that current domestic production levels of carbon fiber and nickel may be lower than the amount required to achieve high levels of wind energy deployment in the United States consistent with a net-zero economy. Additionally, U.S. wind energy demand for rare-earth elements

[14] Demand for chromium, gallium, graphite, lithium, manganese, niobium, tin, and titanium are not compared with domestic production because lithium and titanium production data were withheld from publication and the other materials are not mined domestically (USGS 2021).

(i.e., dysprosium and neodymium) and electrical steel may comprise a large portion of domestic production (if production does not increase beyond 2020 levels). And, globally, production of carbon fiber would need to increase to meet demand for wind energy.

Demand from other sectors of the economy may further limit availability of certain materials. For instance, demand from electric vehicles could further constrain carbon fiber, neodymium, and praseodymium supplies. A strategy to avoid supply issues could be to diversify imports: when considering global production, U.S. wind energy demand for vulnerable materials never exceeds 10% of global production (in either scenario) except for neodymium and carbon fiber in the High Deployment scenario.

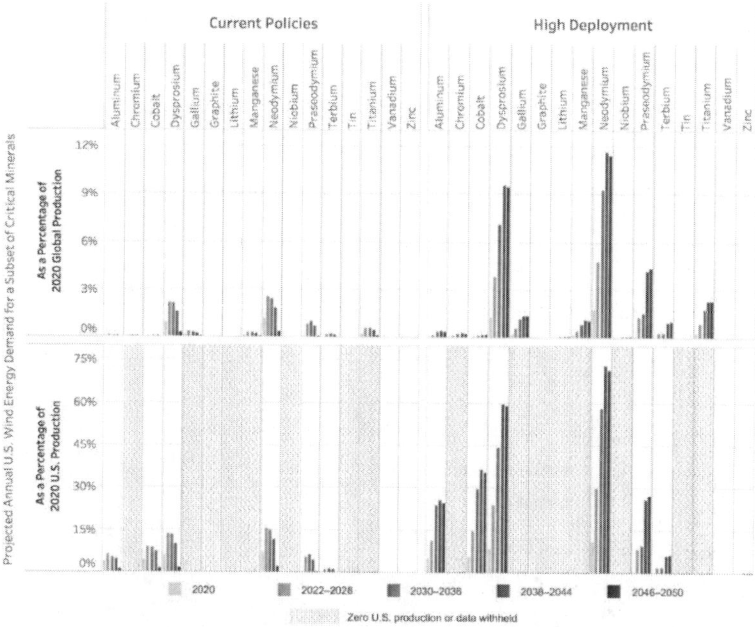

Figure 13. Projected U.S. wind energy demand for a subset of vulnerable materials (excluding nickel, carbon fiber, and electrical steel), as estimated in the Current Policies and High Deployment scenarios as a percentage of 2020 production. Results are presented for global (top row) and United States (bottom row) production. In 2020, gallium, graphite, manganese, and niobium were not produced in the United States; the United States had no primary production of tin or chromium; and the amount of lithium and titanium production were withheld to protect confidential company data (USGS 2021).

3.4. Material Needs for Wind Energy Technologies Compared to Projected Availability

Projected availability of materials depends on many factors, including mineral resources, cost of extraction, and global level of demand from various industries. Current estimates of reserves provide an initial data point from which to estimate projected availability of materials in future years. Reserves are not estimated for electrical steel and carbon fiber, because the concept of reserves does not apply to these highly processed materials in the same way as for critical minerals.

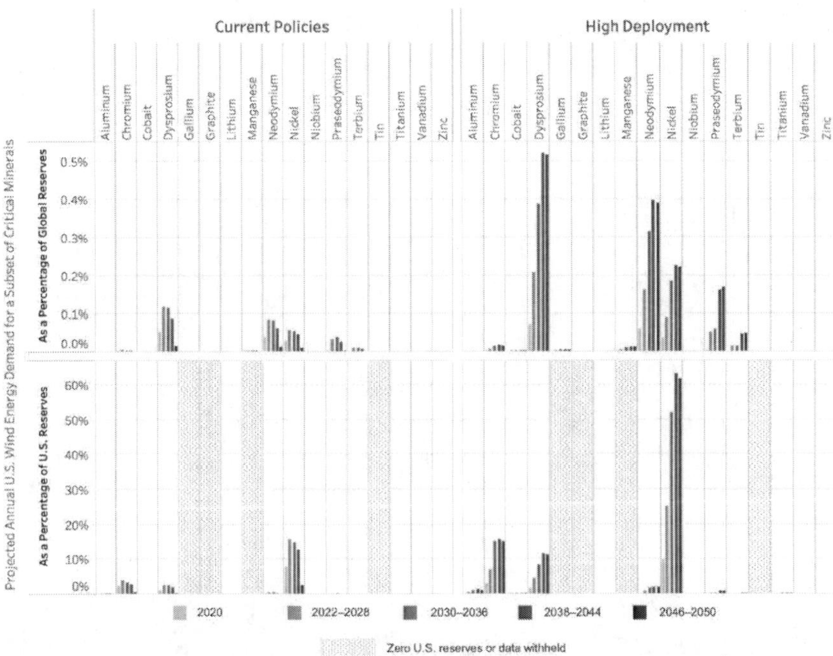

Figure 14. Projected U.S. wind energy demand for critical minerals, as estimated in the Current Policies and High Deployment scenarios as a percentage of reserves. Results are presented for global (top row) and United States (bottom row) reserves (the United States does not have known, economically recoverable reserves of gallium, graphite, manganese, or tin (USGS 2021).

Figure 14 shows how annual U.S. wind energy demand for critical minerals compares to U.S. and global reserves for each material in the Current Policies and High Deployment scenarios. None of the critical mineral needs for U.S. wind energy technologies are expected to exceed 0.5% of global reserves. Most critical mineral requirements are below 2% of U.S. reserves in both the Current Policies and High Deployment scenarios; the three exceptions are nickel, dysprosium, and chromium. In the High Deployment scenario, annual U.S. wind energy demand for nickel, chromium, and dysprosium could reach 63%, 16%, and 11% of U.S. reserves for each material, respectively, from 2038 to 2044.

Overall, Figure 12, 13, and 14 show that global production and reserves are sufficient to meet the U.S. wind energy demand for all vulnerable materials except carbon fiber. The projected demand for carbon fiber, particularly in the High Deployment scenario, will require large increases in both domestic and global supply. In addition, domestic production or imports of nickel will have to increase significantly to meet the material requirements in the High Deployment scenario particularly after 2030. Depending on competing uses, production of electrical steel, cobalt, and rare-earth elements (dysprosium and praseodymium) may also need to increase.

3.5. High-Level Overview of Material Supply Challenges for U.S. Wind Energy

To assess potential material supply challenges for U.S. wind energy, we examined the potential future U.S. wind energy demand for each material in 2050 as a percentage of U.S. and global production in 2020 (illustrations of these results are shown for vulnerable materials in Figure 12 and Figure 13). In this section, we further discuss challenges for both vulnerable and nonvulnerable materials that could exceed 20% of U.S. or global production in 2050. Our results indicate that there are six types of vulnerable materials that exceed this threshold: carbon fiber, electrical steel, aluminum, cobalt, rare-earth elements (i.e., dysprosium, neodymium, and praseodymium), and nickel (Figure 15). (It is important to note that we do not compare demand for chromium, gallium, graphite, lithium, manganese, niobium, tin, and titanium with domestic production because lithium and titanium production data were withheld from publication and the other materials are not mined domestically (USGS 2021); however, in both of our analysis scenarios, U.S. wind energy demand for all of these materials is below 2% of 2020 global production.) There are also three types of nonvulnerable materials that could pose supply challenges: balsa, copper, and glass fiber. We provide a high-level overview of the challenges in this section and discuss potential opportunities for reducing material requirements more generally in the next section.

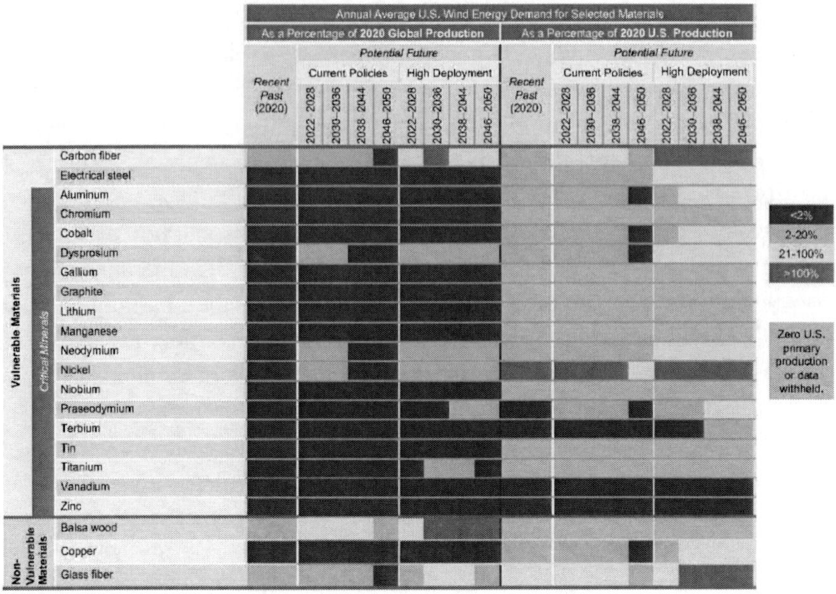

Figure 15. Annual U.S. wind energy demand over time for selected materials as compared to global and U.S. production in 2020.

3.5.1. Vulnerable Materials

3.5.1.1. Nickel

As shown in Table 8, nickel is primarily used in the production of stainless and heat-resisting steels. It is also used to produce other steel alloys, superalloys, other nickel alloys, electroplating, chemicals, batteries, catalysts, ceramics, and coinage. Within a wind plant, nickel is mostly used as an alloy for chromium steel and low-alloyed steel and is therefore accounted for within the background system of the REMPD.[15] The wind energy components that use the most of these types of steel (and therefore the most nickel) are the wind turbine tower, the land-based foundation or offshore substructure, and the wind turbine nacelle.

[15] Refer to Section 2.1.3 for more details about the foreground and background system definitions in the REMPD.

Table 9. Challenges and opportunities for U.S. wind energy demand of selected materials[a]

	Material	Potential Future U.S. Wind Energy Demand for Selected Materials in 2050				Challenges and Opportunities for U.S. Wind Energy
		As a Percentage of U.S. Production in 2020 (%)		As a Percentage of Global Production in 2020 (%)		
		Current Policies	*High Deployment*	*Current Policies*	*High Deployment*	
Vulnerable Materials	Nickel	97	1,600	1	11	• Global production should be sufficient to satisfy future U.S. windenergy demand for nickel, but it could exceed domestic production levels (if production remains unchanged from 2020 levels) • Expanding domestic production and diversifying imports for nickel could help secure domestic supply • Increasing recycling of nickel, identifying alternatives for nickel in steel manufacturing, and identifying alternatives to steel in wind energy • applications could help reduce material requirements
	Carbon fiber	9	440	3	120	• Future wind energy demand for carbon fiber could exceed domestic and global production levels (if production remains unchanged from 2020 levels) • It will be important for the United States to expand domestic production and diversify imports for carbon fiber • Material substitution, recycling, and end-of-life extension could help reduce wind energy material demand for carbon fiber
	Rare-earth elements *Dysprosium:* *Neodymium:* *Praseodymium:*	 3.0 3.8 0.2	 75 91 34	 0.49 0.61 0.03	 12 15 5	• Domestic and global production should be sufficient to satisfy future U.S. wind energy demand for rare-earth elements • Depending on competing uses, it may be important for the United States to expand domestic production and diversify imports of rare-earth elements

Table 9. (Continued)

	Material	Potential Future U.S. Wind Energy Demand for Selected Materials in 2050				Challenges and Opportunities for U.S. Wind Energy
		As a Percentage of U.S. Production in 2020 (%)		As a Percentage of Global Production in 2020 (%)		
		Current Policies	*High Deployment*	*Current Policies*	*High Deployment*	
Vulnerable Materials						• Material substitution, technology substitution, and recycling couldalso help reduce wind energy material demand for rare-earth elements
	Electrical steel	5	94	0.06	1	• Global production should be sufficient to satisfy future U.S. windenergy demand for electrical steel, but it could approach domestic production levels (if production remains unchanged from 2020 levels) • Material substitutions and recycling may not always be viable options for electrical steel • Expanding domestic production and diversifying imports of electrical steel could help secure domestic supply
	Cobalt	3	45	0.01	0.2	• Domestic and global production should be sufficient to satisfy future U.S. wind energy demand for cobalt • Depending on competing uses, it may be important for the United States to expand domestic production and diversify imports of cobalt • Identifying alternatives for cobalt in steel manufacturing and alternatives to steel in wind energy applications could help reduce wind energy material needs for cobalt

	Material	Potential Future U.S. Wind Energy Demand for Selected Materials in 2050				Challenges and Opportunities for U.S. Wind Energy
		As a Percentage of U.S. Production in 2020 (%)		As a Percentage of Global Production in 2020 (%)		
		Current Policies	*High Deployment*	*Current Policies*	*High Deployment*	
Vulnerable Materials	Aluminum	2	31	0.03	0.5	• Domestic and global production should be sufficient to satisfy future U.S. wind energy demand for aluminum • Depending on competing uses, it may be important for the United States to expand domestic production and diversify imports of aluminum • Increasing aluminum recycling and identifying alternatives in windenergy applications could help reduce wind energy material needs foraluminum
Nonvulnerable Materials	Balsa	N/A	N/A	14	520	• Future U.S. wind energy demand for balsa could exceed global production levels (if production remains unchanged from 2020 levels) • Material substitution and end-of-life extension could help reduce wind energy material demand for balsa
	Glass fiber	17	280	2	27	• Global production should be sufficient to satisfy future U.S. wind energy demand for glass fiber, but it could exceed domestic production levels (if production remains unchanged from 2020 levels) • Alternative blade designs, recycling methods, and end-of-life extension could help reduce wind energy material demand for glass fiber

Table 9. (Continued)

	Material	Potential Future U.S. Wind Energy Demand for Selected Materials in 2050				Challenges and Opportunities for U.S. Wind Energy
		As a Percentage of U.S. Production in 2020 (%)		As a Percentage of Global Production in 2020 (%)		
		Current Policies	*High Deployment*	*Current Policies*	*High Deployment*	
Nonvulnerable Materials	Copper	2	30	0.06	1	• Domestic and global production should be sufficient to satisfy future U.S. wind energy demand for copper • Depending on competing uses, it may be important for the United States to expand domestic production and diversify imports of copper • Increasing copper recycling and identifying alternatives in wind energy applications could help reduce wind energy material needs for copper

[a] The materials presented here were selected based on whether the potential future U.S. wind energy demand for them could exceed 20% of U.S. or global production in 2050 under the Current Policies and High Deployment scenarios considered here. Materials not listed here comprise less than 20% of annual U.S. and global production in 2050 (e.g., U.S. wind energy demand for tin is estimated to be less than 0.1% of U.S. and global production in 2050 [if production levels do not change from 2020 levels]).

In 2020, Indonesia produced the most nickel (accounting for 30% of global nickel mine production) and U.S. mine production accounted for less than 1% of global nickel production. If future U.S. production levels remain the same as 2020 levels, the two scenarios analyzed here indicate that future U.S. deployment of wind energy technologies could require 97%–1,600% of current U.S. production of nickel and 5%–63% of U.S. reserves of nickel (see Table 9, Figure 12, and Figure 14). Although the estimated world reserves of nickel are more than sufficient to satisfy U.S. wind energy demand for nickel, current U.S. production will not be sufficient to satisfy the nickel needed for U.S. wind energy deployment. In addition, in the first 3 months of 2022, nickel prices increased by more than 60%, a surge that indicates the possibility of future market volatility and supply disruption. As a result, to enable increased U.S. wind energy deployment, it will be important for the U.S. to secure supply of nickel, which could include expanding domestic production and recycling of nickel, identifying alternatives its use in steel manufacturing, identifying alternatives to steel in wind turbine applications (e.g., through the increased use of concrete in hybrid towers), and diversifying imports.

3.5.1.2. Carbon Fiber

Carbon fiber is stiff, strong, lightweight, and has good fatigue resistance. It is used in high-strength applications, including the manufacture of wind blades. Outside of the wind industry, it is primarily used in industrial applications for automobiles and pressure vessels. It is also used to manufacture aircraft and produce sports equipment (e.g., bicycles, golf equipment, racquets, and marine applications). Although there are substitutes for carbon fiber (e.g., glass fiber), the performance characteristics of these alternatives are generally lower than that of carbon fiber and therefore design modifications (e.g., increased weight) are often necessary to accommodate such a substitution.

In 2020, 76% of the world's carbon fiber was produced by six countries: the United States, Japan, China, Taiwan, and Hungary, with the United States producing more than 25% (refer to Table 8 for more details). The remaining 24% of global production in 2020 came from other countries that produce less than 5% of the global total. Under the High Deployment scenario, U.S. wind energy material needs for carbon fiber could exceed current global production from 2038 to 2044 (Figure 12). Thus, if the United States does not reduce the material requirements for carbon fibers in wind blades (e.g., through recycling, material substitution, or lifetime extension, which are further discussed in Section 4) or increase domestic manufacturing of carbon fiber, it risks relying on other countries to supply a substantial portion of the carbon fiber needed for U.S. wind energy applications under high wind deployment scenarios.

3.5.1.3. Rare-Earth Elements

Rare-earth elements are contained in mineral deposits such as bastnaesite and monazite, and the quality of a mineral deposit is judged on its percentage of total rare-earth oxides or grade (deposits with grades above 5% are considered viable (Alves Dias et al. 2020). Because rare-earth elements are mined together, USGS does not provide individualized information for each element in its mineral commodity summaries (USGS 2021). Data are also scarce, for instance, regarding Chinese rare-earth element reserves (Alves Dias et al. 2020). However, some studies provide some methods to determine production, reserves, and consumption of each rare-earth element, as well as estimates (Alves Dias et al. 2020; Du and Graedel 2011; Hoenderdaal et al. 2013).

The United States is currently the second largest producer of rare-earth elements after China, with U.S. production totaling 16% of global production compared to China's 58% (USGS 2021). However, most ore extracted from U.S. sources is exported for processing and U.S. consumption of rare-earth elements relies heavily on imports (China provides 80% of U.S. rare-earth element imports). Although recycling rare-earth element technologies exist, secondary production has not yet matured, and all commercial-scale U.S. consumption of rare-earth elements comes from primary sources. While the U.S. rare-earth element reserves represent about 1% of global reserves, rare-earth element resources are abundant in North America (e.g., Canada and the United States are estimated to be home to about 15 million tonnes and 2.4 million tonnes of rare-earth element resources, respectively). The main rare-earth element deposits currently exploited in the United States reside in Mountain Pass, California; they are of high grade (around 8%) but do not contain much dysprosium or terbium (Du and Graedel, 2011). As a result, the U.S. shares of global production for those elements are below 3%, whereas the U.S. shares of global production for neodymium and praseodymium are both 6%.

Rare-earth elements are used to produce magnets, ceramics and glass, battery alloys, and catalysts. Within wind power plants, rare-earth elements are primarily used for power generators in the nacelle and the quantity and type of rare-earth element needed depends on the type of generator used. For example, direct-drive permanent-magnet synchronous generators require 180 kilograms (kg)/MW of neodymium, 17 kg/MW of dysprosium, and 7 kg/MW of terbium, whereas gearbox double-fed induction generators only require 12 kg/MW of neodymium, 2 kg/MW of dysprosium, and 0 kg/MW of terbium (European Comission et al. 2020b). Based on the technology configurations and scenario definitions used in this study[16], if future U.S. production levels remain the same as 2020 levels, future U.S. deployment of wind energy technologies could consume 4%–91% of domestic neodymium production and 3%–75% of domestic dysprosium production in 2050. As discussed in greater detail in Section 4, material substitution, technology substitution, and recycling are three viable options for reducing the wind energy demand for rare-earth elements (e.g., a direct-drive generator can be made with a magnet constructed from copper windings, superconducting generators could be used instead of permanent-magnet generators and recycling could enable material recovery).[17] Depending on competing uses, it may be important for the United States to expand domestic production, develop ore processing capabilities, and diversify imports for rare-earth elements, particularly for dysprosium and neodymium.

3.5.1.4. Electrical Steel

Electrical steels (also known as silicon steels, relay steels, and transformer steels) contain iron, silicon, carbon, and some aluminum—thus the constituents are largely not critical by themselves. However, the manufacturing pathway for electrical steels is metallurgically specialized and the technical expertise for manufacturing electrical steels is highly guarded in industry (U.S. Department of Commerce, Bureau of Industry and Security 2021). As a result, the production of electrical steels is vulnerable to supply chain disruptions.

Electrical steels are primarily used to manufacture transformers, motors, and inductors. There are two types of electrical steels: grain-oriented electrical steel (GOES) and nongrain-oriented electrical steel (NOES). Compared to NOES, GOES conducts a magnetic field with a higher degree of efficiency in the direction of rolling and GOES is therefore more efficient at transforming mechanical energy to electrical energy (U.S. Department of Commerce, Bureau of Industry and Security 2021). Due to the highly specialized and advanced manufacturing processes that are required to create its uniformly oriented grain structures, GOES is more challenging to produce than NOES. As a result of its improved properties for transforming energy, GOES is

---

[16] Refer to Section 2.2 for details about the scenario definitions, including the technology configurations.

[17] Refer to Section 4 for more details.

primarily used in medium- and large-sized transformers, whereas NOES is generally used in electric motors, generators, and smaller transformers.

In terms of total electrical steel production (combined production of GOES and NOES), South Korea and China are the two largest exporters of electrical steel (each accounting for 14% of global exports), followed by Japan, Germany, and Russia. The United States comprises just more than 1% of total global exports for all types of electrical steel and relies on Japan, Korea, France, Austria, and China for imports. With respect to GOES, the United States currently consumes about 220,000 t of GOES per year and imports accounted for 21%–37% of U.S. consumption of GOES (by mass) from 2015 to 2019. According to an investigation conducted by the U.S. Department of Commerce, AK Steel, Inc. (a subsidiary of Cleveland Cliffs Inc.) is the only domestic producer of GOES and the company "is unable to meet the domestic demand for the full range of GOES products" (U.S. Department of Commerce, Bureau of Industry and Security 2021).

Within the wind energy industry, electrical steels are used for power generators in the nacelle and for transformers. If future U.S. production levels remain the same as 2020 levels, the High Deployment scenario analyzed here indicates that U.S. deployment of wind energy technologies in 2050 could require up to 94% of 2020 U.S. production of electrical steel. Due to the specialized manufacturing and functions of electrical steel, material substitutions and recycling may not always be viable options for this material. Thus, it will be important for the United States to secure supply of electrical steels by expanding domestic production and diversifying imports.

3.5.1.5. Cobalt

Cobalt is primarily used to produce electrodes in lithium-ion batteries. It is also used to produce other chemicals, superalloys, ceramics, steels, and other alloys. The estimated cobalt resource in the United States is about 1 million tonnes, whereas the world terrestrial resources are about 25 million tonnes. Most cobalt is produced as a byproduct of nickel and copper mines. In 2021, the United States produced about 700 t of cobalt, most of which comes from the nickel-copper Eagle Mine in Michigan. The rest of the cobalt used in the United States comes from imports and scrap materials (USGS 2022). Globally, about 70% of cobalt is mined in the Congo and more than 65% of refined cobalt is processed in China.

Within the wind industry, cobalt is used as an alloy for chromium steel and low-alloyed steel and is therefore accounted for within the background system of the REMPD.[18] The wind energy components that use the most of these types of steel (and therefore the most cobalt) are the wind turbine tower, land-based foundation or offshore substructure, and wind turbine nacelle. If future global production levels remain the same as 2020 levels, the two scenarios analyzed here indicate that global production should be sufficient to satisfy U.S. wind energy demand for cobalt. However, U.S. deployment of wind energy technologies in 2050 could comprise close to 45% of U.S. production of cobalt. As a result, depending on competing uses for cobalt, it may be important for the United States to secure domestic supply of cobalt, which could include expanding domestic production, identifying alternatives for cobalt in steel manufacturing, identifying alternatives to steel in wind turbine applications (e.g., through the increased use of concrete in hybrid towers), and diversifying imports.

3.5.1.6. Aluminum

Aluminum is a relatively lightweight metal—about one-third the weight of steel or copper—with excellent corrosion resistance, strength, and durability. It is used in a wide variety of applications, including the manufacturing of airplanes and automobiles, packaging (e.g., cans and foil), consumer goods such as cooking utensils, construction (e.g., windows and doors), electrical transmission lines, machinery, and appliances. Primary (or virgin) aluminum is manufactured through the electrolysis of alumina, which is refined from bauxite, and the global resources of bauxite are estimated "to be between 55 billion and 75 billion tons and are sufficient to meet world demand for metal well into the future" (USGS 2022). Aluminum can also be recovered from new scrap (manufacturing) and old scrap (discarded aluminum products).

In 2020, China was the world's leading producer of primary aluminum, accounting for 57% of world production (about 37 million tons), followed by Russia (6%), India (5%), Canada (5%), and the United Arab Emirates (4%). In 2020, the United States recovered about 3 million tons of aluminum from scrap, which was equivalent to about 75% of apparent U.S. consumption. The United States. also produced about 1 million tons of primary aluminum in 2020.

Within a wind power plant, aluminum is mostly used in the cables, tower, and nacelle. Our analysis indicates that domestic and global production should be sufficient to satisfy future U.S. wind energy demand for aluminum.

[18] Refer to Section 2.1.3 for more details about the foreground and background system definitions in the REMPD.

However, depending on competing uses for aluminum (e.g., aviation and automotive industries), it may be important for the United States to expand domestic production and diversify imports of aluminum. In addition, increasing aluminum recycling and identifying alternatives in wind energy applications could help reduce wind energy material needs for aluminum.

3.5.2. Nonvulnerable Materials

3.5.2.1. Balsa

Balsa is a lightweight hardwood that is used in a variety of commercial applications, including insulation, packaging, structural reinforcement, and marine applications (e.g., surfboards and boats). Within the wind industry, balsa is used in wind turbine blades to provide increased stiffness through the insertion of relatively light core structure between stiff fiber-reinforced plastic or carbon-reinforced plastic skins.

Ecuador is the top exporter of balsa globally, accounting for about 80% of the world's total balsa exports in 2020, followed by Peru (10% of global exports in 2020) and Brazil (5% of global exports in 2020). If future production levels remain the same as 2020 levels, the High Deployment scenario indicates that U.S. wind energy demand for balsa in 2050 could exceed five times the current level of global production. Material substitution—for example, increasing the use of PVC foam, using fiber-reinforced polymers, or developing alternative bio-based fibers for blade cores—and end-of-life extension could potentially reduce wind energy material demand for balsa.

3.5.2.2. Glass Fiber

Glass fiber is made by heating and drawing silica-based glass into thin strands. The most common type of glass fiber is E-glass, which has low electrical conductivity compared to other types of glass fiber materials (Fitzer et al. 2000). Glass fibers are usually processed into textile glass fabrics (e.g., via weaving, knitting, and braiding) or embedded in a polymer matrix (e.g., epoxy, vinyl ester, polyester) to produce a composite material known as glass-fiber reinforced polymer, which is known for its high strength and low weight. Textile glass fabrics are used to produce circuit boards and thermal insulation for use in the electronics industry and glass fiber-reinforced polymers are often used in structural applications within the construction, transportation, and industrial sectors (Fitzer et al. 2000). Within a wind plant, glass fibers are used within the fiber-reinforced polymers in wind turbine blades.

In 2020, China produced more than 50% of the world's glass fiber, followed by the United States (10%) and Malaysia (9%). If future global production levels remain the same as 2020 levels, the two scenarios analyzed here indicate that global production should be sufficient to satisfy U.S. wind energy demand for glass fiber. However, future U.S. deployment of wind energy technologies could use 20%–300% of 2020 U.S. production of glass fiber annually. Thus, to secure domestic supply of glass fiber for U.S. wind industry applications, it may be important for the country to increase production beyond 2020 levels. Although there are some composite recycling processes (e.g., mechanical recycling, thermal recycling, and chemical recycling) that can be used to recover some of the original glass fiber material, these processes often degrade the quality of the fibers to the extent that they can only be used in applications with less stringent design requirements than are required for wind blades. Alternative wind blade designs (e.g., using thermoplastic resins that can be heated and remolded) could allow for increased recycling of glass fibers within the wind industry. In addition, extending the lifetime of wind blades could also help reduce the quantity of glass fiber that is required for wind blades. However, further research is needed to develop and improve these circular economy approaches for glass fibers.

3.5.2.3. Copper

Copper has excellent electrical conductivity, along with high strength, hardness, and ductility. It is also a good conductor of heat (it conducts heat about 30 times better than stainless steel and 1.5 times better than aluminum). Due to these properties, the metal is largely used in electrical applications, including power transmission and generation, building wiring, telecommunication, electronics, and electronic products.

China is the world's leading producer of refined copper and produced about 10 million tons of refined copper (or 40% of the world's total copper refinery production) in 2020. Refined copper is also produced by Chile, Japan, Congo, Russia, and the United States, each producing 9%, 6%, 4%, 4%, and 4% of the world's refined copper production, respectively, with other countries each producing less than 3%. The recoverable copper content of U.S. mine production was around 1.2 million tons in 2020. Most of the refined copper produced in the United States. comes from primary refinery production of copper from ore; the amount of copper recovered from old and new scrap in the country is only equivalent to about 10% of the apparent U.S. consumption of copper (USGS 2022).

Within a wind power plant, copper is used in the array and export cables, foundation, substation, and turbine. Based on the Current Policies and High Deployment scenarios, domestic and global production should be sufficient to satisfy future U.S. wind energy demand for copper. However, depending on competing uses, it may be important for the United States to expand domestic production and diversity imports. Increasing copper recycling and identifying alternatives in wind energy applications could help reduce wind energy material needs for copper.

4. Opportunities to Reduce Material Requirements

There are many opportunities for innovation to change material requirements for wind energy. Table 10 summarizes innovations that could be relevant to each component and subassembly within a wind power plant. We do not examine their impact and simply provide a high-level overview of how three categories of material innovations could influence wind energy materials. These categories are material substitution, weight reduction, and other circular economy approaches (including recycling).

Table 10. Innovations that could modify material requirements for wind energy technologies

Component, Subassembly	Innovations That Could Modify Material Requirements

Wind turbine, blade	• Material substitution
	  • Bio-based resins and biofibers
	  • Recyclable resin systems (e.g., thermoplastic resin)
	• Weight reduction
	  • Carbon-fiber spar caps
	• Other circular economy approaches
	  • Composite recycling
	  • Lifetime extension

Turbine, nacelle	• Material substitution
	  • Superconducting direct-drive generators
	  • Other generators that use fewer rare-earth elements
	• Weight reduction
	  • High-voltage direct-current generators
	  • Replacing steel with cast iron in the rotor shaft and main bearing
	• Other circular economy approaches
	  • Lifetime extension

Component, Subassembly	Innovations That Could Modify Material Requirements
	• Reuse of permanent magnets in generators • Recycling of rare-earth elements

Turbine, tower	• Material substitution
	  • Alternative materials (e.g., wood, aluminum, concrete, composites), including hybrid designs with steel
	• Weight reduction
	  • Higher-strength steel
	  • Increased tower diameter (would likely require other technology innovations such as spiral welding)
	• Other circular economy approaches
	  • Lifetime extension via partial repowering
	  • Increased recycling
	• Other technology innovations
	  • Land-based: on-site manufacturing (e.g., spiral welding)

Array and export cables and other electrical equipment, all subassemblies	• Material substitution
	  • High deployment conductor materials (e.g., graphene)
	  • Alternatives to sulfur hexafluoride gas in electrical equipment
	• Other circular economy approaches
	  • Dry transformers
	• Other technology innovations
	  • Optimized transmission planning (e.g., high-voltage direct-current backbones)
	  • Increasing cable rating

Turbine, land-based foundation or offshore substructure	• Material substitutions
	  • Hybrid materials (e.g., steel and concrete)
	  • Land-based: steel pile foundation
	• Weight reduction
	  • Offshore: high-alloy steel
	  • Offshore: lattice structures such as jacket designs
	  • Land-based: ribbed or annular concrete foundation
	• Other circular economy approaches
	  • Lifetime extension via partial repowering
	• Other technology innovations
	  • Controls to reduce design loads
	  • Land-based: additive manufacturing
	  • Offshore: floating substructures
	  • Offshore: innovation in post-tensioning for substructure/anchors/tower
	  • Offshore: sustainable geopolymers to replace Portland cement

Table 10. (Continued)

Component, Subassembly	Innovations That Could Modify Material Requirements
	• Offshore: shared mooring and anchors • Offshore: micropile anchors
Substation, all subassemblies	• Other technology innovations • Land-based: increasing plant size • Land-based: gas-insulated switchgear and transformers • Offshore: modular substations that share turbine platform

4.1. Material Substitution

When a material has limited availability, manufacturers may be able to reduce reliance on it by substituting it with another material. The substitution of one material for another can take various forms. In some cases, it is possible to directly substitute one element for another with similar properties. For example, aluminum, cadmium, or plastic-based coatings can replace zinc coatings in some applications (Tolcin 2022). Other substitutions can occur between different technologies that perform the same function (Smith and Eggert 2016). Material substitution is an effective method to reduce dependence on vulnerable materials, but it may increase the mass of other materials required, thereby increasing the total system mass.

Rare-earth elements are vulnerable materials that could be substituted—partially or completely— in wind turbine generators. The sharp rise in rare-earth-element prices in 2011 spurred alternatives to permanent-magnet synchronous generators. Some technologies already existed; for example, induction generators that do not require a permanent magnet. Wind turbines without permanent-magnet generators may use a geared drive with an electromagnetic generator (this remains the dominant configuration for land-based wind turbines) or a direct-drive generator with a magnet made from copper windings (this configuration, known as an electrically excited synchronous generator, eliminates rare-earth magnets but adds weight). Recently, medium-speed generators that use a smaller quantity of rare-earth permanent magnets have begun to be adopted. Although the demand for rare-earth magnets still exists, reducing the size of the magnets allows more generators to be produced from the same quantity of rare-earth elements.

Superconducting generators have also been proposed as an alternative to designs using permanent magnets (Pavel et al. 2017).

The dominant role of steel in wind turbine towers and offshore substructures could also be altered by material substitution. Hybrid towers that replace steel with concrete in the lower sections of the structure were included in the High Deployment scenario (see Table 2). Other alternative materials that have been proposed for wind turbine towers include wood, aluminum, or fiberglass composites with a concrete or foam core (Watson et al. 2019). For offshore wind energy, concrete, gravity-base foundations have been used in some European projects. Concrete is also a possible alternative to steel for floating substructures; one example is the "damping pool" barge design demonstrated in Japan and France (BW Ideol n.d.). Reducing steel requirements would also decrease the wind energy demand for the critical minerals that are used in steel alloying (e.g., nickel).

4.2. Weight Reduction

Material requirements can be reduced by lowering the overall mass of wind energy systems. Reducing weight can have both direct and indirect effects on material use; for example, reducing the weight of the nacelle can lessen loads on the tower, which can then be made lighter as well. Often, weight reduction is achieved by using lightweight, high-performance materials. In some cases, switching to lightweight materials involves increasing the use of critical minerals; for example, adding nickel, molybdenum, and copper to cast iron can reduce the weight of large castings used in the nacelle (International Molybdenum Association 2011). Opportunities for reducing the weight of specific components include:

- *Blades.* Carbon-fiber spar caps enable a weight reduction of approximately 25% compared to fiberglass (Ennis et al. 2019)
- *Tower.* Increasing the diameter of the tower base can decrease the total mass; however, large-diameter towers may require additional innovations such as spiral welding to avoid transportation constraints (Lantz et al. 2019)
- *Generator.* High-voltage direct-current generators could help reduce the weight of electrical equipment compared with alternating-current generators (Watson et al. 2019)

- *Drivetrain.* Replacing steel with cast iron in the rotor shaft and main bearing seat could enable hollow designs with lower mass (Kirsch and Kyling 2021)
- *Substructure.* Lattice structures such as jacket designs could reduce mass requirements for fixed-bottom substructures in water depths of 40 m or more (Damiani et al. 2016).

4.3. Other Circular Economy Approaches

Material requirements for wind energy can also be affected by material handling at end of life. Some approaches that could be used to encourage the development of a more circular economy for wind energy materials include the following:

- *Lifetime extension.* Lifetime extension refers to the continued use of wind turbines and other plant components beyond their design lifetimes. Additional maintenance, monitoring, or refurbishment may be necessary to support the extended operational period, but the overall material requirements are less than would be needed for replacement with new equipment. The impact of life cycle extension on vulnerable material consumption for wind turbine generators was illustrated by Veers, Sethuraman, and Keller (2020), who obtained a 15% decrease in rare-earth element demand when increasing the assumed lifetime from 20 to 25 years. Extending the lifetime of the tower, land-based foundation, or offshore substructure while replacing the nacelle and rotor also reduces material requirements.
- *Reuse.* The large permanent magnets used in wind turbines are relatively straightforward to separate from other materials in the nacelle (compared to other permanent-magnet applications like hard disk drives), which presents an opportunity to reuse the magnets in new turbines after refurbishment, reducing the requirement for virgin rare-earth elements (Yang et al. 2017). The stock of used magnets available for reuse lags installation by 2-3 decades, so there is limited potential for reuse to contribute to reduction in material demand for near-term growth in wind energy.
- *Recycling.* Recycling already makes a significant contribution to the U.S. supply of some metals, with approximately half of the

aluminum, iron, steel, and nickel used in the country coming from secondary sources. Recycling also supplies approximately one-third of copper, chromium, and tin in the United States, and nearly 20% of zinc (Sangine 2022). The scope for increasing recycled content varies by material and depends on the ratio between in-use stocks and material demand as well as the end-of-life recycling rate. For example, there may be room for growth in the recycled content of copper, which has an estimated recycling rate close to 50%, but there is less opportunity to increase lead recycling, which has already reached more than 90% (UNEP 2011). Iron and steel make up the largest portion of wind turbine and offshore substructure mass; globally, these materials are estimated to have an end-of-life recycling rate between 70% and 90% (UNEP 2011). Rare-earth elements are recycled at much lower rates; however, researchers are engaged in developing more effective methods for recycling and material recovery (Jowitt et al. 2018; Lorenz and Bertau 2019; Rademaker et al. 2013).[19] Recycling of wind turbine blade materials is also an active area of research (Deeney et al. 2021; Fonte and Xydis 2021; Sommer and Walther 2021).

5. Conclusion

This report describes how we used the REMPD to determine the quantity of materials that might be needed to satisfy two deployment scenarios for U.S. wind energy. For the two scenarios, the database helped us better understand the constraints and vulnerabilities linked to the availability of physical materials.

5.1. Summary of Approach and Results

Here, we evaluated the types and quantities of materials that are needed to manufacture and construct wind energy plants under two deployment scenarios: Current Policies and High Deployment. We defined these two scenarios using a capacity projection, plant configuration, and technology

[19] Refer to Appendix F for a summary of rare-earth element processing methods and recycling opportunities.

configuration and used the REMPD to determine the wind energy components and computed the total raw and process material requirements (including both foreground and background systems). A moderate wind energy capacity projection, plant configuration, and technology configuration (based on the Standard Scenarios 2022 Mid-case with no new decarbonization policies, limited changes to current plant configurations [e.g., wind turbine size], and no technology innovations) are assumed in the Current Policies scenario. The High Deployment scenario assumes high levels of wind energy deployment (aligned with the Biden administration's goal of carbon-free electricity by 2035 and net-zero economy by 2050) and incorporates significant changes in wind power plant configurations and limited materials-related innovations. The REMPD provided the annual quantity of materials required for each scenario along with material properties (e.g., availability and significant use).

To improve the interpretability of the results, we grouped materials into seven major categories: concrete, road aggregate, steel, composites and polymers, cast iron, other metals and alloys, and other materials. We also performed a more detailed analysis of 19 vulnerable materials, which include 17 critical minerals, along with carbon fiber and electrical steel.[20] Our major findings include:

- In both scenarios, the projected annual U.S. wind energy demand for vulnerable and nonvulnerable materials through 2050 is anticipated to require less than 10% of 2020 global production for all materials except nickel, dysprosium, neodymium, glass fiber, carbon fiber, and balsa. In the High Deployment scenario, from 2030 to 2036, annual average U.S. wind energy demand for nickel, dysprosium, neodymium, glass fiber, carbon fiber, and balsa could account for 8%, 8%, 11%, 23%, 111%, and 268%, respectively, of 2020 global production of those materials.
- In addition, if future U.S. production remains the same as 2020 levels, the material needs for U.S. wind energy could approach or exceed the domestic production of nickel, electrical steel, carbon fiber, and glass fiber. Wind energy facilities also use materials for which there is currently no primary production in the United States, including balsa, chromium, gallium, graphite, manganese, niobium, and tin.

[20] Refer to Table 1 for a list of vulnerable materials included in our analysis.

- Other materials for which demand from wind energy is projected to be high relative to current U.S. production include neodymium, dysprosium, and cobalt.

Overall, we find that U.S. wind energy demand for nickel, balsa, electrical steel, glass fiber, carbon fiber, neodymium, dysprosium, and cobalt could pose challenges under high wind deployment. As a result, it will be important for the United States to secure supply of these materials by increasing domestic production, diversifying imports for these materials, identifying alternative materials, or modifying designs to improve reuse and recycling.

5.2. Opportunities and Vulnerabilities

Using the Current Policies and High Deployment scenarios, we identified several vulnerabilities. Although current global supply chains should be able to meet the U.S. wind energy demand for most materials in both scenarios, it might prove to be more challenging for nickel, balsa, electrical steel, glass fiber, carbon fiber, rare-earth elements (i.e., neodymium and dysprosium), and cobalt. However, there are a variety of opportunities for reducing wind energy requirements for these materials (e.g., increasing domestic production, diversifying import sources, making material substitutions, and increasing other circular economy pathways such as alternative designs and recycling). Our key findings include:

- Nickel and cobalt are primarily used in chromium steel and low-alloyed steel within the wind turbine tower, the land-based foundation or offshore substructure, and the wind turbine nacelle. With more than half of U.S. steel production already coming from scrap, the wind energy industry could benefit from technological solutions that reduce steel requirements. A hybrid wind tower—which uses a large quantity of concrete in the tower and less steel—is one such solution. In addition, the United States could consider identifying alternatives for cobalt and nickel in steel manufacturing, increasing domestic production, and diversifying imports.
- Balsa, carbon fiber, and glass fiber are all used within wind turbine blades. Although the United States is currently one of the top three producers of composite materials, production will need to increase

globally to meet increasing demand from the wind industry and other sectors. If the United States does not increase domestic production of these materials, it risks relying on other countries to supply a substantial portion of the carbon fiber needed for U.S. wind energy applications under high wind deployment scenarios. Lifetime extension and alternative wind blade designs (e.g., foam instead of balsa cores and thermoplastic rather than thermoset resins) could help reduce virgin material requirements for balsa, glass fiber, and carbon fiber. Increased recycling of wind blade composites could also help mitigate material demand for carbon and glass fibers within the wind industry. However, further research is needed to develop and improve circular economy approaches for these materials.

- Electrical steel is used for power generators in the nacelle and for transformers. Due to the specialized manufacturing and functions of electrical steel, material substitution and recycling may not always be viable options. Thus, expanding domestic production and diversifying import sources are likely to be important strategies to ensure a stable supply of electrical steel.
- Rare-earth elements are primarily used within the wind turbine generator. Although some rare-earth ores are mined in the United States, nearly all refining and processing into rare-earth metals occurs in China, causing high supply risks. In the long term, increased domestic production or imports from other countries such as Canada (home to the biggest known rare-earth-element resources) could mitigate this risk. However, in the short term, hybrid generators (with smaller permanent magnets), alternative magnet materials, rare-earth-element recycling, or the reuse of old permanent magnets could be viable solutions.

5.3. Broader Impacts

As a relational database, the REMPD can be extended; for instance, to include additional information on wind energy externalities (e.g., emissions from transportation, material requirements for transmission, impacts on local communities). In future work, the REMPD could be used to compute the life cycle impacts related to a given scenario as the life cycle inventory data used to connect the foreground and background materials contains all the necessary information.

The database taxonomy (see Figure 1) is also designed to allow other technologies or facilities with components and core materials to be stored and analyzed. For example, the REMPD structure has already been used to answer the requirements outlined in Subsection (5) of Section 3004 of the Energy Act of 2020 (i.e., developing "a comprehensive physical property database of materials for use in solar energy technologies, which shall identify the type, quantity, country of origin, source, significant uses, projected availability, and physical properties of materials used in solar energy technologies"). Other renewable energy technologies could also be added.

Additional research could also enable the REMPD to provide a more detailed analysis of potential supply chain risks. For example, three indicators could be added to the database: likelihood of foreign supply disruption (disruption potential), dependency of U.S. manufacturers on foreign suppliers (trade exposure), and the ability of U.S. manufacturers to withstand a supply disruption (economic vulnerability). Such analysis could also highlight mitigation opportunities and trade dynamics (e.g., the role of neodymium substitution with praseodymium). Another option could be to investigate hurdles to increasing material production in supply chains (e.g., for a given material, current U.S. production capacities could increase supply by 10% but a new plant would need to increase it by 15%).

The estimates of total material requirements for wind energy are highly sensitive to the projected deployment of wind energy in the United States. As updated capacity projections data become available, future work will evaluate how the results reported here might change under alternative deployment assumptions. Evolution of the technology used in wind power plants will also affect material requirements. Future work should continue to modify the wind power plant configurations and wind turbine technology to reflect trends in design and material usage.

References

4C Offshore. 2022. Global Offshore Wind Farm Intelligence (Online Database). February 2022. https://www.4coffshore.com/windfarms/.

ABB. 2010. "XLPE Submarine Cable Systems Attachment to XLPE Land Cable Systems -User´s Guide." https://new.abb.com/docs/default-source/ewea-doc/xlpe-submarine-cable-systems-2gm5007.pdf.

Achzet, Benjamin, and Christoph Helbig. 2013. "How to Evaluate Raw Material Supply Risks— an Overview." *Resources Policy* 38 (4): 435–47. https://doi.org/10.1016/j.resourpol.2013.06.003.

Alsaleh, A., and M. Sattler. 2019. "Comprehensive Life Cycle Assessment of Large Wind Turbines in the US." *Clean Technologies and Environmental Policy* 21 (4): 887–903. https://doi.org/10.1007/s10098-019-01678-0.

Alves Dias, P., S. Bobba, S. Carrara, B. Plazzotta. 2020. *The role of rare earth elements in wind energy and electric mobility*. European Commission, Joint Research Centre. EUR 30488 EN. Publication Office of the European Union, Luxembourg. https://data.europa.eu/doi/10.2760/303258.

American Physical Society Panel on Public Affairs and Materials Research Society. 2011. "Energy Critical Elements: Securing Materials for Emerging Technologies." https://www.aps.org/policy/reports/popa-reports/upload/elementsreport.pdf.

Ardani, K., P. Denholm, T. Mai, R. Margolis, E. O'Shaughnessy, T. Silverman, J. Zuboy. 2021. *Solar Futures Study*. Washington, D.C.: U.S. Department of Energy. https://www.energy.gov/sites/default/files/2021-09/Solar%20Futures%20Study.pdf.

Arvesen, Anders, Rasmus Nikolai Nes, Daniel Huertas-Hernando, and Edgar G. Hertwich. 2014. "Life Cycle Assessment of an Offshore Grid Interconnecting Wind Farms and Customers across the North Sea." *The International Journal of Life Cycle Assessment* 19 (4): 826–37. https://doi.org/10.1007/s11367-014-0709-2.

Atwater, Harry, Neil Fromer, and Val Otten. 2011. "Critical Materials For Sustainable Energy Applications." Pasadena, CA: Resnick Institute, California Institute of Technology. https://authors.library.caltech.edu/32727/1/ri_criticalmaterials_report.pdf.

Baars, J., T. Domenech, R. Bleischwitz, H. E. Melin, and O. Heidrich. 2021. "Circular Economy Strategies for Electric Vehicle Batteries Reduce Reliance on Raw Materials." *Nature Sustainability* 4 (1): 71–79. https://doi.org/10.1038/s41893-020-00607-0.

Bauer, D., D. Diamond, J. Li, D. Sandalow, P. Telleen, and B. Wanner. 2010. "U.S. Department of Energy Critical Materials Strategy." U.S. Department of Energy. https://www.osti.gov/biblio/1000846-department-energy-critical-materials-strategy.

Benke, D., M. Fries, M. Specht, J. Wortmann, M. Pabst, T. Gottschall, I. Radulov, K. Skokov, A. I. Bevan, D. Prosperi, C. O. Tudor, P. Afiuny, M. Zakotnik, O. Gutfleisch. 2020. "Magnetic Refrigeration with Recycled Permanent Magnets and Free Rare-Earth Magnetocaloric La–Fe– Si." *Energy Technol.*, 8: 1901025. https://doi.org/10.1002/ente.201901025.

Binnemans, K., P. T. Jones, B. Blanpain, T. Van Gerven, Y. Yang, A. Walton, M. Buchert. 2013. "Recycling of rare earths: a critical review." *J Clean Prod*, 51, pp. 1-22. https://www.sciencedirect.com/science/article/pii/S0959652612006932.

BloombergNEF. 2020. Carbon fiber manufacturing-related data.

Bogart, J., B. Cole, M. Boreen, C. Lippincott, B. Manor, P. Carroll, E. Schelter. 2016. "Accomplishing simple, solubility-based separations of rare earth elements with complexes bearing size-sensitive molecular apertures." *PNAS*, 113 (52) 14887-14892. https://doi.org/10.1073/pnas.1612628113.

BW Ideol. n.d. "Floatgen." Accessed May 6, 2022. https://www.bw-ideol.com/en/floatgen-demonstrator.

Cao, Z., C. O'Sullivan, J. Tan, P. Kalvig, L. Ciacci, W. Chen, J. Kim, G. Liu. 2019. "Resourcing the Fairytale Country with Wind Power: A Dynamic Material Flow Analysis." *Environmental Science & Technology, 53*(19), 11313-11322. doi:10.1021/acs.est.9b03765.

Carrara, S., P. Alves Dias, B. Plazzotta, C. Pavel. 2020b. *Raw materials demand for wind and solar PV technologies in the transition towards a decarbonised energy system.* European Commission, Joint Research Centre, Publication Office of the European Union, Luxembourg. https://data.europa.eu/doi/10.2760/160859.

Connelly, E., M. Penev, A. Milbrandt, B. Roberts, N. Gilroy, M. Melaina. 2020. *Resource Assessment for Hydrogen Production.* https://www.osti.gov/biblio/1660128.

Consolidated Appropriations Act. 2021. https://www.congress.gov/bill/116th-congress/house-bill/133/text.

Cooperman, A., A. Eberle, Eric Lantz. 2021. "Wind turbine blade material in the United States: Quantities, costs, and end-of-life options." *Resources, Conservation and Recycling, 168*, 105439. https://doi.org/10.1016/j.resconrec.2021.105439.

Cooperman, A., A. Eberle, D. Hettinger, M. Marquis, B. Smith, R. Tusing, J. Walzberg. 2023. *Renewable Energy Materials Properties Database: Summary.* Golden, CO: National Renewable Energy Laboratory. NREL/TP-5000-82830. nrel.gov/docs/fy23osti/82830.pdf.

Crawford, R. H. 2009. "Life Cycle Energy and Greenhouse Emissions Analysis of Wind Turbines and the Effect of Size on Energy Yield." *Renewable and Sustainable Energy Reviews* 13 (9): 2653–60.

Damiani, R., K. Dykes, G. Scott. 2016. "A comparison study of offshore wind support structures with monopiles and jackets for U.S. waters." *Journal of Physics: Conference Series, 753,* 092003. doi:10.1088/1742-6596/753/9/092003.

Deeney, Peter, Angela J. Nagle, Fergal Gough, Heloisa Lemmertz, Emma L. Delaney, Jennifer M. McKinley, Conor Graham, Paul G. Leahy, Niall P. Dunphy, and Gerard Mullally. 2021. "End-of-Life Alternatives for Wind Turbine Blades: Sustainability Indices Based on the UN Sustainable Development Goals." *Resources, Conservation and Recycling* 171 (August): 105642. https://doi.org/10.1016/j.resconrec.2021.105642.

Denholm, P., P. Brown, W. Cole, T. Mai, B. Sergi, M. Brown, P. Jadun, J. Ho, J. Mayernik, C. McMillan, R. Sreenath. 2022. *Examining Supply-Side Options to Achieve 100% Clean Electricity by 2035.* Golden, CO: National Renewable Energy Laboratory. NREL/TP-6A40-81644. nrel.gov/docs/fy22osti/81644.pdf.

Du, X., and T. E. Graedel. 2011. "Global In-Use Stocks of the Rare Earth Elements: A First Estimate." *Environ. Sci. Technol., 45*(9), 4096-4101. doi:10.1021/es102836s.

Dunn, Jessica, Margaret Slattery, Alissa Kendall, Hanjiro Ambrose, and Shuhan Shen. 2021. "Circularity of Lithium-Ion Battery Materials in Electric Vehicles." *Environmental Science & Technology* 55 (8): 5189–98. https://doi.org/10.1021/acs.est.0c07030.

Dysinger, D. K., and J. E. Murphy. 1994. *Electrowinning of neodymium from a molten oxide-fluoride electrolyte. Report of investigations/1994*. Reno, NV: Bureau of Mines. https://www.osti.gov/biblio/7168096.

Eberle, A., J. O. Roberts, A. Key, P. Bhaskar, K. L. Dykes. 2019. *NREL's Balance-of-System Cost Model for Land-Based Wind.* Golden, CO: National Renewable Energy Laboratory (NREL). NREL/TP-6A20-72201. https://www.nrel.gov/docs/fy19osti/72201.pdf.

Elshkaki, A., and T. E. Graedel. 2014. "Dysprosium, the balance problem, and wind power technology." *Applied Energy, 136*, 548-559. https://doi.org/10.1016/j.apenergy.2014.09.064.

———. 2015. "Solar cell metals and their hosts: A tale of oversupply and undersupply." *Applied Energy, 158*, 167-177. https://doi.org/10.1016/j.apenergy.2015.08.066.

Energy Act of 2020. 2020. https://www.energy.senate.gov/services/files/847E2D8F-0CED-407E-9D65-C1589C9E1412.

Ennis, B. L., C. L. Kelley, B. T. Naughton, B. Norris, S. Das, D. Lee, D. Miller. 2019. *Optimized Carbon Fiber Composites in Wind Turbine Blade Design.* (SAND2019-14173). https://www.osti.gov/servlets/purl/1592956.

Erdmann, L., and T. E. Graedel. 2011. "Criticality of Non-Fuel Minerals: A Review of Major Approaches and Analyses." *Environmental Science & Technology* 45 (18): 7620–30. https://doi.org/10.1021/es200563g.

Fingersh, L., M. Hand, and A. Laxson. 2006. *Wind Turbine Design Cost and Scaling Model.* Golden, CO: National Renewable Energy Laboratory (NREL). NREL/TP-500-40566. https://www.nrel.gov/docs/fy07osti/40566.pdf.

Fishman, T. and T. E. Graedel. 2019. "Impact of the establishment of US offshore wind power on neodymium flows." *Nature Sustainability*, 2(4), 332-338. doi:10.1038/s41893-019-0252-z.

Fitzer, E., R. Kleinholz, H. Tiesler, M. H. Stacey, R. De Bruyne, I. Lefever, A. Foley et al. 2000. "Fibers, 5. Synthetic Inorganic." In *Ullmann's Encyclopedia of Industrial Chemistry.* Wiley- VCH Verlag GmbH & Co.

Fonte, Rosario, and George Xydis. 2021. "Wind Turbine Blade Recycling: An Evaluation of the European Market Potential for Recycled Composite Materials." *Journal of Environmental Management* 287 (June): 112269. https://doi.org/10.1016/j.jenvman.2021.112269.

Fraunhofer Institute for Systems and Innovation Research ISI, Institute for Energy and Transport (Joint Research Centre), Oakdene Hollins Ltd, Espinoza Luis Tercero, Ruth Bryson, Adrian Chapman, Evangelos Tzimas, et al. 2013. *Critical metals in the path towards the decarbonisation of the EU energy sector: Assessing rare metals as supply-chain bottlenecks in low-carbon energy technologies*. LU: Publications Office of the European Union. https://data.europa.eu/doi/10.2790/46338.

Gaertner, E., J. Rinker, L. Sethuraman, F. Zahle, B. Anderson, G. Barter, N. Abbas, F. Meng, P. Bortolotti, W. Skrzypinski, G. Scott, R. Feil, H. Bredmose, K. Dykes, M. Shields, C. Allen, and Viselli. 2020. *Definition of the IEA 15-Megawatt Offshore Reference Wind.* Golden, CO: National Renewable Energy Laboratory. NREL/TP-5000-75698. https://www.nrel.gov/docs/fy20osti/75698.pdf.

Gagnon, P., M. Brown, Steinberg, D., Brown, P., et al. 2022. *2022 Standard Scenarios Report: A U.S. Electricity Sector Outlook*. Golden, CO: National Renewable Energy Laboratory. NREL/TP-6A40-84327 https://www.nrel.gov/docs/fy23osti/84327.pdf.

Graedel, T. E., and Barbara K. Reck. 2016. "Six Years of Criticality Assessments: What Have We Learned So Far?" *Journal of Industrial Ecology* 20 (4): 692–99. https://doi.org/10.1111/jiec.12305.

Guezuraga, Begoña, Rudolf Zauner, and Werner Pölz. 2012. "Life Cycle Assessment of Two Different 2 MW Class Wind Turbines." *Renewable Energy* 37 (1): 37–44. https://doi.org/10.1016/j.renene.2011.05.008.

Habib, K., P. K. Schibye, A. P. Vestbø, O. Dall, H. Wenzel. 2014. "Material Flow Analysis of NdFeB Magnets for Denmark: A Comprehensive Waste Flow Sampling and Analysis Approach." *Environmental Science & Technology 48*(20), 12229–37. https://doi.org/10.1021/es501975y.

Habib, K., K. Parajuly, and H. Wenzel. 2015. "Tracking the Flow of Resources in Electronic Waste - The Case of End-of-Life Computer Hard Disk Drives." *Environmental Science & Technology, 49*(20), 12441-12449. doi:10.1021/acs.est.5b02264.

Harris, I. R., A. Williams, A. Walton, J. Speight. 2011. Magnet Recycling. https://patents.google.com/patent/US8734714B2/en.

Hitachi. 2010. "Hitachi Develops Recycling Technologies for Rare Earth Metals." https://www.hitachi.com/New/cnews/101206.html.

Ho, Jonathan, Jonathon Becker, Maxwell Brown, Patrick Brown, Ilya Chernyakhovskiy, Stuart Cohen, Wesley Cole, et al. 2021. Regional Energy Deployment System (ReEDS) Model Documentation: Version 2020. Golden, CO: National Renewable Energy Laboratory. NREL/TP-6A20-78195. https://www.nrel.gov/docs/fy21osti/78195.pdf.

Hoenderdaal, Sander, Luis Tercero Espinoza, Frank Marscheider-Weidemann, and Wina Graus. 2013. "Can a Dysprosium Shortage Threaten Green Energy Technologies?" *Energy* 49 (January): 344–55. https://doi.org/10.1016/j.energy.2012.10.043.

Hofmann, Margarethe, Heinrich Hofmann, Christian Hagelüken, and Alessandra Hool. 2018. "Critical Raw Materials: A Perspective from the Materials Science Community." *Sustainable Materials and Technologies* 17 (September): e00074. https://doi.org/10.1016/j.susmat.2018.e00074.

Hydrogen Council. 2020. *Path to Hydrogen Competitiveness: A Cost Perspective*. https://hydrogencouncil.com/en/path-to-hydrogen-competitiveness-a-cost-perspective/.

Ikhennicheu, M., M. Lynch, S. Doole, F. Wendt, M. Schwarzhopf, D. Matha, R. D. Vincente, T. Habekost, L. Ramirez, and S. Potestio. 2020. "Review of the State of the Art of Dynamic Cable System Design." Brussels, Belgium: Corewind. https://corewind.eu/wp-content/uploads/files/publications/COREWIND-D3.1-Review-of-the-state-of-the-art-of-dynamic-cable-system-design.pdf.

International Energy Agency. 2021. "The Role of Critical Minerals in Clean Energy Transitions." World Energy Outlook Special Report, Paris. https://www.iea.org/reports/the-role-of-critical-minerals-in-clean-energy-transitions.

International Molybdenum Association. 2011. *Molybdenum in Irons and Steels for Clean and Green Power Generation*. https://www.imoa.info/download_files/alloy-steel/Clean_and_Green_Energy.pdf.

International Renewable Energy Agency. 2020. *Green Hydrogen Cost Reduction: Scaling up Electrolysers to Meet the 1.5℃ Climate Goal*. https://irena.org/-/media/Files/IRENA/Agency/Publication/2020/Dec/IRENA_Green_hydrogen_cost_2020.pdf.

Iowa Department of Transportation. 2021. *Densities for Use in Estimating Quantities*. https://iowadot.gov/design/dmanual/01b-04.pdf.

Itoh M., K. Miura, K. I. Machida. 2009. "Novel rare earth recovery process on Nd-FeB magnet scrap by selective chlorination using NH4Cl." *J Alloys Compd* 477:484–487. https://doi.org/10.1016/j.jallcom.2008.10.036.

Jakobsson, L. K., M.W. Kennedy, R.E. Aune, G. Tranell. 2016. "Recovery of Rare Earth Elements from the Ferrous Fraction of Electronic Waste." REWAS 2016, Chapter. https://link.springer.com/chapter/10.1007/978-3-319-48768-7_13.

Johnson, N., J. Paquette, P. Bortolotti, N. Mendoza, M. Bolinger, E. Camarena, E. Anderson, and Ennis. 2021. Big Adaptive Rotor Phase I Final Report. Golden, CO: National Renewable Energy Laboratory. NREL/TP-5000-79855. https://www.nrel.gov/docs/fy22osti/79855.pdf.

Jowitt, S. M., T. T. Werner, Z. Weng, G. M. Mudd. 2018. "Recycling of the rare earth elements." *Current Opinion in Green and Sustainable Chemistry* Volume 13, Pages 1-7, https://doi.org/10.1016/j.cogsc.2018.02.008.

Junne, T., N. Wulff, C. Breyer, and T. Naegler. 2020. "Critical Materials in Global Low-Carbon Energy Scenarios: The Case for Neodymium, Dysprosium, Lithium, and Cobalt." *Energy* 211. https://doi.org/10.1016/j.energy.2020.118532.

Kinjal G., G. Ouyang, S. Gupta, V. Kunc, M. P. Paranthaman, I. C. Nlebedim. 2019. "Recycling of additively printed rare-earth bonded magnets." *Waste Management* Volume 90, Pages 94-99, https://doi.org/10.1016/j.wasman.2019.04.040.

Kinjal G., I. C. Nlebedim, V. Kunc, E. Lara-Curzio, R. Fredette, M. P. Paranthaman. 2020. "Additive manufacturing of highly dense anisotropic Nd–Fe–B bonded magnets." *Scripta Materialia*, Volume 183, 2020, Pages 91-95, https://doi.org/10.1016/j.scriptamat.2020.03.012.

Kirsch, J., and H. Kyling. 2021. "Optimized Cast Components in the Drive Train of Wind Turbines and Inner Ring Creep in the Main Bearing Seat." *Forschung Im Ingenieurwesen* 85 (2): 199–210. https://doi.org/10.1007/s10010-021-00458-x.

Lantz, E., O. Roberts, J. Nunemaker, E. DeMeo, K. Dykes, G. Scott. 2019. *Increasing Wind Turbine Tower Heights: Opportunities and Challenges*. Golden, CO: National Renewable Energy Laboratory (NREL). NREL/TP-5000-73908. https://www.nrel.gov/docs/fy19osti/73629.pdf.

Lawrence Livermore National Laboratory. 2021. "Americans move to more solar and wind power in 2021." https://www.llnl.gov/news/americans-move-more-solar-wind-power-2021.

Li, L., A. Tirado, I. Nlebedim, O. Rios, B. Post, V, Kunc, R. R. Lowden, E. Lara-Curzio, et al. 2016. "Big Area Additive Manufacturing of High Performance Bonded NdFeB Magnets." *Sci Rep* 6, 36212. https://doi.org/10.1038/srep36212.

Li, J., K. Peng, P. Wang, N. Zhang, K. Feng, D. Guan, J. Meng, W. Wei, Q. Yang. 2020. "Critical Rare-Earth Elements Mismatch Global Wind-Power Ambitions." *One Earth, 3*(1), 116-125. https://doi.org/10.1016/j.oneear.2020.06.009.

Lorenz, T., and M. Bertau. 2019. "Recycling of Rare Earth Elements from FeNdB-Magnets via Solid-State Chlorination." *Journal of Cleaner Production* 215 (April): 131–43. https://doi.org/10.1016/j.jclepro.2019.01.051.

Lyman, J. W., G. R. Palmer. 1993. "Recycling of Rare Earths and Iron from NdFeB Magnet Scrap." *High Temp Mater Process* 11(1–4):175–187. https://doi.org/10.1515/HTMP.1993.11.1-4.175.

Martínez, E., F. Sanz, S. Pellegrini, E. Jiménez, and J. Blanco. 2009. "Life-Cycle Assessment of a 2-MW Rated Power Wind Turbine: CML Method." *The International Journal of Life Cycle Assessment* 14 (1): 52–63. https://doi.org/10.1007/s11367-008-0033-9.

McQueen, S., J. Stanford, S. Satyapal, E. Miller, N. Stetson, D. Papageorgopoulos, N. Rustagi, Vanessa, et al. 2020. *Department of Energy Hydrogen Program Plan*. United States. https://www.osti.gov/biblio/1721803.

Musial, W., P. Spitsen, P. Beiter, P. Duffy, M. Marquis, A. Cooperman, R. Hammond, M. Shields. 2021. *Offshore Wind Market Report: 2021 Edition.* U.S. Department of Energy, Washington, D.C. https://www.energy.gov/sites/default/files/2021-08/Offshore%20Wind%20Market%20Report%202021%20Edition_Final.pdf.

Nassar, N., D. Wilburn, and T. Goonan. 2016. "Byproduct Metal Requirements for U.S. Wind and Solar Photovoltaic Electricity Generation up to the Year 2040 under Various Clean Power Plan Scenarios." *Applied Energy* 183 (December): 1209–26. https://doi.org/10.1016/j.apenergy.2016.08.062.

Nassar, N. T., and S. M. Fortier. 2021. *Methodology and technical input for the 2021 review and revision of the U.S. Critical Minerals List*. U.S. Geological Survey Open-File Report 2021–1045, 31 p., https://doi.org/10.3133/ofr20211045.

National Research Council. 2008. *Minerals, critical minerals, and the US economy.* Washington, DC: National Academies Press.

National Renewable Energy Laboratory. 2022. *2022 Annual Technology Baseline.* https://atb.nrel.gov/.

———. 2023. "Renewable Energy Materials Properties Database." Version 1.0. Washington, D.C.: U.S. Department of Energy. https://apps.openei.org/REMPD/.

Negro, V., J. López-Gutiérrez, M. D. Esteban, P. Alberdi, M. Imaz, and J. Serraclara. 2017. "Monopiles in Offshore Wind: Preliminary Estimate of Main Dimensions." *Ocean Engineering* 133 (March): 253–61. https://doi.org/10.1016/j.oceaneng.2017.02.011.

Ozoemena, M., W. M. Cheung, and R. Hasan. 2018. "Comparative LCA of Technology Improvement Opportunities for a 1.5-MW Wind Turbine in the Context of an Onshore Wind Farm." *Clean Technologies and Environmental Policy* 20 (1): 173–90. https://doi.org/10.1007/s10098-017-1466-2.

Paranthaman, M. P., I. C. Nlebedim, F. Johnson, S. K. McCall. 2016. "Additive Manufacturing of Permanent Magnets." *Material Matters*, 11.4. https://www.sigmaaldrich.com/US/en/technical-documents/technical-article/materials-science-and-engineering/nanoparticle-and-microparticle-synthesis/additive-manufacturing-of-permanent-magnets.

Parker, J. G. and Baroch, C. T. 1971. *Rare-earth elements, yttrium, and thorium. A materials survey.* Bureau of Mines Information Circular, 8476, U.S. Department of Interior, Washington, D.C. https://www.osti.gov/biblio/7111745.

Paulick, H., and Machacek, E. 2017. "The global rare earth element exploration boom: An analysis of resources outside of China and discussion of development perspectives." *Resources Policy, 52*, 134-153. https://doi.org/10.1016/j.resourpol.2017.02.002.

Pavel, C. C., R. Lacal-Arántegui, A. Marmier, D. Schüler, E. Tzimas, M. Buchert, W. Jenseit, D. Blagoeva. 2017. "Substitution strategies for reducing the use of rare earths in wind turbines." *Resources Policy, 52*, 349-357. https://doi.org/10.1016/j.resourpol.2017.04.010.

PostgreSQL Global Development. 2022. "PostgreSQL." PostgreSQL. March 8, 2022. https://www.postgresql.org/.

Prosperi, D., C. O. Tudor, A. I. Bevan, E. De Leon Quiroz, H. Peh, G. Furlan, R. Skomski, D. Suvorov, M. Zakotnik. 2019. "Thermal-stability and microstructure comparison between recycled M2M and conventionally produced sintered neodymium-iron-boron magnets, *Journal of Cleaner Production*, Volume 239, 118091. https://doi.org/10.1016/j.jclepro.2019.118091.

Rademaker, Jelle H., René Kleijn, and Yongxiang Yang. 2013. "Recycling as a Strategy against Rare Earth Element Criticality: A Systemic Evaluation of the Potential Yield of NdFeB Magnet Recycling." *Environmental Science & Technology* 47 (18): 10129–36. https://doi.org/10.1021/es305007w.

Rajaei, M., and J. M. Tinjum. 2013. "Life Cycle Assessment of Energy Balance and Emissions of a Wind Energy Plant." *Geotechnical and Geological Engineering* 31 (6): 1663–70. https://doi.org/10.1007/s10706-013-9637-3.

Ren, K., X. Tang, X., P. Wang, J. Willerström, M. Höök. 2021. "Bridging energy and metal sustainability: Insights from China's wind power development up to 2050." *Energy*, 227, 120524. https://doi.org/10.1016/j.energy.2021.120524.

Riddle, M. E., E. Tatara, C. Olson, B. J. Smith, A. Bennett Irion, B. Harker, D. Pineault, E. Alonso, D. J. Graziano. 2021. Agent-based modeling of supply disruptions in the global rare earths market. *Resources, Conservation and Recycling, 164*, 105193. https://doi.org/10.1016/j.resconrec.2020.105193.

Ruth, M. F., P. Jadun, N. Gilroy, E. Connelly, R. Boardman, A. J. Simon, A. Elgowainy, J. Zuboy. 2020. *The Technical and Economic Potential of the H2@Scale Hydrogen Concept within the United States*. Golden, CO: National Renewable Energy Laboratory (NREL). NREL/TP-6A20-77610. https://www.nrel.gov/docs/fy21osti/77610.pdf.

Saito T., H. Sato, S. Ozawa, J. Yu, T. Motegi. 2003. "The extraction of Nd from waste Nd-Fe-B alloys by the glass slag method." *J Alloys and Compd* 353:189–193. https://doi.org/10.1016/S0925-8388(02)01202-1.

Sangine, E. S. 2022. "Recycling—Metals." In *2018 Minerals Yearbook*. United States Geological Survey. https://pubs.usgs.gov/myb/vol1/2018/myb1-2018-recycling.pdf.

Schreiber, A., J. Marx, and P. Zapp. 2019. "Comparative life cycle assessment of electricity generation by different wind turbine types." *Journal of Cleaner Production, 233*, 561-572. https://doi.org/10.1016/j.jclepro.2019.06.058.

Schrijvers, D., A. Hool, G. A. Blengini, W.-Q. Chen, J. Dewulf, R. Eggert, et al. 2020. "A review of methods and data to determine raw material criticality." *Resources, Conservation and Recycling, 155*, 104617. https://doi.org/10.1016/j.resconrec.2019.104617.

Sharma, R. A. 1987. "Neodymium Production Processes." *Journal of Metals*, Volume 39, pp 33-37. https://doi.org/10.1007/BF03259468.

Shirayama, S., and T. H. Okabe. 2018. "Selective Extraction and Recovery of Nd and Dy from Nd-Fe-B Magnet Scrap by Utilizing Molten MgCl2." *Metallurgical and Materials Transactions B* 49, 1067-1077. https://doi.org/10.1007/S11663-018-1176-0.

Smith, B. J., and R. G. Eggert. 2016. "Multifaceted Material Substitution: The Case of NdFeB Magnets, 2010–2015." *JOM 68*(7), 1964–71. https://doi.org/10.1007/s11837-016-1913-2.

Sommer, Valentin, and Grit Walther. 2021. "Recycling and Recovery Infrastructures for Glass and Carbon Fiber Reinforced Plastic Waste from Wind Energy Industry: A European Case Study." *Waste Management* 121 (February): 265–75. https://doi.org/10.1016/j.wasman.2020.12.021.

Sprecher, Benjamin, Ichiro Daigo, Shinsuke Murakami, Rene Kleijn, Matthijs Vos, and Gert Jan Kramer. 2015. "Framework for Resilience in Material Supply Chains, With a Case Study from the 2010 Rare Earth Crisis." *Environmental Science & Technology* 49 (11): 6740–50. https://doi.org/10.1021/acs.est.5b00206.

The White House. 2021. "Executive Order on Catalyzing Clean Energy Industries and Jobs Through Federal Sustainability."https://www.whitehouse.gov/briefing-room/presidential-actions/2021/12/08/executive-order-on-catalyzing-clean-energy-industries-and-jobs-through-federal-sustainability/. Tolcin, Amy. 2022. "Zinc." In *Mineral Commodity Summaries*, by the United States Geological Survey. https://pubs.usgs.gov/periodicals/mcs2022/mcs2022-zinc.pdf.

Tse, P.-K. 2011. *China's rare-earth industry*. U.S. Department of the Interior, United States Geological Survey. https://pubs.usgs.gov/of/2011/1042/of2011-1042.pdf.

Uda, T. 2002. "Recovery of Rare Earths from Magnet Sludge by FeCl2." *Master Trans*, 43:55–62. https://www.jstage.jst.go.jp/article/matertrans/43/1/43_1_55/_pdf.

Ueberschaar, M., J. Geiping, M. Zamzow, S. Flamme, V. S. Rotter. 2017. "Assessment of element-specific recycling efficiency in WEEE pre-processing." *Resour Conserv Recycl*, 124, pp. 25-41. https://doi.org/10.1016/j.resconrec.2017.04.006.

United Nations Environment Program. 2011. "Towards a green economy: Pathways to sustainable development and poverty eradication." http://www.unep.org/greeneconomy.

———. 2016. "Global Material Flows and Resource Productivity." UNEP International Resource Panel. https://www.resourcepanel.org/reports/global-material-flows-and-resource-productivity-database-link.

U.S. Department of Commerce, Bureau of Industry and Security. 2021. *Publication of a Report on the Effect of Imports of Transformers and Transformer Components on the National Security: An Investigation Conducted Under Section 232 of the Trade Expansion Act of 1962*. 64606 Federal Register, Vol. 86, No. 220. https://www.govinfo.gov/content/pkg/FR-2021-11-18/pdf/2021-24958.pdf.

U.S. Department of Energy. 2021a. "Secretary Granholm Launches Hydrogen Energy Earthshot to Accelerate Breakthroughs Toward a Net-Zero Economy." https://www.energy.gov/articles/secretary-granholm-launches-hydrogen-energy-earthshot-accelerate-breakthroughs-toward-net.

———. 2021b. *Solar Futures Study*. Office of Energy Efficiency and Renewable Energy. https://www.energy.gov/sites/default/files/2021-09/Solar%20Futures%20Study.pdf.

———. 2022. *America's Strategy to Secure the Supply Chain for a Robust Clean Energy Transition.* https://www.energy.gov/sites/default/files/2022-02/America%E2%80%99s%20Strategy%20to%20Secure%20the%20Supply%20Chain%20for%20a%20Robust%20Clean%20Energy%20Transition%20FINAL.docx_0.pdf.

U.S. Department of State and the United States Executive Office of the President. 2021. *The Long-Term Strategy of the United States, Pathways to Net-Zero Greenhouse Gas Emissions by 2050.* https://www.whitehouse.gov/wp-content/uploads/2021/10/US-Long-Term-Strategy.pdf.

United States Environmental Protection Agency. 2012. "Rare Earth Elements: A Review of Production, Processing, Recycling, and Associated Environmental Issues." Washington, D.C., Report No.: EPA 600/R-12/572. https://cfpub.epa.gov/si/si_public_record_report.cfm?dirEntryId=251706&Lab=NRMRL.

United States Geological Survey. 2021. *Mineral Commodity Summaries 2021.* https://pubs.usgs.gov/periodicals/mcs2021/mcs2021.pdf.

———. 2022. "2022 List of Critical Minerals." https://www.usgs.gov/news/national-news-release/us-geological-survey-releases-2022-list-critical-minerals.

Veers, P., L. Sethuraman, and J. Keller. 2020. "Wind-Power Generator Technology Research Aims to Meet Global-Wind Power Ambitions." *Joule 4*(9), 1861–63. https://doi.org/10.1016/j.joule.2020.08.019.

Vestas. 2013. *Life Cycle Assessment of electricity production from an Onshore V90-3.0 MW Wind Plant*. Vestas Wind Systems A/S, Hedeager 44, Aarhus N, 8200, Denmark. https://www.vestas.com/en/sustainability/reports-and-ratings#lcadownload.

———. 2017a. *Life Cycle Assessment of electricity production from an Onshore V136-3.45 MW Wind Plant*. Vestas Wind Systems A/S, Hedeager 42, Aarhus N, 8200, Denmark. https://www.vestas.com/en/sustainability/reports-and-ratings#lcadownload.

———. 2017b. *Life Cycle Assessment of electricity production from an Onshore V126-3.45 MW Wind Plant*. Vestas Wind Systems A/S, Hedeager 42, Aarhus N, 8200, Denmark. https://www.vestas.com/en/sustainability/reports-and-ratings#lcadownload.

———. 2017c. *Life Cycle Assessment of electricity production from an Onshore V117-3.45 MW Wind Plant*. Vestas Wind Systems A/S, Hedeager 42, Aarhus N, 8200, Denmark. https://www.vestas.com/en/sustainability/reports-and-ratings#lcadownload.

———. 2017d. *Life Cycle Assessment of electricity production from an Onshore V112-3.45 MW Wind Plant*, Version 1.1. Vestas Wind Systems A/S, Hedeager 42, Aarhus N, 8200, Denmark. https://www.vestas.com/en/sustainability/reports-and-ratings#lcadownload.

———. 2017e. *Life Cycle Assessment of electricity production from an Onshore V105-3.45 MW Wind Plant*, Version 1.1. Vestas Wind Systems A/S, Hedeager 42, Aarhus N, 8200, Denmark. https://www.vestas.com/en/sustainability/reports-and-ratings#lcadownload.

———. 2018a. *Life Cycle Assessment of electricity production from an Onshore V116-2.0 MW Wind Plant*. Vestas Wind Systems A/S, Hedeager 42, Aarhus N, 8200, Denmark. https://www.vestas.com/en/sustainability/reports-and-ratings#lcadownload.

———. 2018b. *Life Cycle Assessment of electricity production from an Onshore V120-2.0 MW Wind Plant*. Vestas Wind Systems A/S, Hedeager 42, Aarhus N, 8200, Denmark. https://www.vestas.com/en/sustainability/reports-and-ratings#lcadownload.

———. 2019. *Life Cycle Assessment of electricity production from an Onshore V117-4.2 MW Wind Plant*. Vestas Wind Systems A/S, Hedeager 42, Aarhus N, 8200, Denmark. https://www.vestas.com/en/sustainability/reports-and-ratings#lcadownload.

———. 2022a. *Life Cycle Assessment of electricity production from an Onshore V150-4.2 MW Wind Plant*. Vestas Wind Systems A/S, Hedeager 42, Aarhus N, 8200, Denmark. https://www.vestas.com/en/sustainability/reports-and-ratings#lcadownload.

———. 2022b. *Life Cycle Assessment of electricity production from an Onshore V136-4.2 MW Wind Plant*. Vestas Wind Systems A/S, Hedeager 42, Aarhus N, 8200, Denmark. https://www.vestas.com/en/sustainability/reports-and-ratings#lcadownload.

Watari, T., K. Nansai, K. Nakajima. 2020. "Review of critical metal dynamics to 2050 for 48 elements." *Resources, Conservation and Recycling, 155*, 104669. https://doi.org/10.1016/j.resconrec.2019.104669.

Watson, S., A. Moro, V. Reis, C. Baniotopoulos, S. Barth, G. Bartoli, F. Bauer, et al. 2019. "Future emerging technologies in the wind power sector: A European perspective." *Renewable and Sustainable Energy Reviews, 113,* 109270. https://doi.org/10.1016/j.rser.2019.109270.

Wellens, S., B. Thijs, and K. Binnemans. 2012. "An environmentally friendlier approach to hydrometallurgy: highly selective separation of cobalt from nickel by solvent extraction with undiluted phosphonium ionic liquids." *Green Chem* 14:1657–1665. https://doi.org/10.1039/C2GC35246J.

Wilburn, David R. 2011. "Wind Energy in the United States and Materials Required for the Land-Based Wind Turbine Industry from 2010 through 2030." *Scientific Investigations Report*. United States Geological Survey. https://pubs.usgs.gov/sir/2011/5036/sir2011-5036.pdf.

Wiser, R., M. Bolinger, B. Hoen, D. Millstein, J. Rand, G. Barbose, N. Darghouth, W. Gorman, S. Jeong, A. Mills, B. Paulos. 2021. *Land-Based Wind Market Report: 2021 Edition.* U.S. Department of Energy, Washington, D.C. https://www.energy.gov/eere/wind/articles/land-based-wind-market-report-2021-edition-released.

World Bank Group. 2017. "The Growing Role of Minerals and Metals for a Low Carbon Future." World Bank, Washington, DC. https://doi.org/10.1596/28312.

Yang, Y., A. Walton, R. Sheridan, K. Guth, R. Gauß, O. Gutfleisch, et al. 2017. "REE Recovery from End-of-Life NdFeB Permanent Magnet Scrap: A Critical Review." *J. Sustain. Metall.* 3, 122–149. https://doi.org/10.1007/s40831-016-0090-4.

Yang, J., L. Zhang, Y. Chang, Y. Hao, G. Liu, Q. Yan, Y. Zhao. 2020. "Understanding the material efficiency of the wind power sector in China: A spatial-temporal assessment."

Resources, Conservation and Recycling, 155, 104668. https://doi.org/10.1016/j.resconrec.2019.104668.

Zakotnik, M., I. R. Harris, A. J. Williams. 2008. "Possible methods of recycling NdFeB-type sintered magnets using the HD/degassing process." *J Alloys Compd* 450:525–531. https://doi.org/10.1016/j.jallcom.2007.01.134.

Appendix A. Biden Administration Objectives

A.1. Carbon-Free Power Sector by 2035 and Irreversible Path to a Net-Zero Economy by 2050

The "Long-Term Strategy of the United States: Pathways to Net-Zero Greenhouse Gas Emissions by 2050" describes the Biden administration's decarbonization goals for the country and outlines an approach for achieving them (United States Department of State and the United States Executive Office of the President 2021). The primary goal is to achieve net-zero greenhouse gas emissions by no later than 2050. This goal is supported by five key transformations:

1. Decarbonizing electricity (i.e., achieving 100% clean electricity by 2035)
2. Electrifying most of the economy and switching to other clean fuels
3. Cutting energy waste
4. Reducing methane (and other non-carbon-dioxide) emissions
5. Scaling up carbon dioxide removal.

The capacity projection used in the High Deployment scenario in this study is aligned with the goal of achieving a carbon-free power sector by 2035 and supports the goal of putting the United States on an irreversible path to a net-zero economy by 2050. Refer to United States Department of State and the United States Executive Office of the President 2021 for more details about the administration's objectives and strategy.

A.2. Role of Renewable Wind Energy in Hydrogen Shot Goal

Hydrogen is a versatile energy carrier that can be stored in liquid, gas, or as a chemical compound and converted to energy through traditional combustion methods or electrochemical processes via fuel cells (McQueen et al. 2020). If produced from clean energy sources, it will help decarbonize some of the most energy-intensive U.S. sectors: transportation, electricity generation, buildings' heat and power, and manufacturing.

In June 2021, the U.S. Department of Energy (DOE) launched the first Energy Earthshots Initiative—the Hydrogen Shot—which aims to reduce the cost of clean hydrogen (produced without emitting greenhouse gas emissions) by 80% in the next decade (DOE 2021a). This reduction would bring the costs of clean hydrogen from its current $5/kilogram (kg) to $1/kg, which would make clean hydrogen competitive with blue hydrogen (produced from fossil fuels with carbon capture and storage) (International Renewable Energy Agency 2020) and help reach the goal of a net-zero economy by 2050.

Reaching the Hydrogen Shot goal requires reducing the cost of electrolyzers and increasing their efficiency as well as inexpensive and abundant access to renewable energy resources (International Renewable Energy Agency 2020). The "Solar Futures Study" estimated that deep decarbonization of the United States supported by hydrogen would almost triple the demand for renewable energy in 2050 (DOE 2021b). Furthermore, according to a National Renewable Energy Laboratory study, wind energy represents the second-highest technical potential—after solar—for hydrogen production from renewable resources (Connelly et al. 2020). Specifically, the technical potential from land-based and offshore wind is 700 million metric tonnes per year — enough to supply the current hydrogen demand and serviceable consumption potential of 10 tonnes and 106 tonnes, respectively (Ruth et al. 2020).

Hence, while contributing to laying a path toward a net-zero economy, the addition of wind power to reach the carbon-free power sector, net-zero economy, and Hydrogen Shot goals also implies that considerable amounts of materials will be needed to manufacture new wind turbines.

Appendix B. Brief History of Critical Materials Research

The term "critical material" was first used by the U.S. government in 1939, when it referred to material stockpiles needed to supply the nation during wartime. Subsequently, these materials came to include those required for civilian and industrial uses in addition to national defense (National Research Council 2008). In 2010–2011, a trade dispute between China and Japan caused prices of rare-earth elements[21] to increase up to tenfold and drew attention to the potential for disruptions in the supply of critical materials, thereby impacting the clean energy transition (Sprecher et al. 2015). Since then, many studies on material criticality have been published, with six articles providing summaries (Achzet and Helbig 2013; Erdmann and Graedel 2011; Graedel and Reck 2016; Habib, Parajuly, and Wenzel 2015; Schrijvers et al. 2020; Watari et al. 2020). For example, Watari et al. (2020) identified 88 studies exploring the projected long-term status of 48 critical materials with respect to their use in 10 emerging technologies.

Following the 2010–2011 China-Japan trade dispute and its ripple effects on the pricing of rare-earth metals, several studies dedicated to the supply and demand of rare-earth metals were published (Du and Graedel 2011; Paulick and Machacek 2017; Tse 2011). Riddle et al. (2021) highlighted that out of the 10 rare-earth metals included in the analysis, dysprosium is the most vulnerable to supply disruption, showing the highest increases in prices. The study also shows that supply disruptions may foster more—and earlier—development of new mining projects outside China, which is in line with the rare-earth metal deposits exploration boom that followed the 2010–2011 trade dispute (Paulick and Machacek 2017). Due to supply restrictions, higher prices of rare-earth metals may trigger reductions in demand for neodymium magnets, threatening wind energy development (especially offshore projects).

Another challenge posed by critical minerals is that they are often byproducts of host metals and, therefore, depend on the demand for the latter (Watari et al. 2020). For example, production of indium, gallium, and dysprosium relies on demand for their respective host metals of zinc, aluminum, and yttrium (Riddle et al. 2021; Watari et al. 2020). Dysprosium oxide production is constrained by the co-production of other rare-earth metals with lesser demand, which suggests that dysprosium production may fall short of demand (Riddle et al. 2021). While constraints from host metal mining may limit the supply of critical materials, removing those constraints may also create issues. Elshkaki and Graedel (2015) estimated that greenhouse gas emissions from mining oversupplies of the host metals zinc and copper to obtain tellurium, indium, and germanium for solar cells might, in some cases, exceed the greenhouse gas emission reductions from replacing coal with solar photovoltaics. The production of dysprosium is likely to increase the

[21] The rare-earth elements are scandium, yttrium, and the 15 elements of the lanthanide series of chemical elements: lanthanum, cerium, praseodymium, neodymium, promethium, samarium, europium, gadolinium, terbium, dysprosium, holmium, erbium, thulium, ytterbium, and lutetium.

production of thorium (Elshkaki and Graedel 2014). This oversupply of thorium may be costly to deal with because of the metal's radioactivity.

Appendix C. Data Sources for Life CycleInventories

As described in the main text, the authors use tools from life cycle assessment (LCA) to identify and characterize background material quantities. LCA allows researchers to model the inputs (e.g., materials and fossil-fuel consumption) and outputs (air pollutant emissions, discharges to water) of a product or process's life cycle from raw material extraction through manufacturing, use, and disposal. In the Renewable Energy Materials Properties Database (REMPD), we use life cycle inventory data generated for the life cycle assessment to estimate the background material for the foreground system (refer to Section 2.1.3 for a definition of foreground and background systems and Figure 3 for an illustration of the system boundary used in the REMPD). This approach positions future work on the REMPD to calculate other types of resource requirements and impacts (e.g., global warming potential, emissions to water, energy consumption) using the background material data available in the database.

Whenever possible, we sourced life cycle inventory data from publicly available data sources using the material that most closely aligned with each foreground material. When a specific material was unavailable, we used a proxy to model life cycle inputs and outputs. For example, for the unique fiberglass/epoxy materials used in wind turbine blades, our team modeled life cycle assessment results for epoxy and fiberglass separately, and combined the results based on the allocation of materials by mass. Table C-1 lists the database foreground materials modeled in the REMPD along with the associated life cycle assessment proxy material and data source. Our team used U.S. supply chain scopes whenever possible in modeling these results for all materials. If U.S. scopes were unavailable, we used global supply chain assumptions. European supply chain assumptions were used if neither U.S. nor global scopes were available.

Table C-1. Data sources and proxy materials used for background life cycle inventory data

REMPD Material Type	Life Cycle Assessment Proxy Material	Data Source
Road aggregate	aggregate	Gursel, P. and C. Custodio. 2012. Green Concrete LCA Webtool. https://greenconcrete.berkeley.edu/
Aluminum	aluminum	U.S. Life Cycle Inventory Database. 2012. National Renewable Energy Laboratory (NREL). https://www.nrel.gov/lci/
Balsa	veneer hardwood	U.S. Life Cycle Inventory Database. 2012. NREL. https://www.nrel.gov/lci/
Carbon fiber	polyacrylonitrile fibers	European Reference Life Cycle Database of the Joint Research Center – Version 3.2. 2015. European Commission Joint Research Center. https://eplca.jrc.ec.europa.eu/ELCD3/
Cast iron	cast iron	U.S. Life Cycle Inventory Database. 2012. NREL. https://www.nrel.gov/lci/
Casting steel	low-alloyed steel	Frischknecht R., N. Jungbluth, H.-J. Althaus, G. Doka, R. Dones, T. Heck, S. Hellweg, R. Hischier, T. Nemecek, G. Rebitzer, M. Spielmann. 2005. "The ecoinvent database: Overview and methodological framework." *International Journal of Life Cycle Assessment* 10, 3–9. https://link.springer.com/article/10.1065/lca2004.10.181.1
Chromium steel	chromium steel	Frischknecht R., N. Jungbluth, H.-J Althaus, G. Doka, R. Dones, T. Heck, S. Hellweg, R. Hischier, T. Nemecek, G. Rebitzer, M. Spielmann. 2005. "The ecoinvent database: Overview and methodological framework." *International Journal of Life Cycle Assessment* 10, 3– 9. https://link.springer.com/article/10.1065/lca2004.10.181.1
Concrete	concrete	Yang, Y., W. Ingwersen, T. Hawkins, M. Srocka, and D. Meyer. 2017. "USEEIO: a New and Transparent United States Environmentally Extended Input-Output Model." *Journal of Cleaner Production*. Elsevier Science Ltd, New York, NY, 158:308-318. https://doi.org/10.1016/j.jclepro.2017.04.1 50

REMPD Material Type	Life Cycle Assessment Proxy Material	Data Source
Copper	copper	Yang, Y., W. Ingwersen, T. Hawkins, M. Srocka, and D. Meyer. 2017. "USEEIO: a New and Transparent United States Environmentally Extended Input-Output Model." *Journal of Cleaner Production*. Elsevier Science Ltd, New York, NY, 158:308-318. https://doi.org/10.1016/j.jclepro.2017.04.1 50
Electrical steel	low-alloyed electric steel	Frischknecht R., N. Jungbluth, H.-J Althaus, G. Doka, R. Dones, T. Heck, S. Hellweg, R. Hischier, T. Nemecek, G. Rebitzer, M. Spielmann. 2005. "The ecoinvent database: Overview and methodological framework."
		International Journal of Life Cycle Assessment 10, 3–9. https://link.springer.com/article/10.1065/lc a2004.10.181.1
Epoxy	epoxy	Keoleian, G., S. Miller, R. D. Kleine, A. Fang, J. Mosley. 2012. *Life cycle material data update for GREET model*. Report No. CSS12-12. Argonne National Laboratory, Argonne, IL (United States). https://greet.es.anl.gov/publication-greet2-lca-update
Fiberglass_e	fiberglass E	U.S. Life Cycle Inventory Database. 2012. NREL. https://www.nrel.gov/lci/
Galvanized steel	galvanized steel, low alloy	Internal life cycle assessment by NREL for this report.
Glass-fiber-reinforced polymer	glass-fiber-reinforced plastic	Keoleian, G., S. Miller, R. D. Kleine, A. Fang, J. Mosley. 2012. *Life cycle material data update for GREET model*. Report No. CSS12-12. Argonne National Laboratory, Argonne, IL (United States). https://greet.es.anl.gov/publication-greet2-lca-update
Glass-reinforced plastic	glass-fiber-reinforced plastic	Keoleian, G., S. Miller, R. D. Kleine, A. Fang, J. Mosley. 2012. *Life cycle material data update for GREET model*. Report No. CSS12-12. Argonne National Laboratory, Argonne, IL (United States). https://greet.es.anl.gov/publication-greet2-lca-update
Graphite	graphite	Dunn, J. B., C. James, L. Gaines, K. Gallagher, Q. Dai, J. C. Kelly. 2015. *Material and energy flows in the production of cathode and anode materials for lithium ion batteries* (No. ANL/ESD-14/10 Rev). Argonne National Laboratory, Argonne, IL (United States). https://doi.org/10.2172/1224963
Iron	iron	U.S. Life Cycle Inventory Database. 2012. NREL. https://www.nrel.gov/lci/

Table C-1. (Continued)

REMPD Material Type	Life Cycle Assessment Proxy Material	Data Source
Lead	lead	GHGs Regulated Emissions, and Energy use in Transportation 2 Model. 2021. Argonne National Laboratory. https://greet.es.anl.gov/
Low-carbon steel	low-alloyed steel	Frischknecht R., N. Jungbluth, H.-J. Althaus, G. Doka, R. Dones, T. Heck, S. Hellweg, R. Hischier, T. Nemecek, G. Rebitzer, M.Spielmann. 2005. "The ecoinvent database: Overview and methodological framework." *International Journal of Life Cycle Assessment* 10, 3– 9. https://link.springer.com/article/10.1065/lca2004.10.181.1
Magnetic steel	galvanized steel, low alloy	Internal life cycle assessment by NREL for this report.
PET foam	polyethylene terephthalate	U.S. Life Cycle Inventory Database. 2012. NREL. https://www.nrel.gov/lci/
Polyester	polyester	U.S. Life Cycle Inventory Database. 2012. NREL. https://www.nrel.gov/lci/
PVC foam	polyvinyl chloride	U.S. Life Cycle Inventory Database. 2012. NREL. https://www.nrel.gov/lci/
Reinforcing steel	low-alloyed steel	Frischknecht R., N. Jungbluth, A.-J. Althaus, G. Doka, R. Dones, T. Heck, S. Hellweg, R. Hischier, T. Nemecek, G. Rebitzer, M. Spielmann. 2005. "The ecoinvent database: Overview and methodological framework." *International Journal of Life Cycle Assessment* 10, 3–9. https://link.springer.com/article/10.1065/lca2004.10.181.1
Thermoplastic	HDPE	U.S. Life Cycle Inventory Database. 2012. NREL. https://www.nrel.gov/lci/
Plastic	polyurethane, rigid	Idemat https://www.ecocostsvalue.com/data/download-of-databases/
Silicone	silicone	Frischknecht R., N. Jungbluth, H.-J Althaus, G. Doka, R. Dones, T. Heck, S. Hellweg, R. Hischier, T. Nemecek, G. Rebitzer, M. Spielmann. 2005. "The ecoinvent database: Overview and methodological framework." *International Journal of Life Cycle Assessment* 10, 3– 9. https://link.springer.com/article/10.1065/lca2004.10.181.1

Abbreviations: PET = polyethylene terephthalate; PVC = polyvinyl chloride; HDPE = high-density polyethylene.

Appendix D. ScalingRelationships

To capture changes in material intensity that might occur as wind power plants and wind turbines increase in size, the authors multiply the fractional contribution of materials by type (e.g., % concrete, % steel, % carbon fiber) by a scaling relationship related to the plant configuration (e.g., the plant size, number of turbines, rotor diameter). These scaling relationships are summarized in Table D-1. The following sections provide more detail about each scaling relationship.

Table D-1. Scaling relationships used to compute material requirements over time

Component, Subassembly	Scaling Relationship	Data Source
Roads, all subassemblies	$mass = 12{,}208(N_t - 1) * D_r + 15{,}256{,}640$	Eberle, A., J. O. Roberts, A. Key, P. Bhaskar, K. L. Dykes. 2019. *NREL's Balance-of-System Cost Model for Land-Based Wind.* Golden, CO: NREL. https://www.nrel.gov/docs/fy19osti/72201.pdf.
Foundation, all subassemblies	$mass = 0.67042 * h * \frac{\pi}{4} * D_r^2 + 444{,}740$	Linear regression performed for this report
Substructure, all subassemblies	$mass \propto C$	Mass is assumed to scale linearly with turbine capacity
Substation, all subassemblies	$mass \propto P_c$	Mass is assumed to scale linearly with plant size
Array and export cables, all subassemblies	Array cable: $mass \propto Pc$ Export cable: • Offshore: mass scales linearly with export cable length (assumed to be 70 kilometers (km) based on project announcements) • Land-based: mass scales linearly with distance to interconnection (assumed to be 8 km)	Array cable mass is assumed to scale linearly with plant size; export cable length varies based on relevant export distances
Turbine, nacelle	$mass = 108813.2 - 642.5C - 76252.4D_r + 2920.3h$	Linear regression performed for this report

Table D-1. (Continued)

Component, Subassembly	Scaling Relationship	Data Source
Turbine, hub	$mass = (0.954 M_{Singleblade} + 5680.3)$	Fingersh, L., M. Hand, A. Laxson. 2006. *Wind Turbine Design Cost and Scaling Model.* Golden, CO: NREL. NREL/TP-500-40566. https://www.nrel.gov/docs/fy07osti/40566.pdf.
Turbine, blade	$m_{Blade} = L_{Blade}{}^{x}$, where x varies based on the facility type and technology configuration (see Section D.7 for details)	Derived based on expert input
Turbine, tower	$\ln(mass) = 1.206 + 2.432 * \ln(h)$	Linear regression performed for this report

Abbreviations: N_t = number of turbines (unitless); D_r = rotor diameter (m); C = turbine rated capacity (MW); P_c = plant rated capacity; h = hub height (m); $M_{singleblade}$ = mass of a single blade (kg); R = rotor radius (m).

D.1. Land-Based Roads

The total material volume needed to construct roads for land-based wind plants, W_{rr}(in loose cubic yards), is calculated based on the length of the road, LL_{rr}, (in m) and is given by

$$V_r = (L_r * W_n * t_r) * 1.39 \tag{D-1}$$

$$L_r = (N_t - 1) * D_r * S_t + l_{adder} \tag{D-2}$$

Where N_t is the number of wind turbines, D_r is the rotor diameter in meters, S_t is the spacing of the turbines relative to the rotor diameter (unitless; assumed to be 4), and l_{adder} is the excess roads for access to road strings from existing public roads and/or a highway (in meters (m); assumed to be 5,000 m), W_n is the width of the road (in m; assumed to be 4.9 m), tt_{rr} is the thickness of the road (in m; assumed to be 0.2 m), and 1.39 is a factor from RSMeans that converts embankment cubic yards to loose cubic yards (Eberle et al. 2019).

Substituting in the assumed values for road width, thickness, turbine spacing, and road adder, provides:

$$V_r = ((4(N_t - 1) * D_r + 5000) * 4.9 * 0.2) * 1.39 \tag{D-3}$$

which further simplifies to:

$$V_r = 5.45\ (N_t - 1) * D_r + 6811 \tag{D-4}$$

To compute the mass of material (in this case, aggregate) needed to construct roads, we multiply by the density, such that:

$$M_r = d_a\,(5.45(N_t - 1) * D_r + 6811 \qquad \text{(D-5)}$$

Where d_a, is the density of road aggregate, which we assume to equal 2,240 kg/m^3 (Iowa Department of Transportation 2021).

D.2. Land-Based Foundation

The mass scaling model for wind turbine foundations assumes a linear relationship between the foundation mass and the product of the rotor swept area and hub height. Empirical data were combined with modeled data from NREL's Land-based Balance of System Systems Engineering (LandBOSSE) model (Eberle et al. 2019) to create a data set of 25 turbines covering 2 megawatts (MW) to 15 MW capacity, 80-m to 315-m rotor diameter, and 72-m to 200-m hub height designs. We regressed the concrete mass against the rotor swept area ($\frac{\pi}{4} D_r^2$) and the hub height, h, obtaining an R^2 of 0.989, such that:

$$mass = 444740 + 0.67042 * \text{h} * \frac{\pi}{4} D_r^2 \qquad \text{(D-6)}$$

Based on industry data for spread-foot foundations, the total mass of a spread-foot wind turbine foundation is assumed to be 98.5% concrete and 1.5% steel.

D.3. Offshore Substructure

For offshore wind energy, the mass of the substructure is assumed to scale linearly with wind turbine size.

D.4. Substation

The mass of materials required for substations is assumed to scale linearly with plant size.

D.5. Array and Export Cables

For offshore wind energy, export cables are assumed to scale linearly with distance to shore. For land-based wind energy, export cables are assumed to scale linearly with the distance to interconnection. Because we do not vary the distance to shore or the distance to interconnection in this analysis, the material requirements for export cables remains constant for all scenarios.

Array cables are assumed to vary based on the number of wind turbines and the rotor diameter.

D.6. Wind Turbine: Nacelle

The nacelle mass scaling model was developed in R (V.4.1.2) using the base linear regression function with leave-one-out cross validation. We regressed the nacelle mass against the hub height, h, rated turbine capacity, C, and rotor diameter, D_r, obtaining an R^2 of 0.987 and F-statistic P-value < 0.001, such that:

$$mass = 108813.2 - 642.5 * C - 76252.4 * D_r + 2920.3 * h \qquad \text{(D-7)}$$

D.7. Wind Turbine: Hub

The mass scaling relationship for the hub is based on Fingersh et al. (2006), such that:

$$mass = (0.954 M_{singleblade} + 5680.3 + 0.954 * M_{singleblade}) \qquad \text{(D-8)}$$

where $M_{singleblade}$ is the mass of a single blade.

D.8. Wind Turbine: Blade

Wind turbine blades primarily comprise fiber-reinforced composite materials that combine a polymer resin (e.g., epoxy) with glass or carbon fibers, and a balsa wood or polymer foam core. Fiber-reinforced composites are an engineered material in which a polymeric resin is combined with reinforcement fibers and then cured to become a solid composite. The technology configurations used in this analysis rely on different combinations of a reinforcement fiber and resin system, as well as two different composite manufacturing processes. An E-type fiberglass is the predominant reinforcement fiber (by mass) in each of the designs, being used in all parts of the blade, except for the spar cap in some cases. Carbon-fiber reinforcement is only used in the spar cap, wherein the high relative costs and mechanical performance of carbon fiber are best utilized.

The traditional manufacturing process for a wind turbine blade uses a vacuum-assisted resin transfer molding (VARTM) infusion, wherein dry fabric (reinforcement fibers) and the core material are laid up in molds, covered with vacuum bags, and a pressure vacuum is applied to infuse the resin system into the mold and blade materials. The VARTM process is standard across the wind energy industry and results in low-cost, structurally efficient composites. A slight modification to this process occurs when using pultruded composites for the spar cap. A pultrusion manufacturing process is where reinforcement fibers are wetted by the resin (through a resin bath or spray, based on resin system) and then pulled through a heated die, resulting in a fully cured fiber-reinforced composite with a constant cross section. Pultrusion manufacturing is an intermediate step for wind turbine blades and the only composite form being considered for use of carbon fiber in spar caps, due to the high consistency and control of the manufacturing process.

Wind turbine blade mass, m_{blade}, scales exponentially with blade length, l_{blade}, as given by:

$$m_{blade} = L_{blade}^{x} \tag{D-9}$$

where x is a value greater than 2. The precise exponent depends on several design factors, including the wind speed classification, structural reinforcement materials used, power rating for a given blade length, blade segmentation, and the aerodynamic design. Due to the exponential relationship of blade mass with swept area and energy capture, cost relationships favor longer blades to achieve lower levelized cost of energy. Thus, current designs may continue to push toward longer blade lengths to enable continued reductions in the levelized cost of energy.

For the purposes of this analysis, we vary the blade mass scaling exponent as outlined in Table D-2. To simplify the analysis to fewer levels of discretization, the scaling exponents are specified based only on two parameters: (1) spar-cap reinforcement materials (fiberglass or carbon fiber) and (2) siting conditions (land-based or offshore). The mass scaling exponents used in this analysis are a function of spar-cap reinforcement fiber (carbon fiber or fiberglass) and vary slightly for offshore versus land-based machines, mostly a result of higher wind speed sites offshore. The scaling exponent varies further within this discretization for specified design configurations and wind classification, so the values used are averages based on recent design models (Ennis et al. 2019; Johnson et al. 2021; Gaertner et al. 2020) and based on blade mass for relevant commercial wind turbine blades.

Table D-2. Wind turbine blade mass scaling exponents

Spar Reinforcement	Land-Based	Offshore
Fiberglass, VARTM infusion	2.39	Not studied
Carbon, pultrusion	2.31	2.34
Fiberglass, pultrusion	Not studied	2.39 (reduced blade mass calculated by database)

When using pultruded spar material, fully cured pultruded planks are added in the blade mold with the dry fabric and core material for the VARTM infusion to adhere the pultruded planks to the rest of the blade materials. As a result of the distinction in the manufacturing processes, the resin system for the VARTM blade infusion and for the pultruded members do not have to be the same. For example, even when using an epoxy resin throughout the blade, there are slight distinctions for the pultrusion resin and the infusion resin. The resin systems included in the analysis are summarized in Table D-3, including the associated manufacturing process.

Table D-3. Study resin systems, composite manufacturing processes, and representative densities

Resin System	Manufacturing Method	Density [kg/m^3]
Epoxy	VARTM infusion	1,150
Epoxy	Pultrusion	1,225
Vinylester	Pultrusion	1,100
Polyurethane	Pultrusion	1,130
Reversible epoxy	VARTM infusion	1,150
Thermoplastic	VARTM infusion	1,035

To determine the fundamental material requirements for a wind turbine blade, we divide the blade mass into the major components, or regions. These components serve various purposes during the operation of a wind turbine, both structural and to produce the aerodynamic shape of the blade. The component mass is calculated using a mass fraction of the total blade weight based on trends that have been generalized for this analysis. The component mass fractions are distinct based on the spar-cap reinforcement material (carbon fiber or fiberglass) but are not treated independently for land-based versus offshore applications.

The mass fraction values used in this analysis for the major underlying components in a wind turbine blade are shown in Tables D-4 and D-5, representing current material and manufacturing approaches. This component list for materials in a wind turbine blade does not include lightning protection (primarily using aluminum), leading-edge protection, or paint, as these materials represent an insignificant portion of the wind turbine blade or are considered abundant and do not require assessment. It is worth noting that a blade with a fiberglass spar cap is around 25% more massive than one using a carbon-fiber spar cap (Ennis et al. 2019), so keep this in mind when comparing mass percentages directly between Tables D-3 and D-4.

Table D-4. Blade component mass breakdown for blades with a fiberglass-reinforced spar cap

Blade Mass Breakdown — Fiberglass Spar (Percentage of Blade Weight)	
Glass-fiber-reinforced polymer spar	30%
Root buildup	15% (80% triaxial, 20% root fasteners)
Trailing-edge reinforcement	10%
Shell panel	33% (80% biaxial, 20% core)
Shear web	8% (67% biaxial, 33% core)
Gelcoat	2%
Adhesive	2%

Table D-5. Blade component mass breakdown for blades with a carbon-fiber-reinforced spar cap

Blade Mass Breakdown — Carbon Spar (Percentage of Blade Weight)	
Carbon-fiber-reinforced polymer spar	12%
Root buildup	15% (80% triaxial, 20% root fasteners)
Trailing-edge reinforcement	8%
Shell panel	51% (80% biaxial, 20% core)
Shear web	10% (67% biaxial, 33% core)
Gelcoat	2%
Adhesive	2%

There are various material and design approaches used for commercial wind turbines beyond the spar cap material type. These variations include resin systems currently used or considered in the infusion and pultrusion composite manufacturing processes, core materials used, and blade segmentation. Tables D-4 and D-5 are used for each technology approach being considered, with minor modifications. For fiber-reinforced polymers, many of the mechanical properties are determined by the volume of the composite (through the reinforcement fiber amount) and not its mass. For composites using various resin systems, we assume that the composites will have a fixed volume fraction for each resin system, but the mass of the composite will vary based on the density of the resin. The mass fractions mentioned earlier are modified by the relative composite weight ratio for the non-epoxy composites studied. This approach assumes there is the same mass of reinforcement fibers and the same volume of the various resin systems, which causes a modified component mass fraction due to minor differences in the resin densities.

Commercial wind turbine blade designs vary based on design decisions related to siting and energy considerations, as well as underlying materials. Designs will continue to develop and use different materials based on future technology approaches, advances, and constraints. As outlined in this report, two technology configurations were developed for land-based and offshore wind turbines, which are used in the analysis study scenarios to represent expected advancements in wind turbine blade design.

The blade component technology configurations used in this report are based on both material and design choices or advances and are subject to fundamental differences between land-based and offshore applications. Land-based wind turbines are mostly constrained by the capital cost of the wind turbine and transportation limits. Offshore wind turbines are primarily constrained by mass of the turbine and the implications it has on system costs, in addition to technologies that assist with long blade lengths. One distinction this constraint causes for the Low Innovation technology configuration is that the land-based wind turbine is designated to have an infused fiberglass spar cap due to the higher relative costs of commercial carbon-fiber systems, whereas the offshore turbine uses carbon-fiber spar caps to control the mass and resist large deflections. Because of the system mass sensitivities, offshore wind turbines are not expected to have infused fiberglass spar caps; yet, would have a slight mass penalty. However, due to the variety of design approaches for blade joints in the industry, the additional mass and material requirements for the joint are not quantified in this analysis. The added mass of a blade joint is small relative to the total blade mass; therefore, given the averaging of blade mass scaling exponents in the analysis this approach is justified.

The Moderate Innovation technology configuration for land-based and offshore wind turbines assumes the same fundamental materials as the Low Innovation configuration, but the land-based blades are segmented with an additively manufactured (three-dimensional-printed) tip portion of the segmented blade. Additional innovations could include printed tips through ongoing research efforts by several national laboratories and industry partners where value has been identified in using printed materials for this reduced stress portion of the blade.

The resulting Low Innovation and Moderate Innovation technology configurations for land-based and offshore wind turbines are summarized in Table D-6. This table also shows the market share distribution for core material and resin systems at the different levels, where applicable.

Table D-6. Wind turbine blade technology configurations used for the study scenarios

	Current Policies Scenario	High Deployment Scenario
	Technology Configuration: Low Innovation	*Technology Configuration: Moderate Innovation*
Land-based wind	• Fiberglass/thermoset shell (glass fibers and epoxy resin using VARTM infusion) • Fiberglass/thermoset spar cap using VARTM infusion • Core: 50% balsa, 50% polyvinyl chloride (PVC)	• Fiberglass/thermoset shell with epoxy resin for VARTM infusion • Segmented blade tip • Carbon-pultruded spar cap (85% epoxy, 15% vinylester) • Core: 50% balsa, 50% PVC
Offshore wind	• Fiberglass/thermoset shell with epoxy resin for VARTM infusion • Carbon-pultruded spar (85% epoxy, 15% vinylester) • Core: 50% balsa, 50% PVC	• Fiberglass/thermoset shell with epoxy resin for VARTM infusion • Carbon-pultruded spar (85% epoxy, 15% vinylester) • Core: 50% balsa, 50% PVC

The REMPD combines the technology configurations (Table D-6) with the mass calculation assumptions listed in Tables D-2 through D-5 to estimate the fundamental material requirements for various wind energy development scenarios.

For each of the reference turbines, we used Eq. D-9 to calculate two reference blade mass values for a fiberglass and carbon-fiber spar cap with the baseline exponents from Table D-2. These baseline mass values are not equal to the resulting blade mass value calculated by the database for each of the technology configurations. The baseline blade mass values are inputs to the database, which then recalculate the resulting blade mass based on the technology configurations and their inherent assumptions listed in Table D-6. The resulting blade mass for the two technology configurations is shown in Table D-7 as a percentage of the baseline blade mass when using fiberglass or carbon-fiber reinforcement in the spar cap.

Table D-7. Relative blade mass values for the study of technology configurations (for land-based and offshore wind turbine designs)

	Land-Based Turbine		Offshore Turbine	
Fiber reinforcement	Low Innovation	Moderate Innovation	Low Innovation	Moderate Innovation
Fiberglass spar	100%	n/a	n/a	n/a
Carbon-fiber spar	n/a	100%	100%	100%

The study-defined technology configurations, mass percentages, and blade mass scaling exponents help calculate the constituent materials required for the reference wind turbines used in the study scenarios. The underlying blade materials are divided into four fundamental categories of reinforcement fiber, resin systems, core materials, and assembly materials. The set of materials for a given technology configuration is not the same as for other technology configurations, as shown in Table D-6, but the full set in the analysis includes the following materials:

- Reinforcement fibers:
    - Fiberglass used in unidirectional, biaxial, or triaxial composites
    - Carbon fiber used in unidirectional pultrusions only
- Resin systems:
    - Traditional epoxy; slight variations for pultruded spar caps or VARTM infusion
    - Vinylester
    - Polyurethane
- Core materials:
    - Balsa wood
    - Polyvinyl chloride (PVC) foam
    - Polyethylene terephthalate foam
- Assembly materials:
    - Steel root fasteners
    - Epoxy adhesive
    - Methyl methacrylate adhesive.

As mentioned previously, the REMPD's calculated blade mass accounts for different component mass fraction values (compared to Tables D-4 and D-5) resulting from minor variations in the underlying composite mass based on the resin system used. Fiber-reinforced polymers combine reinforcement fiber (in various orientations) and resin, and an important property is the relative portion of fibers in a given volume, known as the fiber volume fraction. For wind turbines with VARTM infusion manufacturing, the fiber volume fraction varies between 49% and 57% based on the orientation of adjacent fiber layers (laminae) in a composite (unidirectional, biaxial, triaxial). Modern carbon-fiber pultrusions are moving to a fiber volume fraction of around 68%, which is used in this study to represent pultruded composites using carbon or glass fiber. To calculate the constituent material mass, the fiber mass fraction is used in the database for the various fiber-reinforced polymers that depend on the fiber volume fraction and fiber and resin system. Tables D-8 and D-9 list the fiber mass fractions for the various composites defined for use in the Low Innovation and Moderate Innovation technology configurations. The tables list the mass fractions for glass-fiber-reinforced polymers and for carbon-fiber-reinforced polymers, respectively.

Table D-8. Glass-fiber-reinforced polymer mass fractions for various volume fractions and resin systems using pultrusion on infusion manufacturing

	Composite Fiber Volume Fraction and Manufacturing Method			
Resin systems	68% pultrusion	57% unidirectional infusion	54% biaxial infusion	49% triaxial infusion
Traditional epoxy	0.816	0.746	0.722	0.681
Thermoplastic	n/a	0.766	0.743	0.703
Reversible epoxy	n/a	0.746	0.722	0.681

Table D-9. Carbon-fiber-reinforced polymer mass fractions for various volume fractions and resin systems using pultrusion manufacturing

	Composite Fiber Volume Fraction and Manufacturing Method			
Resin systems	68% pultrusion	57% unidirectional infusion	54% biaxial infusion	49% triaxial infusion
Epoxy	0.752	n/a	n/a	n/a
Vinylester	0.772	n/a	n/a	n/a
Polyurethane	0.767	n/a	n/a	n/a

The preceding approach enables a representative calculation of the mass of constituent materials in future wind turbine blades, with blade lengths between 70 to 128 m, assuming material mechanical properties consistent with commercial materials being used today. There are alternative material approaches being developed for use in fiber-reinforced polymers for a variety of reasons, including reduced cost, recycling and reuse, and reduced carbon emissions.

Additional future technology configurations for wind turbine blades could consider circular economy considerations for the fiber-reinforced polymers to enable materials to be downcycled into other markets or reprocessed to aid in the production of future wind turbine blades. Such a configuration would assume improvements in manufacturing, inspection, and/or design of wind turbine blades (e.g., the blade components might only require 90% of the baseline mass fractions or blade composite materials might use reversible epoxies). However, we do not consider any such configuration in this analysis.

D.9. Wind Turbine: Tower

The tower mass scaling model was developed in R (V.4.1.2) using the base linear regression function with leave-one-out cross validation. We regressed the tower mass against the hub height and the rated turbine capacity. Based on the results, the rated capacity is not considered a significant explanatory variable at the $p < 0.001$ level of significance. Data were then log-transformed, and the regression was re-run, regressing tower mass on hub height, h, obtaining an R^2 of 0.931 and F-statistic P-value < 0.001, such that:

$$\log(mass) = 1.206 + 2.432 * \log(h). \qquad \text{(D-10)}$$

Appendix E. Material Intensities for Vulnerable Materials

Tables E-1 and E-2 provide the underlying data the authors used to generate Figure 7. These tables summarize material intensities for vulnerable materials in current and future wind energy technologies. Current technology is represented using results from the Current Policies scenario in the year 2020. Potential future technology is represented using results from the High Deployment scenario in the year 2050. Table E-3 documents the data sources that were used to calculate the material intensities in Tables E-1 and E-2.

Table E-1. Average material intensity for vulnerable materials in current wind energy technology

	Land-Based Wind Material Intensity (kg/MW)				Offshore Wind Material Intensity (kg/MW)			
	Cables	Foundation	Substation	Turbine	Cables	Substructure	Substation	Turbine
Aluminum	2,400	1.3	4.1	680	242			680
Carbon fiber				590				580
Chromium	2.9	170	7.2	1,000	6.4	72	5.3	150
Cobalt	5.4E-07	0.42	0.08	2.1	0.01	3.2	0.14	1.1
Dysprosium		6.6E-06	5.3E-06	1.9	7.9E-07	6.2E-05	2.2E-06	6.6
Electrical steel			510	1,300			200	2,600
Gallium		8.0E-03	5.9E-04	0.04	2.7E-04	0.04	1.6E-03	0.01
Graphite				3.4				4.7
Lithium		2.8E-03	4.5E-06	0.68	6.3E-07	5.3E-05	1.9E-06	0.93
Manganese	0.64	38	14	1,600	11	3,700	130	830
Neodymium				40				110
Nickel		40	9.1	1,800	12	2,800	100	680
Niobium		0.07	1.5E-03	0.24	1.7E-03	0.57	0.02	0.12
Praseodymium		0.10	2.7E-03	0.42	3.0E-03	1.0	0.04	43
Terbium		3.9E-06	3.2E-06	1.9E-05	4.7E-07	3.7E-05	1.3E-06	0.38
Tin	1.1E-04	0.03	0.01	0.13	9.2E-04	0.25	0.01	0.26
Titanium	0.01	9.7	0.28	40	0.31	94	3.4	21
Vanadium	1.4E-05	1.6E-05	7.8E-07	1.3E-04	5.2E-05	1.4E-04	5.0E-06	3.7E-05
Zinc	0.09	0.35	0.22	27	11	31	1.1	7.7

Table E-2. Average material intensity for vulnerable materials in potential future wind energy technology

	Land-Based Wind Material Intensity (kg/MW)				Offshore Wind Material Intensity (kg/MW)			
	Cables	Foundation	Substation	Turbine	Cables	Substructure	Substation	Turbine
Aluminum	2,400	1.4	4.1	680	240			890
Carbon fiber				2,300				1,100
Chromium	2.9	190	7.2	1,000	6.4	80	5.3	200
Cobalt	5.4E-07	4.6E-01	0.08	1.9	0.01	3.5	0.14	1.5
Dysprosium		7.2E-06	5.3E-06	1.9	7.8E-07	6.9E-05	2.3E-06	6.5
Electrical steel			510	1,300			200	2,700
Gallium		8.7E-03	5.9E-04	0.04	2.7E-04	0.04	1.6E-03	0.02
Graphite				1.8				8.8
Lithium		3.0E-03	4.5E-06	0.35	5.9E-07	1.4E-04	1.9E-06	1.8
Manganese	0.64	410	14	1,300	11	4,100	130	1,200
Neodymium				44				110
Nickel		430	9.1	1,600	12	3,100	100	1000
Niobium		0.06	1.5E-03	0.20	1.7E-03	0.63	0.02	0.19
Praseodymium		0.10	2.7E-03	0.36	3.0E-03	1.1	0.04	43
Terbium		4.3E-06	3.2E-06	1.6E-05	4.7E-07	4.1E-05	1.4E-06	0.37
Tin	1.1E-04	0.03	0.01	0.11	9.2E-04	0.28	0.01	0.29
Titanium	0.01	11	0.28	34	0.31	110	3.4	31
Vanadium	1.4E-05	1.7E-05	7.8E-07	1.0E-05	5.2E-05	1.5E-04	5.1E-06	5.4E-05
Zinc	0.09	3.9	0.22	22	11	34	1.1	11

Table E-3. Data sources used to compute wind energy material quantities in the renewable energy materials properties database (REMPD)

Facility Type	Component	Data Source(s) Used to Compute Wind Energy Material Quantities in the REMPD
Land-based wind	Array and export cables	• Proprietary data from original equipment manufacturers (OEMs)
	Foundation	• Proprietary data from OEMs • Selected Vestas life cycle assessments[a] • Crawford (2009) • Eberle et al. (2019) • Schreiber, Marx, and Zapp (2019)
	Roads	• Eberle et al. (2019)
	Substation	• Proprietary data from OEMs • Alsaleh and Sattler (2019)
	Turbine	• Proprietary data from OEMs • Scaling relationships documented in Appendix D • Crawford (2009) • Alsaleh and Sattler (2019) • Martínez et al. (2009) • Ozoemena, Cheung, and Hasan (2018) • Rajaei and Tinjum (2013) • Guezuraga, Zauner, and Pölz (2012)
Offshore wind	Array and export cables	• Proprietary data from OEMs • ABB (2010) • Arvesen et al. (2014) • Ikhennicheu et al. (2020)
	Substructure	• 4C Offshore (2022) • Negro et al. (2017)
	Substation	• Proprietary data from OEMs • Arvesen et al. (2014)

Table E-3. (Continued)

Facility Type	Component	Data Source(s) Used to Compute Wind Energy Material Quantities in the REMPD
	Turbine	• Proprietary data from OEMs • Scaling relationships documented in Appendix D • Crawford (2009) • Guezuraga, Zauner, and Pölz (2012)

[a] Selected Vestas LCAs include Vestas 2013, 2017a, 2017b, 2017c, 2017d, 2017e, 2018a, 2018b, 2019, 2022a, and 2022b.

Appendix F. Opportunities to Recycle Rare-Earth Elements

The rare-earth elements that are most used in the wind energy industry are neodymium and dysprosium, plus small quantities of praseodymium and terbium. Alloys of these four elements are key constituents of the powerful neodymium iron boron permanent magnets used in synchronous generators employed in some wind turbines. Compared to other applications, like vehicles, the generators of modern wind turbines are relatively large (1,000–2,000 kilograms). Supply and demand challenges for rare-earth elements are reviewed in Alves Dias et al. (2020a).

The primary processing steps in the rare-earth permanent-magnet supply chain are:

1. *Mineral Production.* Rare-earth ores (monazite, bastnaesite, and xenotime) are discovered and mined.
2. *Mineral Processing.* Rare-earth ores are crushed and milled. A flotation process is typically used to separate the tailings from the rare-earth elements.
3. *Chemical Extraction.* The most common extraction process is leaching, which involves dissolution of the valuable metals into an aqueous solution. Monazite, which contains neodymium, is often processed with sulfuric acid (Parker and Baroch 1971).
4. *Separation.* This step may be the most challenging due to the nearly identical chemical properties of the rare-earth elements. Common separation techniques include leaching, solvent extraction, and ion exchange followed by precipitation.
5. *Reduction.* Reduction is an energy-intensive process that can be carried out using metallothermic (Sharma 1987) or electrochemical (Dysinger 1994) methods.
6. *Magnet Production.* Refined rare-earth metal(s) are alloyed with other metals that will produce the desired microstructure and magnetic properties in the final product. The alloy is heated in a vacuum furnace and a stream of molten metal is forced under pressure onto a cooled drum where it is rapidly cooled to produce very small grains of metal. Next, a jet mill (a high-speed stream of cyclonic inert gas) grinds the alloy into powder. The cyclone automatically classifies the particles by size as they go through the system, so a narrow— and very favorable—particle size distribution is maintained. The powder enters a mold and is pressed between plates

while under a strong magnetic field, forming a block of material. The magnetic field orients the grains so that the magnetic domains remain aligned in the designed direction. It is then machined and ground. The final step is applying a coating to avoid corrosion.

Currently, there is no effective method of recycling rare-earth magnets. Only 3% to 7% of rare-earth magnets are recycled from end products because current processes are uneconomical and/or generate huge amounts of toxic waste. While current processing methods are still at various research and development stages, it is estimated that in the coming 10 to 15 years, the recycled rare-earth elements from end-of-life permanent magnets will play a significant role in the total supply of rare-earth elements for the magnet sector (Yang 2017). As a result, there is a need to develop and implement efficient recycling technologies.

A significant amount of waste is produced during machining and grinding, generating what is known as "swarf" (Yang 2017). Many tonnes of rare-earth-element-rich grinding slurry are accumulated each year, representing nearly one-third of the rare-earth-elements' input to the manufacturing process (Binnemans et al. 2013; Jowitt et al. 2018). This slurry is a valuable resource, yet an appropriate, cost-effective, and clean recycling process is still a challenge. The swarf can potentially be inserted in the following steps in the manufacturing supply chain (see Figure F-1).

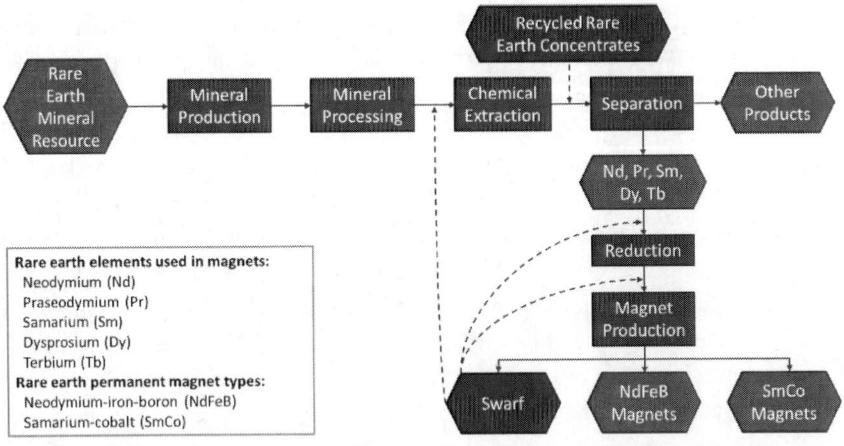

Figure F-1. Opportunities to insert recycled materials (swarf and recycled rare-earth concentrates) into the manufacturing supply chain of rare-earth permanent magnets.

Waste from electrical and electronic equipment (notably from computer hard-disk drives and mobile phones) can be used as a more valuable source of rare-earth elements from end-of-life magnets. Dismantling and separation are mostly performed by manual and mechanical processes, and their methods are applied depending on the type of waste products generated. The preprocessing waste from electrical and electronic equipment involves the automatic sorting of neodymium magnets, potentially resulting in crushed magnetic components being attracted to ferrous metal scrap. Although this method can effectively separate the rare-earth element into a single output stream, the resulting mass concentrations are typically low. Further, in most cases, this approach can contaminate this stream (Ueberschaar et al. 2017), potentially meaning that the recovery of rare-earth elements from this recycling stream is not possible. Hitachi developed the process using a rotating drum for disassembling hard-disk drives, and the National Institute for Materials Science developed a small-scale electronic crushing device and three-dimensional ball mill, resulting in a shortened work time (United States Environmental Protection Agency 2012; Hitachi 2010).

Some opportunities for recycling the rare-earth elements found in rare-earth magnets include:

- Traditional hydrometallurgical recovery techniques where magnets are dissolved in acids (or potentially in the future using ionic liquids) before the rare-earth elements are precipitated out of the solution [e.g., Lyman 1993, Wellens et al. 2012, Jakobsson 2016].
- A hydrometallurgical approach (Bogart 2016) that enables fast, efficient separations through a simple leaching process.
- Pyrometallurgical recovery techniques where rare-earth-element alloys are re-melted, separated from alloyed transition melts in a liquid metallic state, refined in anelectroslag process, or dissolved out of alloys by reaction with a molten flux, with the rare-earth element then supercooling with the flux to form a glass. The approach used depends on the nature of the rare-earth-element alloys within the magnets (e.g., Saito 2003). Molten salts have also been explored (Shirayama 2018).
- Gas phase extraction methods wherein the rare-earth elements are transferred to a volatile chloride phase and separated based on differences in volatility (Itoh 2009; Uda 2002).

- Hydrogen decrepitation has been tried, wherein hydrogen is intentionally induced into the alloy to promote embrittlement to produce powers (Zakotnik 2008).
- An innovative method, big area additive manufacturing, has been developed to fabricate isotropic and anisotropic near-net-shape neodymium-bonded magnets (Li et al. 2016; Kinjal et al. 2020; Paranthaman 2016). Potentially, this method could allow the swarf to be reused at the last stage for magnet production (Figure 1), thereby eliminating the need for solvents and other materials used in earlier stages (Kinjal et al. 2019).

F.1. Industrial Efforts

Several North-American-based industrial efforts are on their way to utilizing secondary rare-earth-element feedstock. However, assessing their technoeconomic viability is beyond the scope of this study. The following provides a brief description of these efforts, based on what is openly available.

- *Rare Earth Salts* is based in Beatrice, Nebraska. The company claims to have developed a methodology for separating and refining all rare-earth elements to high purity from ore-based and recycled feedstock. They have a proprietary, patented modular technology that efficiently separates the rare-earth element in a cost-effective and environmentally friendly process. A critical component of the separations process is based on chemical electro-winning. Furthermore, the process can be performed at a range of temperatures, (including ambient), operated with low-cost consumables, and allows for continuous operations.
- *Geomega* is based in Quebec, Canada, and claims to be able to process 1.5 tonnes of magnet waste per 8-hour day. Their technology, developed by their subsidiary Innord, seems to be based on a reusable reagent. The separation was originally based on electrophoresis; the current method (Innord's Separation of Rare Earths) is not revealed but achieves separate elements and uses zero organic solvents.
- *Noveon Magnetics* (formerly Urban Mining Company) is based in San Marcos, Texas, and claims that its M2M technology uses waste or recycled magnetic material to support its neodymium-magnet

manufacturing process. Their approach is to recover rare-earth particulate material by exposing a rare-earth magnet to hydrogen gas to effect hydrogen decrepitation of the rare-earth magnet and produce a rare-earth particulate material that can be separated from the rest of the assembly (Harris et al. 2011). The company has demonstrated their method using recycled feedstock (e.g., Benke et al. 2020 and Prosperi 2019).

Chapter 6

American WAKE Experiment (AWAKEN)*

Patrick Moriarty[1]
Nicholas Hamilton[1]
Mithu Debnath[1]
Tommy Herges[2]
Brad Isom[3]
Julie K. Lundquist[1,4]
David Maniaci[2]
Brian Naughton[2]
Rebecca Pauly[1]
Jason Roadman[1]
Will Shaw[3]
Jeroen van Dam[1]
and Sonia Wharton[5]
[1]National Renewable Energy Laboratory
[2]Sandia National Laboratories
[3]Pacific Northwest National Laboratory
[4]University of Colorado Boulder
[5]Lawrence Livermore National Laboratory

* This is an edited, reformatted and augmented version of a Technical Report of the national laboratory of the U.S. Department of Energy; Office of Energy Efficiency & Renewable Energy, Publication No. NREL/TP-5000-75789, dated May 2020.

In: Wind Energy
Editor: Mehmet Landrey
ISBN: 979-8-89113-528-4

Notice

This work was authored in part by the National Renewable Energy Laboratory, operated by Alliance for Sustainable Energy, LLC, for the U.S. Department of Energy (DOE) under Contract No. DE-AC36-08GO28308. Funding provided U.S. Department of Energy Wind Energy Technologies Office. The views expressed herein do not necessarily represent the views of the DOE or the U.S. Government.

This work was performed under the auspices of the U.S. Department of Energy by Lawrence Livermore National Laboratory under Contract DE-AC52-07NA27344. This report is also published as Lawrence Livermore National Laboratory Technical Report LLNL-TR-806278.

Pacific Northwest National Laboratory is operated by Battelle Memorial Institute for the DOE under Contract DE- AC05-76RL01830.

Sandia National Laboratories is a multimission laboratory managed and operated by National Technology & Engineering Solutions of Sandia LLC, a wholly owned subsidiary of Honeywell International Inc., for the U.S. Department of Energy's National Nuclear Security Administration under contract DE-NA0003525. This report is also published as Sandia National Laboratories Report SAND2020-4296 O.

This report is available at no cost from the National Renewable Energy Laboratory (NREL) at www.nrel.gov/publications.

U.S. Department of Energy (DOE) reports produced after 1991 and a growing number of pre-1991 documents are available free via www.OSTI.gov.

Executive Summary

The national laboratories, directed by the U.S. Department of Energy Wind Energy Technologies Office, will organize, design, and execute a landmark international wake observation and validation campaign known as the American WAKE experimeNt or AWAKEN. This document describes the vision and purpose for AWAKEN and describes some of the organizational activities that have taken place to date. The driving need for this experimental campaign is that wake interactions are among the least understood physical phenomena in wind plants today, leading to unexpected power and financial losses. New observation data gathered will be used to further validate wind plant models and lead to both improved layout and more optimal operation of wind farms with greater power production and improved reliability, ultimately leading to lower wind energy costs.

This campaign will occur in the U.S. Midwest, where the largest concentration of wind farms in the U.S. is located and where significant future growth is most likely. The field campaign will provide a data set that is unique among wake studies, based on location, scope, and observational fidelity: most wake studies have largely focused on offshore wind farms, which are not currently critical to the U.S. wind energy supply, and with a limited number of observations. While AWAKEN is focused on land-based wind farms in flat terrain, many of the observations will help researchers understand fundamental wake behavior applicable in both offshore and complex terrain environments.

Project organization has begun with initial kickoff meetings to stimulate interest in the community and prioritize science goals that a new wake field campaign can and should accomplish. The design of the experimental campaign will rely on the successful collaborative model based on one used within the New European Wind Atlas. Leading up to the experiment, U.S.-Department-of-Energy-funded researchers will assist interested participants with developing proposals for individual external funding agencies to build a large collaborative team funded through multiple sources. The AWAKEN team will include industry members, such as wind farm owners, who wish to learn more about wake impacts within their wind farms, turbine manufacturers who want to study turbine response in wind farm environments, and observational measurement companies with a desire to demonstrate their own sensing technology applied to the wind energy wake problem. The national laboratories are uniquely suited to assemble the international team to perform the observational and model validation campaign envisioned.

Researchers will narrow their experimental design and select a site that will fulfill the critical science goals identified by the international group of experts. Once a site has been identified, numerical studies of the potential observation campaign will be conducted to give researchers a sense of quantities that can be observed and to what resolution in the weather situations typical of this location. These simulations will also assist researchers in identifying optimal placement of instruments for the greatest potential benefit toward validation of numerical models. Concurrently, researchers will develop a roadmap for advanced microscale sensing technologies to improve observational capabilities of complex flow fields within wind plants. The expert-driven roadmap will guide the multiyear effort to advance remote sensing technologies specifically for wind energy applications. New sensing technologies will be simulated and potentially prototyped for use in the AWAKEN campaign, once sufficient measurement quality and reliability are achieved.

Researchers will then identify the optimal instrumentation to acquire through purchase, rental, or collaboration and then commence setup of the experiment. With all partners in place, the experiment will be executed through an iterative process of observation and model validation focused primarily on the important wake science goals outlined in planning meetings. Researchers will develop new model validation benchmarks that will be used internationally to improve a wide range of wind plant models worldwide. Data gathered from the field campaign and a select set of validation simulations will be made publicly available through the Atmosphere to Electrons Data Archive and Portal. A series of publications of the field campaign and validation studies will be prepared and presented at relevant academic and industry events.

1. Introduction

The American WAKE experimeNt (AWAKEN) is an international multi-institutional wind energy field campaign to better understand wake losses within operational wind farms. Wake interactions are among the least understood physical interactions in wind plants today, leading to unexpected power and profit losses. For example, Ørsted, the world's largest offshore wind farm developer, recently announced (Ørsted 2019) a downward revision in energy estimates across their energy generation portfolio, primarily caused by underprediction of energy losses from wind farm block- age and wakes. In their announcement, they noted that the standard industry models used for their original energy estimates were inaccurate, and this was likely an industry-wide issue. To help further improve and validate wind plant models across scales from individual turbines as well as interfarm interactions between plants, new observations, such as those planned for AWAKEN, are critical. These model improvements will enable both improved layout and more optimal operation of wind farms with greater power production and improved reliability, ultimately leading to lower wind energy costs.

The average wind power plant in the United States loses 10% of its potential energy production to wake interactions between turbines and wind farms (Clifton, Smith, and Fields 2016), some of which may be recoverable through advanced wind farm design and operation. The models used to predict these losses have uncertainty levels of 20%– 50%, which lead to excessive wind project financial costs. The reduction in overall plant power output and the mechanical stresses associated with transient turbine-to-turbine wake interaction are among the major causes of poor wind plant performance as well as a contributor to degraded turbine reliability. Addressing these issues requires detailed knowledge about the complex fluid dynamic interactions throughout a plant and well-validated models to predict future performance and impacts. The ability to predict wake evolution, merging, and dissipation, along with plant inflow conditions and turbine control strategies, will improve optimization of plant layout, power output, and turbine reliability. More accurate modeling capabilities can also lead to revolutionary turbine and plant designs that will support the renewable-energy-centered grid of the future (Veers et al., 2019). Because of their impact on wind plant profitability and unresolved physical understanding, wakes have been a significant focus of the U.S. Department of Energy's (DOE's) Atmosphere to Electrons (A2e) program (Cline, Haupt, and Shaw 2017) since its inception, and AWAKEN represents a culmination of the work done within this program to date.

Numerous data sets gathered from around the world have been used in initial validation studies of increasing sophistication with the passage of time. Some of the first data sets consisted primarily of coarse observations of wakes using supervisory control and data acquisition (SCADA) data available from offshore wind farms (e.g., Barthelmie et al., 2010; Hansen et al., 2012). More recently, new data sets have directly measured the wake and its impact. The advent of scanning lidar and scanning radar technology led to remote-sensing studies of individual wakes. Some of these campaigns took place at DOE facilities (Smalikho et al., 2013; Aitken et al., 2014a), research sites (Iungo, Wu, and Porté-Agel 2013; Vollmer et al., 2015; Machefaux et al., 2016; Kumer et al., 2016; Trujillo et al., 2016; Krish- namurthy et al., 2017), or at commercial facilities (Käsler et al., 2010; Hirth et al., 2012; Iungo, Wu, and Porté-Agel 2013; Aitken and Lundquist 2014; Bodini, Zardi, and Lundquist 2017; Menke et al., 2018; Wildmann, Vasiljevic, and Gerz 2018; Bromm et al., 2018). Recent field campaigns have also included two measurement campaigns currently taking place under the A2e Wake Dynamics program: the Sandia National Laboratories (Sandia) Scaled Wind Farm Technology (SWiFT) wake measurement and validation campaign (Herges et al., 2017; Doubrawa et al., 2019) and the NextEra Peetz Table wake steering campaign (Fleming et al., 2019; Murphy, Lundquist, and Fleming 2019; Fleming et al., 2020). Experiments at the SWiFT facility on smaller-scale V27 turbines have produced the highest resolution data to date (~1.25 m) (Herges et al., 2017), but questions remain as to what physics are sufficiently represented that are relevant to modern utility-scale turbines. The ongoing test at Peetz Table has primarily utilized SCADA data output, with a small effort to directly observe wakes using scanning lidars.

To beyond smaller-scale and ungeneralizable tests that examine the wakes from a handful of turbines and capture full-plant interaction physics, a multinational wakes experiment is needed. This new experiment can leverage the experience, instrumentation, and combined funding of multiple institutions to create the most highly detailed, comprehensive wind energy wake experiment to date is needed.

Investment in augmented observational technology is also required to complement high-fidelity simulations to improve the systematic validation of wind plant modeling tools. Validation has not advanced as quickly as model development, mainly because of the lack of observations at a similar range of fidelities, particularly for full-scale wind plants. Recent advances in remote-sensing technology, such as Ka-band and X-band radars, have been promising (Nygaard and Newcombe 2018), but further development across a range of technologies is necessary to improve data availability, temporal and spatial resolution, and the reliability of observations of quantities of interest related to wind plant performance.

2. Previous Wake Observation Campaigns

Wind energy wake experiments have taken place since the 1970s and can generally be separated into two categories: wind tunnel testing and field testing in the atmosphere. Both approaches are helpful for improved understanding of wind plant atmosphere interactions, but have limitations leading to knowledge gaps to be addressed by AWAKEN. Wind tunnel tests are executed in a controlled environment where flow conditions are well quantified but lack large-scale atmospheric behavior, which can only be addressed through full-scale testing. Early and more recent field tests at smaller than utility scale have provided valuable information about single wake formation and evolution, but lack the physics of larger farms, such as deep array effects and farm blockage. Utility-scale observations have been limited to lower-fidelity observations and an inability to control the driving physical processes. A large-scale field campaign such as AWAKEN will fill many of the knowledge gaps of previous observations by providing higher fidelity measurements across a wider range of scales, particularly at the mesoscale where data are limited.

Researchers have performed wind tunnel tests of single turbines (e.g., Krogstad and Adaramola 2012, some of which include the impacts of atmospheric stability (e.g., Chamorro and Porté-Agel 2010. There have also been studies of scaled full wind farms within wind tunnels (e.g., Cal et al., 2010; Hamilton and Cal 2015. Such studies have been useful for calibration of wind farm models and studies of wind farm layout impacts.

At full scale, measurements of older and smaller turbines have produced valuable data sets for the early wind industry. For example, in a series of observations at Sexberium (Cleijne 1992, 1993), a wind farm containing eighteen 310- kW turbines and three meteorological masts proved useful for model intercomparison and validation (Moriarty et al., 2014). In this data set, tower-based wind speed measurements were taken over a period of 6 months and wake impacts measured as the wind changed directions for multiple rotors and downstream spacings.

Some years later, operational data from full-scale offshore wind farms have been used for better understanding of wind farm flows, such as in the ENDOW project (Rados et al., 2001) and UpWind project (Barthelmie et al., 2011). SCADA data taken from Horns Rev, a wind farm in the Danish North Sea, have been popular and serve as a base validation data set for many industry and research wind farm wake models (Barthelmie et al., 2009; Hansen et al., 2012). The focus on offshore experiments has been because lower turbulence levels result in longer propagation of wakes, which are thought to have greater impact offshore than wakes from land-based wind farms. These validation cases are based on meteorological observations coupled with turbine supervisory control and SCADA data, binned according to wind direction, atmospheric stability, and turbulence.

SCADA data have also been useful to diagnose large-scale interactions between wind plants (Nygaard 2014; Nygaard and Hansen 2016). These studies highlighted the importance of accounting for spatial variations in inflow conditions that must be observed as wind farm size increases. The disadvantages of these offshore data sets are that the atmospheric conditions, such as wind direction and stability, as well as detailed turbine characteristics and operating behavior, are not always observed or reported. In any full-scale test, turbine design information, particularly control system behavior, is often proprietary, making scientific analysis and comparisons to models more difficult.

More high-fidelity observations with direct measurement of wakes using remote sensing are now being used to augment data taken from wind farm SCADA systems. Remote-sensing measurements made in land-based wind farms have provided direct observation of wakes (e.g., Barthelmie et al., 2006; Krishnamurthy et al., 2017, and similar validation studies were also done for land-based scale turbines (Machefaux et al., 2016), including the DOE Crop Wind Energy Experiment campaign, as published in Rajewski et al., 2013; Rajewski et al., 2014; Bodini, Zardi, and Lundquist 2017; Lee and Lundquist 2017; Muñoz-Esparza et al., 2017; Sanchez Gomez and Lundquist 2020. Researchers have also found value in airborne and unmanned aerial vehicle observations that can be used to augment remote sensing (Kocer et al., 2011; Reuder and Jonassen 2012; Wildmann et al., 2014; Lundquist and Bariteau 2015).

Remote-sensing instruments are also valuable in gathering wind field data of an atmospheric boundary layer profile needed for wind energy. For example, there have been field campaigns using remote-sensing instruments in land-based (Rhodes and Lundquist 2013; Klein et al., 2015; Wilczak et al., 2015; Bianco et al., 2017; Lundquist et al., 2017; Mann et al., 2017; Wilczak et al., 2019) and offshore (Pichugina et al., 2012) locations to examine the inflow and thermodynamic conditions within potential wind energy areas. Radiometers used in many of these campaigns have been valuable for determining atmospheric stability, which can substantially influence turbine wake behavior.

New remote-sensing measurements are becoming available that will serve as validation data sets for the near future. New observations with scanning lidars (Bingöl, Mann, and Larsen 2010; Käsler et al., 2010; Trujillo et al., 2011; Aitken et al., 2014b; Iungo and Porté-Agel 2014; Aitken and Lundquist 2014; Banta et al., 2015) are proving valuable; for example, Sandia National Laboratories have the SWiFT facility, and researchers there have performed detailed observations of a single turbine wake using a high-resolution scanning lidar (Herges et al., 2017). Remote-sensing techniques will continue to be important as turbines grow in size, and more traditional meteorological towers will become cost prohibitive at larger heights. New technologies, such as Ka- and X-band radars, are making real-time measurements of entire wind farm interactions and wind farm wakes possible (Hirth et al., 2012; Nygaard and Newcombe 2018). Additionally, using an offshore synthetic aperture radar is a promising method for investigating wind farm wakes (Christiansen and Hasager 2005; Ahsbahs et al., 2018; Jagdish et al., 2018; Hasager et al., 2019). While combining satellite with aircraft observations has been shown to be useful (Emeis et al., 2016), larger scale instrumented aircraft can measure wind speed, wind direction, turbulence, and temperature data within the wind farm or wake from wind farms (Platis et al., 2018; Tucker et al., 2018). In complex terrain, recent observations (Menke et al., 2018; Barthelmie et al., 2018; Wildmann, Kigle, and Gerz 2018) of single-turbine wakes show that local features of the surface topology can have significant effects on wake evolution.

2.1. Future Directions for Observational Technology

The importance of gathering quality data within the atmosphere is not limited to the wind energy community. Researchers in the atmospheric, aeronautical, and even planetary exploration communities are exploring new technologies for atmospheric observation. The quantities of interest for each community may be diverse, but the AWAKEN project and wind energy industry as a whole can benefit from what is being done in other sectors.

A major task within AWAKEN is the development and execution of an instrumentation development roadmap that will advance the state of the art in observational technologies for wind energy applications. The AWAKEN team will organize a group of researchers and industry partners early in the project to create this roadmap. They will build on similar roadmaps developed in other communities, such as recent ones published by the National Academy of Sciences (Board, Sciences, and Medicine 2018) and the National Center for Atmospheric Research (Vömel et al., 2018; Geerts et al., 2018).

Given the growing size of turbines and potential cost of more traditional tower-based observation systems, much focus will be on remote-sensing technologies based on the ground, in the air, and possibly in space. Note that AWAKEN will serve as a starting point for other field campaigns; much of the instrumentation deployed will also be deployed for a future DOE-planned offshore wind farm field experiment in the United States, known as the Wind Forecast Improvement Project 3 (WFIP3). Of particular importance for AWAKEN is gathering observations of the sort of mesoscale impacts on wind turbine wakes that have been observed offshore (e.g., Emeis et al., 2016; Platis et al., 2018), as land-based data are not currently available.

3. Project Objectives and Outcomes

The overall objective of this project is to gather high-fidelity observations of wind turbines and power plants operating in representative atmospheric conditions and to use the data to advance the understanding of wind plant physics. The data gathered as part of this project will also be used to validate computational simulation tools developed under the A2e initiative and by project partners. Advanced instrumentation for microscale observation will be developed and deployed to enable higher-resolution observations beyond present capabilities. Validation studies will be undertaken for both high-performance and engineering models of wind plant operation and design, as well as publicly shared, to ensure dissemination of this work to the wind research and development and industry community. Specific multiyear project objectives are to:

1) Acquire high-quality, high-resolution wind plant wake data sets suitable for the validation of wind turbine and wind plant high-fidelity modeling tools.
2) Demonstrate validation and uncertainty quantification of computational modeling tools developed under the A2e initiative and those of project partners.
3) Advance the fundamental understanding of wind turbine and wind plant physics.
4) Enhance measurement instrumentation and methods as needed to support validation data requirements.
5) Create an international team to observe critical wind plant phenomena and advance wind plant models.
6) Leverage funding opportunities outside of DOE and partnerships to design and collect the highest-quality wind plant observational data set to date.

Advanced observation of wind turbine wakes at utility scale will produce a unique data set for validating high-fidelity models, reduce the uncertainty of a range of wake models, and lead to better overall design and operation of wind plants. This data set will be unique as a U.S. wind energy resource and therefore provide validation data for U.S. developers and manufacturers. By advancing observational technology, researchers will be in a better position to gather observational data with quantified uncertainty at the temporal and spatial resolution required to properly confirm new theories and validate models for wind farm performance. New technology will help enhance the fundamental understanding of wind plant flow physics by research and industry communities alike. Technology transfer will benefit industry by providing more advanced observational tools for understanding the wind resource, wake interactions, and other phenomena of interest that drive wind plant performance and industry profits.

4. Project Tasks

The key tasks for the AWAKEN project are: organize an international team to perform new wind farm observations, leverage funding and partnership opportunities external to DOE, design the experiment, advance the state of the art in microscale observational technologies, execute the experiment, perform useful validation studies to advance wind plant models, and disseminate validation data and publish research results.

4.1. Test Planning

One of the first subtasks in the project will be to finalize the science goals of the AWAKEN field campaign. These goals are aligned with the A2e verification and validation documents (Hills, Maniaci, and Naughton 2015) and presented in detail through phenomena importance and ranking tables (Maniaci 2016). These goals were further prioritized during an international expert meeting held in 2018 to begin defining the motivation and objectives of the new field campaign, which is explained in more detail in Section 5 and Appendix A. The science goals identified by National Renewable Energy Laboratory (NREL) researchers and the international community range from turbulence breakdown of individual wakes and their interactions with downstream turbines to large-scale wind plants and their mesoscale interactions with the atmosphere and neighboring wind plants. Figure 1 provides a view of the rankings of important phenomena to be studied in the AWAKEN field campaign, according to expert meeting breakout groups. Research topic areas listed on the left side of the graph were deemed most important by all the groups. The final set of science goals will synthesize items from this list into a few overarching topics, many of which can be observed simultaneously using similar instrumentation. An overview of potential instrumentation and their link to science goals is provided in Appendix A.1.

As for site selection, the site must first have representative physical behavior commensurate with the science goals. The site will be located somewhere in the central U.S., where the majority of wind farms are located. The willingness of both the wind farm owner and landowners to share access to the wind farm turbines, as well as the openness of the turbine manufacturer to share turbine data, will all be considered for the selection of the site. Researchers will work with local landowners to identify locations for instrument placement and begin land lease procurement, which is often a long lead-time item in field studies.

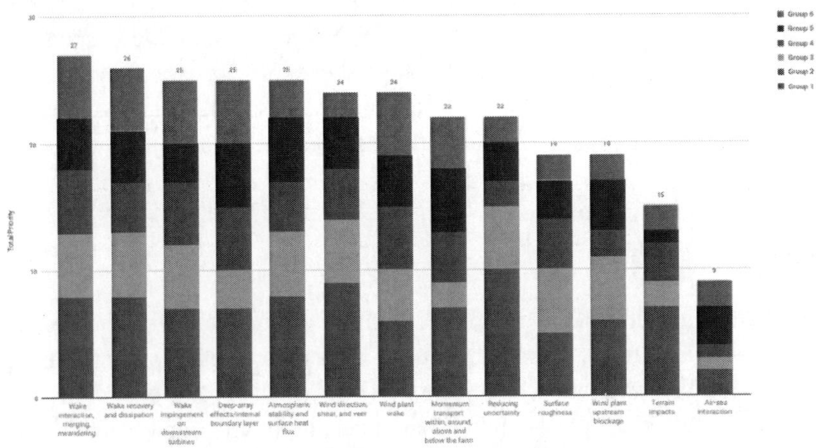

Figure 1. Prioritization of science goals from expert meeting feedback sessions. Color bars are ranking of individual feedback groups. A larger bar indicates a higher priority score. Goals on left hand side of chart were highest priority for all groups combined.

4.2. Partnership Engagement

Because the wind farms and turbines used in the AWAKEN experiment will not be DOE-owned, partnerships with industry are critical. With work and research partnerships comes the need for agreements on specific issues (e.g., site access and data sharing) that may require considerable negotiation between project partners. A fully open data set that is publicly available would be ideal for DOE goals, but certain data may need to be protected to ensure participation by industry members. Often, protections such as normalizing data and granting editorial power on publications to industry partners are useful compromises that can strengthen partnerships between collaborating entities and open paths forward for the project.

DOE laboratories have experience in complex partnership arrangements and commercial data sharing through the Wind Forecasting Improvement projects (Wilczak et al., 2015; Shaw et al., 2019; Wilczak et al., 2019), which can be applied to the AWAKEN project. This includes data sharing with researchers before, during, and after the field campaigns as well as commercially sensitive data protected in the A2e Data Archive and Portal (DAP) (Sivaraman et al., 2014).

Following the example of the New European Wind Atlas Perdigão experiment (Mann et al., 2017; Fernando et al., 2019), external funding sources will be pursued to leverage existing DOE funds and enable the largest possible field and validation campaign. The New European Wind Atlas consortium consisted mainly of European entities, but also included three teams funded by the U.S. National Science Foundation. Two National-Science-Foundation-funded teams were consortia of multiple universities, and the other was the National Center for Atmospheric Research's Earth Observing Laboratory, all of which added considerably to the instrumentation suite available for the project. While the key science goals of AWAKEN will be achieved by the core DOE-funded team, the acquisition of external funding will enhance the team and value of the overall data set. Potential sources for AWAKEN within the United States include government agencies like the National Science Foundation, National Oceanic and Atmospheric Administration, and U.S. Department of Interior, who may be willing to fund their traditional partner institutions. The team will also investigate international funding cost sharing from industry in the form of staff time and access to equipment and instrumentation.

4.3. Microscale Instrumentation Development

The current state of the art for instrumentation to observe complex wind farm flows is generally of much lower fidelity than the current range of simulation tools in terms of temporal and spatial resolution. Current instrumentation can provide real-time data with higher temporal frequency but with lower spatial resolution. To cover spatial variability of a wind field within the atmospheric boundary layer and wind farm, the accuracy of the measurements needs to be compromised. Therefore, validation of numerical models based on the field observation data is not at the required or targeted fidelity, and the quality of the assessment of the numerical models is often obstructed by the limitations of measurements. To narrow the gap between observation and simulation and improve confidence in validation exercises, researchers involved in this task will seek to develop next-generation instrumentation that will improve the spatial and temporal fidelity of microscale observations. Researchers will also quantify the uncertainty of experimental data and the propagation of that uncertainty to the quantities of interest.

The national laboratories will develop and execute an advanced wind plant observation instrumentation development strategy. The strategy will begin with a summary of the state of the art of remote-sensing technology for wind plant applications, which begins with an instrumentation atlas (Lundquist et al., 2019) and will be augmented through a meeting of international experts. The outcome of the experts' meeting will be a document to detail and compare currently available technology and make recommendations and estimate potential costs for the most promising technologies to develop. During this meeting, prioritization of advanced instrumentation will be informed by the A2e high-fidelity modeling (HFM) verification and validation (V&V) roadmap (Hills, Maniaci, and Naughton 2015; Maniaci and Naughton 2019) and AWAKEN science goals (see Appendix A). Potential technologies of interest include X-band radars, acoustic tomography, the Sandia Wake Imaging System (Herges et al., 2016), unmanned aerial vehicle swarms, multistatic lidar, and so on. Using this document as a basis, an International Energy Agency task on remote- sensing technology will be created to allow for international collaboration on advanced instrumentation during the period of the project. Starting with the feedback from the experts' meeting, national laboratory researchers will conduct observing system studies experiments using numerical simulations of different technologies to better understand their ability to measure wind plant quantities of interest.

Results from these studies will highlight the temporal and spatial resolutions possible across a range of technologies for the highest-priority quantities of interest. From these studies, laboratory researchers will identify promising technologies for future development. Assuming technologies are at a sufficient technology readiness level, laboratory researchers will collaborate with private industry to help develop and validate the performance of early prototypes in the field. Prototype testing will occur at DOE field testing sites with atmospheric observation capabilities, such as NREL's Flatirons Campus, SWiFT, and DOE Atmospheric Radiation Measurement (ARM) sites. Testing may also occur at operational wind farms as needed for validation. Finally, new technologies will be compared with current industrial and research technologies to help quantify the improvement in observational capability.

An important task will be the creation of an instrumentation development roadmap, starting from the current state of the art and proceeding through the end of the AWAKEN project. In 2019, an instrumentation atlas (Lundquist et al., 2019) was developed that provided a current state of the art of microscale instrumentation and measurement capability gaps. The instrumentation atlas has been updated and ported to the AWAKEN wiki website (https://openei.org/wiki/AWAKEN), where it can be maintained throughout the AWAKEN project. This summary will be useful for providing current capabilities of existing observational equipment as a baseline for improvements made within this task. The primary purpose of the instruments to be developed will be those to provide high-resolution validation data within AWAKEN following the A2e HFM V&V roadmap. However, we envision that such technology will also be useful for industry in advancing their observational methods. Therefore, this subtask will involve close collaboration with industry remote-sensing manufacturers and others. The roadmap will initially be written by laboratory researchers but then distributed to outside partners and finalized after feedback from an expert elicitation meeting in the beginning of 2020. The roadmap will highlight the potential capabilities of new instruments, matching observational requirements for AWAKEN validation goals and projected timelines and funding requirements for development. Researchers will modify the roadmap during the project as more accurate performance estimates and realistic development timelines become available, and the AWAKEN experimental design becomes clearer.

During expert elicitation meetings, promising technologies will be identified that show the greatest promise for investment in further development. The different technologies, depending on their level of maturity and needs for future growth, will first be modeled in high-fidelity simulation tools such as A2e-developed Nalu-Wind (Lawson et al., 2019). Assuming these simulations show promise and significant potential impact for better microscale observation, prototypes will be built and tested in a laboratory setting before being deployed and validated in a field location. Go/no-go decisions for each technology will be made prior to prototyping based on the potential ability to fill measurement gaps and estimated development timeline. Researchers will ensure technology testing can be completed within the project time window and, if not available for the AWAKEN experiment, be available for future observational campaigns. Facilities at the NREL Flatirons

Campus, Sandia, SWiFT, and DOE ARM sites will be used for development and testing of new instrumentation. Depending on the pace of development, these new technologies could also be deployed during the actual AWAKEN test.

4.4. Experimental Design

A major task following site selection is the planning of the experiment itself. Planning will involve developing multiple timelines, incorporating the availability and applicability of different sorts of instrumentation, and understanding project partner constraints. Logistical considerations, such as instrument power supply, spare parts, and estimated repair times, will also be included. We envision a field campaign of at least 18 months to capture seasonal variability of the atmosphere. Such a long field campaign can be very challenging, and lessons learned regarding instrumentation monitoring and support, real-time data visualization, and so on, from the 18-month Wind Forecast Improvement Project 2 and Perdigão field campaigns will be incorporated into the planning. The range of instruments to be deployed in the field will be simulated with high-fidelity models, and the overall test plan will be optimized to gather the best possible data set with the available instruments based on previous work to reduce the measurement errors caused by instrument coordination or their placement (Stawiarski et al., 2013; Simley et al., 2016; Choukulkar et al., 2017; Debnath et al., 2017; Newsom et al., 2017; Peña and Mann 2019). The combination of instruments with different spatial and temporal resolutions can modify measurement error. Therefore, the instrument setup and measurement strategy will be tested and optimized to reduce the error within the measurements prior to and during the experiment. Figure 2 provides an example of some instrumentation that may be deployed and its relative location to wind turbines.

Using the instrumentation atlas, science goals, and site properties, researchers will develop and prioritize a list of instrumentation needed for the campaign. See Appendix A.1 for an initial matching of potential instruments to science goals. A portion of the project budget will fund purchases and rental of critical systems. Partnerships will augment DOE funding with other participants providing their own equipment for microscale measurement. As stated in Section 4.2, these externally funded partnerships will not be required to achieve the key AWAKEN science goals but will increase the amount and value of the overall data gathered. Researchers will work with partners to develop timelines for deployment. In some cases, such as with the National Center for Atmospheric Research Earth Observing Laboratory and DOE ARM programs, team members will submit proposals for usage of the institutions' equipment over specific stages of the field test. These proposals require significant lead time and have been initiated. Another consideration in the selection process is the time period each instrument will be deployed; some instruments will be observing complex flow for the duration of an 18-month test, whereas others will be deployed for shorter durations of 6 months or less.

High-fidelity simulation tools such as Nalu-Wind will be used to identify the errors associated with a particular instrument and its applications to achieve the related science goals. Such observing system simulation experiments will provide researchers with an idea of the types of data the AWAKEN experiment will produce and the advantages and disadvantages of different types of instrumentation. The experiments will also serve as the basis for the first validation studies. A prerequisite for such studies is representations of each observational technology within HFM codes. Models for profiling lidars, scanning lidars, and Ka/X-band radar are currently in development. Other technologies will be added as needed. Researchers will use these tools to inform the sensor layout of the AWAKEN experiment during the different phases of the campaign.

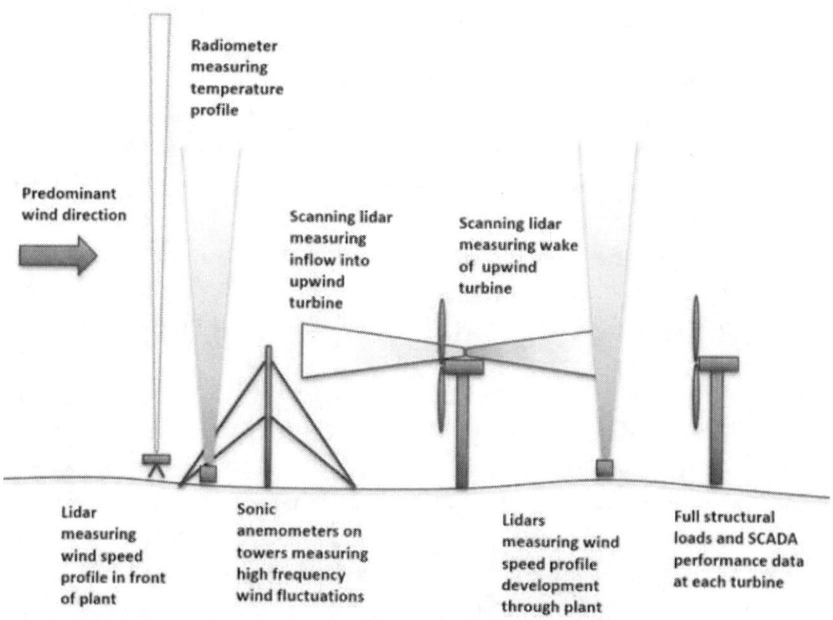

Figure 2. Examples of instrumentation for the AWAKEN project.

4.5. Experimental Execution

Execution of the AWAKEN field campaign at minimum will occur over a period of 18 months to ensure capture of a full set of seasonal variations with some overlap to identify a wide range of atmospheric conditions at a specific site. We anticipate that not all equipment will be available during the entire campaign, particularly those from externally funded collaborators and those instruments that require significant labor to operate. However, we will seek to maximize the amount of data gathered to ensure operation of a critical mass of instruments during the different phases of the campaign.

A month prior to test execution, researchers will begin to deploy the instrumentation at the test site. Instruments will initially be deployed according to the experimental design studies, and adjustments will be made to meet real-world constraints (e.g., insufficient power available, terrain or access complications, and so on). All instruments will be calibrated and ready to use prior to the first simultaneous day of testing.

Long-term observations will be recorded over the duration of the test. Researchers will seek to leverage existing instrumentation in the area, such as locally owned meteorological towers or automated surface observing system stations or state-supported mesonets (Brock et al., 1995; Schroeder et al., 2005) (or DOE ARM facilities [Sisterson et al., 2016]), and incorporate their data streams into the overall data set. These existing data sets will likely be most useful for observing larger-scale physical phenomena. DOE-owned/rented instrumentation, such as meteorological towers, vertically profiling lidars, some scanning lidars, sodars, and radiosonde launches, will be deployed and kept operational for the duration of the experiment. Time and budget will be planned for instrumentation maintenance, data handling, and quality control. Data will be uploaded to the A2e DAP (Sivaraman et al., 2014) in a timely, reasonable manner.

Researchers will perform more highly detailed measurements over short targeted time periods using instrumentation that cannot be easily deployed over longer periods. As possible, data will be made available in real-time to assist with episodic intensive measurement periods and deployments of airborne instrumentation. Examples include unmanned aerial vehicles for measuring turbine wakes, aircraft to measure wind plant impacts, and mobile radar systems that have short deployment windows. The actual deployment of each of these instruments is highly dependent on the instrument itself and the data gathered. Researchers using instruments that are best for observational campaigns lasting less than a month will plan for multiple deployments during the test to capture some seasonal variability and provide repeatability of observations.

4.6. Data Analysis and Model Validation and Archival

Model validation is one of the key goals of the AWAKEN project, and while the primary focus is on high-fidelity simulation tools such as Nalu-Wind, other models will be used in a multifidelity validation approach. The process for this validation is being developed in a related verification & validation and uncertainty quantification project being led by Sandia researchers and currently being applied in other A2e tasks. Many researchers overlap between the AWAKEN project and the verification and validation/uncertainty quantification project and thus will be well educated as to the best practice for validation using observations from the test. The process of validation requires careful processing, analysis, and understanding of observations, which will lead to new insights of wind plant behavior and driving physical phenomena.

It is important that during execution of the AWAKEN field campaign, researchers perform validation tests to ensure that the data being gathered are valuable. These iterative tests provide feedback to the researchers in the field, who may have time to adjust the observational campaign and improve validation results during the test. Prior to gathering and analyzing data, team members will identify key quantities of interest on which to focus initial validation studies based on the A2e V&V roadmap. Validation metrics for simulation tools will also be devised to measure overall performance of the simulation tools, identify areas of improvement, and help steer ongoing measurements in the field.

International Energy Agency Task 31, also known as Wakebench (Sanz Rodrigo et al., 2017), is a critical venue for validation exercises with a large representative group of researchers interested in wind farm model validation. The group has been active for 6 years and is a well-established international working group. Various new wind plant model benchmarks from the field campaign will be created and led by AWAKEN team members. International researchers will use their own models following the prescribed validation process, beginning with a blind comparison between observations and simulations, followed by a gradual release of observational results. The iterative validation process provides insights into which models perform best under a range of conditions. Guidelines for "best practices" for each of the different models will then be developed, following the recommendations produced in the validation study. The principal investigator of AWAKEN is a co-lead of Wakebench and will ensure seamless interaction between the international efforts.

All observations will be uploaded to the A2e DAP on an ongoing basis during the AWAKEN test, including a capability for real-time monitoring of instrument performance. A dedicated web page for the AWAKEN field campaign will be created on the DAP site. AWAKEN researchers have experience with archiving with DAP and uploading data in near real time, so the process is well vetted. The Wind Forecast Improvement Project 2 18-month experiment highlighted the importance of real-time instrument monitoring and local technical support to quickly identify and repair instrumentation problems. During the long-term observational campaign, weekly weather and turbine event discussions will enable the creation of an event log that will also be archived at the DAP to identify notable events of interest to the modeling community. An event discussion coordinator will be identified early in the project to lead regular meetings during the field campaign. Researchers will work with the DAP team to ensure commonality of data formats that will make it easier for future data users to analyze. Both raw data from observations and processed data from analysis will be uploaded and maintained. As validation studies proceed, researchers will upload simulation results of interest for the wider community. Data storage limits prevent storage of all simulations, and criteria will need to be developed to determine what data will be kept over what time periods. Analysis scripts used by team members will be uploaded to enable reproducible research results and easier continued analysis after the AWAKEN project is complete. In general, researchers will strive to follow FAIR (Findable, Accessible, Interoperable, and Reusable) data principles (Wilkinson et al., 2016) to ensure lasting value from the data gathered.

5. Expert Elicitation Meeting

The first AWAKEN meeting to gauge interest in collaborative wind farm field observations and to begin the formation of a central repository for past experiences was convened by NREL at the University of Colorado in Boulder, Colorado, March 21, 2018. The agenda for the meeting can be found in Appendix C. More than 50 participants attended the meeting, representing the interests of academics, wind energy industry professionals, government agencies, and national laboratories. The agenda for the meeting is found in

Appendix C, and names of the participants and their affiliations are listed in Appendix D. Throughout the meeting, researchers offered their perspectives on past experience in utility-scale wind farm observation campaigns and provided insights on lessons learned about international collaborative work and emerging technologies and instrumentation; they also contributed to the prioritization of scientific objectives of the forthcoming study.

5.1. Key Points by Speakers

Some of the key points emphasized by the speakers include the following:

- Models, physics, experiments, and applications must all work together. Improving one benefits each of the other elements, as described by the"virtuous cycle" introduced by David Maniaci from Sandia and repeated by several other speakers. The importance of model validation cannot be overstated. Validation needs to outstrip observational capabilities, and remote-sensing technology is improving, but not as fast a model's technology. Therefore, comparisons between models and data may be most effective when reconciled against performance or other averaged turbine data.
- Multiple forms of observations are important. Co-locating and correlating observations reduce uncertainty and increase reliability. Duplicate measurements are important for redundancy, reduction of uncertainty, and corre- lated observations. Long-term turbine SCADA data are necessary to get a complete picture of the operation of a wind farm. SCADA data include using turbines as anemometers through power output. Turbine anemometer and unsteady responses correlate to turbulence from performance and loads observations. Common data formats and time stamping across observational platforms are vital for consistent access to data and method development.
- Long-term project planning and careful partnership engagement are essential. Three to four years' planning before a campaign begins might be necessary. The success of the project will depend on long-term relationships. Landowner/wind farm owner/operator relationships are crucial to the success of a project. Service

relationships with instrument manufacturers are important during long campaigns.

– The added value of collaboration in large projects is attractive to a variety of potential partners. There are many funding sources that could be leveraged to benefit the group effort. Within DOE, collaboration frequently occurs between laboratories and sometimes across programs. One question discussed was: How will other funding agencies interact with DOE? Collaboration may be difficult because of the proposed timeline alone, and establishing funding could potentially require years of planning. Funding collaborations between sources will be important if the scope will be on the order of the Perdigão campaign.

5.2. Feedback Sessions

During the meeting, feedback from the attendees was solicited through three short surveys completed in group breakout sessions, and results were tabulated into a spreadsheet. The meeting attendees were divided into working groups of 8–10 people, with representatives from each sector of the wind energy research community. Each survey was designed to collect input on the relevance of the supplied science goals, attendees' respective priority of study, and what should be added or removed from the preliminary test plan. Below, the feedback solicited from the small work groups during the meeting is summarized. The aggregated data from each of the feedback sessions have been consolidated into a summary that is located in Appendix B.

5.2.1. Which Science Goals Can Be Removed or Combined?

Although most of the small work groups suggested some sort of reorganization and/or combination of the science goals (Appendix A and priorities, a single recommended hierarchy or structure did not emerge from the feedback suggested by the groups. Rather, most groups suggested identifying the two or three overarching topic areas for the campaign, under which all other science questions were nested within.

Some other comments made about the science goals were:

- While not a science goal per se, instrumentation and experimental design should be given equal standing in terms of planning and consideration. Questions that should be considered carefully are: How will measurements be made to address science goals? How well can the quantities of interest be measured? Does an optimal arrangement of sensors exist? Can we reach it?

- Add depth to the large-scale phenomena/wind farm effects science goal. Mesoscale effects on a wind farm are important. To capture these effects, ensure instruments are movable and dispersed instead of localized at a fixed point. Expand the instrumentation plan to include observations of mesoscale recirculation effects. Also, some observations should seek to measure the correlation between large coherent structures' effects on wind turbines and structural fatigue loads.
- Additional thinking is needed around terrain impacts and how boundary conditions affect science goals. Be sure to include observations that can capture the homogeneity (or heterogeneity) of the surface (trees, roughness, terrain, and so on). It is useful to start with flat terrain to eliminate areas of uncertainty. It is also easy to start with areas or data sets with consistency of wind direction. Layout spacing is similar; large idealized wind farms exist on flat terrain, but not very many are placed on regular gridded layouts.
- How do the science goals tie into the desired outcomes of the campaign? Consider which goals are achievable in terms of observations and for model validation. Should goals be tuned to match needs for model development or validation? What data do wind farm control designs need?
- More detail could be added to the inflow and operating conditions science goals. Inflow and atmospheric conditions are equally important as measuring wakes: shear, veer, wind direction, wind direction variability, speed, turbulence, and stability. If turbine structural loads are important, then so are turbulence and variations in flow parameters along the blade. Can the freestream wind speed be measured? It is important to measure the flow around the wind farm, in addition to inside it. Lastly, atmospheric stability has a big impact on the ability to control turbine wakes. For example, in highly turbulent/unstable conditions, flow wakes will be difficult to control, but also create less energy loss due to faster wake dissipation.
- Selection of instrumentation is a priority. Try to get wind direction variability from SCADA data immediately to begin analysis. Use the

Richardson number or Obukhov length for stability; use temperature profile from the meteorological mast or the sonic anemometer close to the ground. The preferred measurement depends on the surface homogeneity. Some planning is required for dissipation measurements and how best to make measurements with tethered lift systems and/or hot-wire anemometry. Strongly consider how to best instrument blades with strain measurements along blades, accelerometers, and so on.

5.2.2. Instrumentation

The next question posed of the breakout groups was: What instrumentation should be added to the testing plan and why? Participants suggested a substantial list of instrumentation they would like to see deployed and used during the testing campaign. In addition to a list of instruments, many attendees offered recommendations on specific models of sensors and measurement devices for a wide range of reasons. Next is a brief summary of recommendations. Additional suggestions and their relation to the science goals are discussed further in Appendix A.1.

- Meteorological masts and in-situ point measurements
  - Sonic anemometers
  - Pressure-temperature-humidity
- Remote sensing
  - Sodar and radar radio acoustic sounding system)
  - Ka-band and X-band radar
  - A diversity of lidars (e.g., ground-based, nacelle-mounted, scanners, profilers)
  - Field-particle-image-velocimetry-type systems
  - Satellite imaging
- Other instruments
  - Radiosondes and sounding systems (difficult with the Federal Aviation Administration; labor-intensive)
  - Eddy covariance (surface flux stations) both upwind and throughout the wind plant
  - Ceilometer for planetary boundary layer heights
  - SCADA (performance/operational data, loads)
  - Upwelling/downwelling radiation sensors

    - Sensors for surface properties (e.g., soil moisture/temperature, surface albedo,âA˘ e̦)
    - Tethered lifting systems
    - Airborne measurement systems and unmanned aerial vehicles
- Other advice
    - Plan for backup units, (re)calibration of instruments, maintenance/downtime, data formatting, and time synchronization.

5.2.3. Recommendations for Test Site

The breakout groups were asked about a potential site for the AWAKEN campaign. Most groups recommended sites that have a wide range of atmospheric conditions and constrained terrain complexity. It was also recommended that the potential test site be one that has already been the subject of previous research experience, as forging relationships with landowners and wind farm operators can be a significant hurdle in any field campaign. A short list of potential sites includes the following.

- Wind Forecast Improvement Project site (Columbia River Gorge): too complex terrain for this test, but instrumentation already exist
- ARM climate research facilities
- Story County, Iowa: there are existing, maintained tall towers, and the site of the successful CWEX campaigns
- North of Mason City, Iowa: there are three farms close together
- NextEra pre/postconstruction test site in the Midwest: data sharing may be sensitive
- Texas wind farms (Roscoe, Loraine, Champion).

5.2.4. Additional Thoughts on Test Planning

Lastly, breakout groups were asked about preliminary thoughts on the initial AWAKEN test plan. Potential testing plans will be compiled in an open-source test planning document, which will soon be open to public scoping. Input requested included specific strategies that would be beneficial to the AWAKEN project as well as potential locations for the campaign, instrumentation to be considered, and wind turbine models that would be good test candidates. The main suggested additions to the testing document fall into the categories of:

- Preassessment
- Flowfield measurements
- Turbine operational data/site characterization
- Logistics (points of contact, responsible parties, scheduling, and so on)
- Concurrent modeling.

Please refer to the aggregated results in Appendix B for a complete list of suggestions.

6. Next Steps

With the vision of the AWAKEN project summarized in this document, the next steps are execution of the tasks outlined earlier and according to a timeline described in Section 7. The initial year of the project will focus on partnership management, such as working with interested parties on agreements for wind farm and data access. There will be considerable time spent working with research partners submitting proposals to augment DOE funding to increase the overall scope and potential impact of the project. Site selection will be an early goal as well, once strong partnerships are established. While negotiations proceed, instrument selection can proceed concurrently, identifying key instruments and matching them to science goals. The identification of instruments will also include a gaps analysis of the current state of the art and formulation of an instrumentation development roadmap. This roadmap will be executed throughout the AWAKEN project. Once a site is selected, detailed simulation studies will be done to design the experiment for optimal observation of phenomena related to the science goals. Once the design is complete, instruments will be acquired, then deployed and calibrated, and the observational campaign will begin. Throughout the campaign, validation studies against models will be executed and the observation plan adjusted accordingly. Validation studies will continue past the completion of the field observations until the end of the project. All data and some simulations will be uploaded to the A2e DAP for further use by the wind energy community.

As part of the ongoing AWAKEN planning efforts, organizers at NREL will continue to solicit feedback from the wind energy community as a whole through monthly update meetings and sporadic expert meetings. Given the early stage of the campaign, participants are encouraged to communicate with any potential participants or collaborators and to share the public AWAKEN-related literature.

7. Timeline and Milestones

Milestones and events key to the success of the AWAKEN project are described in the AWAKEN timeline, reproduced below. Note that the proposed timeline here will develop as the number of participants increases, the scope of the campaign comes into focus, and opportunities arise for funding and collaboration. Meetings and working group sessions will be added to the timeline as they are scheduled.

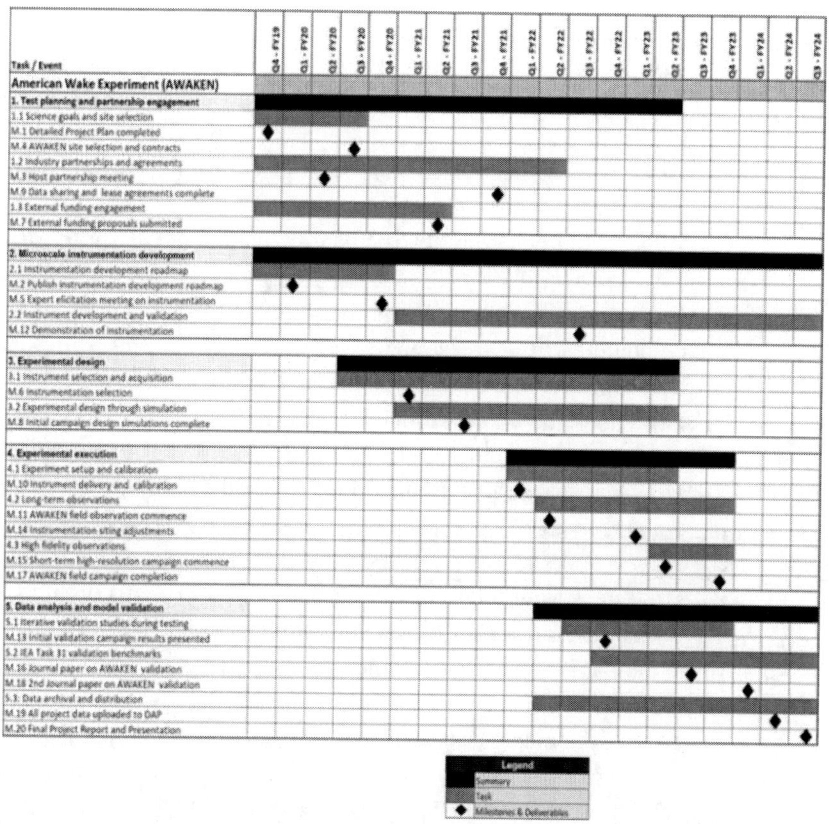

Figure 3. Timeline and milestones for AWAKEN project, as of publication.

A. Science Goals

The science goals for the Atmosphere to Electrons project have been identified at a high level during previous strategic workshops (Hammond et al., 2015) under a process using Phenomena Identification Ranking Tables (Hills, Maniaci, and Naughton 2015). The major phenomena important for wind plant flow physics are identified in this section, and many have been identified as high priority (Maniaci 2016; Maniaci and Naughton 2019; Maniaci et al., 2020). The experts who attended the first AWAKEN expert elicitation meeting helped refine the critical scientific goals and questions to be answered through a new test based on their experience. The science goals and associated questions to be answered are as follows:

1. Wake recovery and dissipation
   A) How does the wake recovery rate change with atmospheric stability throughout the wind plant? How does stability within the wind farm evolve over the diurnal and seasonal cycles in contrast to concurrent ambient stability?
   B) How does wake-added and ambient turbulence affect the recovery rate and expansion?
   C) What is the best way to calculate or parameterize and observe wake recovery?
   D) How and why is turbulent dissipation in a wake modified compared to the freestream, and how can it be measured and modeled? How does dissipation rate affect wake recovery?
   E) How does the wake width change with background conditions, especially the stratification and shear?
   F) How does available momentum above the wind farm impact wake recovery?
   G) How is wake recovery related to wake meandering?
   H) How does yaw angle influence wake recovery and dissipation under different stratification conditions?
2. Wake interaction, merging, meandering
   A) How do wakes merge under different atmospheric conditions?
   B) How is the meandering affected by turbine separation distance?
   C) What are the dominant contributing factors to wake meandering? How do atmospheric conditions con- tribute to wake meandering? How does turbine design contribute to wake meandering? How does turbine operation contribute to wake

meandering? Can these various effects be separated in a measurement campaign?

D) How does turbine yaw misalignment influence meander and merger?

E) How does meandering change between the upstream wakes (for the front row of turbines) and wakes deep in the plant?

F) Is meandering driven only by the large scales in the flow, or does turbulence kinetic energy redistribution within the wake (and therefore its expansion) also contribute to wake motion? What is the influence of topography?

3. Wake impingement on downstream turbines

A) How do wakes influence blade, tower, and nacelle loading and acceleration on downstream turbines?

B) How is wake influence impacted by different wake mixing scenarios (single wake vs. mixed wakes from multiple turbines)? How does that change in partial-wake scenarios?

C) How do yawed wakes impinge differently than standard wakes?

D) At what wind speeds and other atmospheric conditions do wakes have the greatest loading impact on downstream turbines?

E) Is the half-wake situation the most detrimental for loads?

F) What are the important loads for wake interaction? Can loading impacts be best isolated in a wake experiment, in combination with simulation or with simulation alone?

G) How can the inflow—wind turbine—wake system be modeled effectively to predict loads on downstream turbines?

4. Deep-array effects/internal boundary layer

A) How would a deep-array effect be seen and detected in a land-based wind farm?

B) What are the dominant physical processes required to model the deep-array effect for a land-based wind farm?

C) How does the deep-array effect change with atmospheric stability?

D) How does the deep-array effect change with different turbine spacing and layout?

E) How does the deep-array effect change with turbine operation?

F) What is the shape of internal boundary layer growth above and on the sides of a wind farm? And how is this related to the ambient boundary layer depth?

5. Atmospheric stability and surface heat flux

A) How do the dynamics of a single wake and of the wind plant (intensity/recovery/meandering) change under clear-sky, radiative forcing-driven atmosphere vs. under specific weather events?
B) What is the impact of lateral coherence (and its height dependence) in the freestream atmosphere on an unwaked turbine vs. a waked turbine (e.g., in terms of yaw loading), and how does it vary with turbine separation?

6. Momentum transport within, around, above, and below the farm
   A) How does boundary layer height (and the development of a wind farm internal boundary layer) affect momentum entrainment above the farm? How does this depend on stability conditions?
   B) Can the influence of the wind farm on momentum be separated from the atmospheric dynamics? In simulation or observation?
   C) Does the speed reduction of the wind farm lead to surface convergence and enhanced upward motion?
7. Wind direction, shear, and veer
   A) How does the wake behavior change under changing shear and veer?
   B) How does the wake move under sudden changes in wind direction? What is a quantitative measure of direction change vs. a turbulent gust? Is the International Electrotechnical Commission standard a sufficient definition?
8. Surface roughness
   A) How do seasonal changes in roughness affect power performance and plant aerodynamics in general?
   B) How important is it to accurately model roughness in our wind plant aerodynamics simulations? How dependent is the surface roughness impact to other local features such as terrain? How can individual roughness and terrain impacts be isolated in measurements?
   C) How is the levelized cost of energy affected by roughness management (e.g., corn vs. soy, dormant season roughness management)?
9. Wind plant wake
   A) How far downstream does a wind plant wake persist? How can this be defined if the background wind field is changing?

B) How does the intensity/fetch of the plant wake vary with atmospheric conditions/terrain/obstacles/plant operating conditions?

C) How is the plant wake defined? Are there near and far wakes with different physical descriptions and influences as with individual turbine wakes?

D) Are there downstream turbine loading impacts from wind plant wakes, or is the dominant effect power loss?

E) How does the wind plant wake affect the local climate downstream?

F) Does a wind plant wake meander?

G) How can a wind plant wake be best observed?

10. Terrain impacts

A) How does terrain affect wake recovery and trajectory (of a single wake vs. the entire plant)?

B) How does terrain affect loads? If increased shear/turbulence reflects in increased fatigue loads, do terrain-induced speedup effects outweigh the decreased fatigue lifetime of the components/turbine?

C) At which resolution does the terrain need to be modeled so that its speedup effect and its effect on wake dynamics are reproduced?

11. Wind plant upstream blockage

A) How far upstream is the wind speed changed by presence of the wind farm?

B) How is the blockage effect influenced by turbine size, spacing, operation, and layout?

C) How does the wind accelerate around the wind farm?

12. Air-sea interaction (American WAKE experimeNt [AWAKEN] is envisioned for the U.S. Midwest, but if offshore is prioritized highly, these science goals can be incorporated.)

A) Relationship between wave height and period on tower loads (and platform/anchor system loads for floating offshore wind turbines), on the flow within the plant, on wake recovery.

B) How far off are we when we use similarity theory to estimate surface fluxes (from buoy or satellite or model data), and how does this propagate to our extrapolation to hub heights?

C) Should we revisit how we model (or not model) offshore roughness?

D) What are the consequences of using equivalent neutral winds (satellite product) instead of stability-dependent winds when estimating vertical wind profiles and extrapolating low wind speeds to nominal hub heights? Corrections accounting for stability are often based on similarity/models.
E) How do swell conditions affect the atmospheric surface layer, wake recovery, and the turbine power/loads?

13. Reducing Uncertainty
    A) How will measurement sites be selected to construct ensembles of multiple experiments (e.g., single wakes, merged wakes, wind plant wakes, wind plant-to-wind-plant interactions)?
    B) How will the selection of sites help reduce measurement uncertainty?

A1. Instrumentation Matched to Science Questions

In addition to the feedback sessions conducted at the experts meeting, an online participation survey was conducted to gauge interest in participating in AWAKEN. Feedback was collected about what instrumentation may be available to AWAKEN for data collection. Between the participation survey and the expert meeting feedback, the following instruments are the primary instrument types identified to meet the science questions in Appendix A.

- Lidar: Lidars are remote-sensing instruments widely used in the wind energy community to remotely measure wind speed and direction in and around wind farms through the Doppler effect.
  - Scanning lidar: A scanning lidar was identified as a key instrument for the AWAKEN project, and numerous instruments may be available for deployment for the AWAKEN campaign.
  - Profiling lidar: A profiling lidar can be used to measure the profile of wind, sheer, and turbulence. Based on the participation survey, multiple profiling lidars may be available for the AWAKEN project.
- Radar: Radars are remote-sensing instruments that use long-wave radiation and can be used to measure wind speed and direction at long ranges from the instrument, using the Doppler effect. Several radars,

both mobile and fixed, may be available for the AWAKEN field measurements.

- Ceilometers: Ceilometers are remote-sensing instruments that can be used to measure vertical profiles of particulates in the air. They can primarily be used for determining the boundary layer and cloud heights.
- Sodar: Sodar use sound waves to measure wind speed and direction.
- Temperature/humidity profilers: Temperature/humidity profilers measure the profile of temperature and humidity is important for understanding atmospheric stability, which is important for setting boundary condition for model runs.
- Radiometers: Radiometers are used to measure radiative flux.
- Pitot tubes and turbine surface pressure measurement systems: Pitot tubes and other pressure measurement systems are used to measure the speed of air flow near the turbine surface. Several different systems may be available for AWAKEN measurements.
- Tethersondes: Tethersondes are balloon systems that are tethered to the ground, allowing for repeated profile measurements, or stationary measurements at a singular elevated location. Measurements of wind, temperature, pressure, humidity, and other atmospheric quantities are possible, depending on the instruments in the payload. Several tethersonde systems may be available for the AWAKEN project.
- Radiosondes: Radiosondes are single-use, balloon-borne instruments used to measure profiles of wind, temperature, pressure, and humidity.
- Towers and surface meteorological stations: Towers and surface meteorological stations can be used to measure wind, pressure, temperature, humidity, and other atmospheric parameters at set heights using in-situ instruments. Numerous towers and their associated instruments (i.e., sonic anemometers) may be available for the AWAKEN measurements campaign.
- Turbine loads and alignment measurements: Measuring the loads and alignment of turbines will be critical to understanding how the wakes impact turbine performance as well as how turbine operation impacts the creation of wakes.

Table A.1 matches the previously mentioned instrument type to science questions listed in Appendix A. Instruments in each class of instrument can be used to address the questions listed.

There are certainly other instruments available for use in the AWAKEN project, such as those that can be mounted on aerial vehicles that will be deployed. This is considered an initial list of ideas based on feedback from meeting experts as to what is immediately available.

Table A.1. Instruments match to science questions

Instrument	Science Questions
Scanning lidar	1.A, 1.B, 1.C, 1.D, 1.E, 1.G, 1.H, 2.A, 2.B, 2.E, 3.B, 3.C, 4.A, 5.A, 7.A, 7.B, 9.A, 9.B, 9.F, 11.A, 11.C
Profiling lidar	1.B, 1.D, 1.E, 1.F, 2.C, 3.D, 4.A, 4.B, 4.C, 4.F, 5.A, 5.B, 6.A, 6.C, 7.A
Radar	1.A, 1.G, 1.H, 2.A, 2.B, 2.E, 3.B, 3.C, 4.A, 5.A, 7.A, 9.A, 9.B, 9.F, 11.A, 11.B
Ceilometer	2.A, 2.C, 4.F, 5.A, 6.A, 9.B
Sodar	1.A, 1.B, 1.E, 1.F, 1.H, 2.A, 3.D, 4.A, 4.B, 4.C, 4.F, 5.A, 5.B, 6.A, 6.B, 7.A
Temperature/ humidity profilers	1.A, 1.B, 2.A, 2.C, 4.B, 4.C, 4.F, 5.A, 6.A, 6.B, 9.B
Radiometer	1.A, 2.B, 2.C, 5.A, 9.A
Pitot probes	All
Tethersondes	1.E, 1.F, 1.H, 2.1, 3.D, 4.B, 4.C, 6.A, 7.A, 9.A
Radiosondes	1.E, 1.F, 1.H, 2.1, 3.D, 4.B, 4.C, 6.A, 7.A, 9.A
Tower-based systems	1.A, 1.B, 2.A, 2.C, 3.D, 4.B, 4.C, 6.B, 6.C
Turbine loads and alignment	1.H, 2.C, 2.D, 3.A, 3.D, 3.F, 3.G, 9.D

B. Feedback Sessions

Here are the aggregated notes from feedback sessions that took place at the American WAKE experimeNt expert elicitation meeting that took place on March 21, 2018.

B.1. Feedback Session #1 - Science Goals

B.1.1. Question .1#1 - Which Science Goals Can Be Removed or Combined?

Reduce the length of the list of science goals. While there are many important topics listed, it seems ambitious to try and address all of them in a single campaign. In addition, several goals or topics seem nested or interdependent. For instance, while characterizing the atmospheric conditions (wind direction, wind speed, stability, shear, veer) is necessary to drawing consistent and reproducible conclusions from the observations; they are not in and of themselves high-priority goals for wind energy.

By reorganizing the science goals, the main purpose of the campaign can be more clearly identified. The other topics can be nested within the main purpose as subgoals or additional topics of interest that could be answered along the way.

The main goals appear to combine points 1 through 4:

- Goal 1. Turbine interaction with the near wake, correlation between near- and far-wake interactions. Wake vortex startup and dissipation. Wake contraction based on turbine operation
- Goal 2. How do wakes from individual turbines combine to form a wind farm wake? Does that affect our wake management strategy from one place to another?
- Goal 3. How do we characterize the transition to fully developed internal boundary layers? How does wake steering affect transition to or development of the internal boundary layer?

Within the main goals, other points from the list could be combined:

- Combine 5, 7, 8, and 10 as atmospheric boundary conditions.
  - Atmospheric stability and surface heat flux
  - Wind direction, shear, and veer
  - Surface roughness
  - Terrain impacts.
- Combine 6, 9, and 11 as transport and plant-scale phenomena.
  - Momentum transport within, around, above, and below the farm
  - Wind plant wake
  - Wind plant upstream blockage.

Point #13 (reducing uncertainty) as written is important but is not a science goal by itself. It is more about experiment design. However, there should be a goal related to quantifying uncertainties of the full range of models we use, with situational dependence.

B.1.2. Question #2 - Which Science Goals Are Missing?

While not a science goal, instrumentation and experimental design should be give equal standing in terms of planning and consideration.

- How do we make measurements in order to address science goals?
- How well can we measure quantities of interest?
- Does an optimal arrangement of sensors exist? Can we reach it?
- Unmanned aerial vehicles? Thermal imaging? Add to large-scale phenomena/wind farm effects.
- Mesoscale effects on a wind farm, maybe not localizing instruments at one point (for example, recirculation from a nearby mountain range) or maybe expand item 6 to include these mesoscale recirculation effects.
- Large coherence structure of wind turbines for fatigue load assessment. Add to terrain impacts/boundary conditions.
- Homogeneity (or heterogeneity) of surface (e.g., trees, roughness, terrain). How do the science goals tie into the desired outcomes of the campaign?
- Should goals be tuned to match needs for model development or validation?
- What data do wind farm controllers need?

B.1.3. Question #3 - Additional Comments on Science Goals?

Reorganize priorities, including:

- Goals 3 and 4 seem to be the overarching goals of the campaign, and the other points seem to feed into those high-level concepts. Those are the points that are most salient on wind plant controls. Does the wake management strategy for a single wind turbine scale work for a full plant?
- A lot of these topics are interconnected (e.g., stability greatly impacts wake recovery).

- Some reorganization might be needed, as the goals are at different levels, perhaps inputs (turbine parameters, boundary conditions) vs. outputs (e.g., wake).
- Misalignment between academia/industry on goals. Single validation/case study is not enough for industry.
- Goals we want to see/view as important:
  - Atmospheric stability vs. wake controllability
  - Freestream flow; need to have an idea of what happens around the wind farm
  - Wind-farm-to-wind-farm interaction
  - Catch the dynamics of the wake
  - Wake-to-wake interaction
  - Wind plant blockage/induction zone.
- Active control of wind turbines for wake control—industry hesitant to adopt. For example:
  - Open-loop strategies will not work
  - As an owner: never a chance to increase production after installation; wake steering could change that.
  - Need additional measurements on the turbines
  - Misalignment between academia/industry on goals; single validation/case study not enough for industry. Idealize/control conditions
- Large idealized wind farm on flat terrain, but not very many regular gridded layouts
- Validation purposes: it is useful to start with a flat surface to eliminate areas of uncertainty and consistency of wind direction
- We must consider which goals are achievable in terms of observations and for model validation as we plan the campaign and identify goals.

Inflow/operating conditions

- Inflow is equally important as measuring the wakes, such as shear, veer, wind direction, wind direction variability, speed, turbulence, and stability
- If loads are important, then turbulence is important and the change of inflow parameters along the blade.
- Freestream flow: need to have an idea of what happens around the wind farm

- Atmospheric stability vs. wake controllability. Instrumentation is a priority
- Think about potential new ways to instrument a site, but limited advancement by 2020
- Try to get wind direction variability from supervisory control and data acquisition (SCADA) data today; also use Richardson number or Obukhov length for stability; use temperature profile from meteorological tower or sonic anemometer close to ground; depends on homogeneity of the surface.
- How will dissipation be measured (e.g., tethered lift systems, hot-wire anemometry)?
- Possibility to instrument blades, strain measurements along blades; different (scales of) models require different measurements (e.g., long-term statistics for wind farm layout models vs. high-resolution measurements for high-fidelity models). Suggest scanning techniques to answer multiple science questions (e.g., different scanning strategies [faster scans for wake meandering vs. multiple wakes at the same time]).

Catalog "lessons learned" from previous work

- Do we have lessons learned from other measurement campaigns? Do we know at this stage?
- Researchers have a list of what they wish they did differently in Wind Forecast Improvement Project 2 (WFIP2). Offshore
- In air-sea interactions (gulf-stream interaction—sea surface temperature and winds)
- There is a great deal of interest (professionally and scientifically) about offshore, for most of the science goals identified.
- Offshore is important. Meteorology and underlying surface are simpler. But instrumentation is more challenging.

B.2. Feedback Session #2 - Science Goal Prioritization

Please see the prioritized list of science goals and questions in Appendix A.

B.3. Feedback Session #3 - Testing Plan Feedback

B.3.1. Question #1 - What Should Be Added to the Testing Plan?

Preassessment

- Lidar aerial survey prior (and perhaps after) to experiment helps in planning, and useful for modeling
- Strong surface characterization.

Flowfield measurements

- Long-range lidar/radar for multiple wakes
- Multiple lidars for multi-Doppler retrievals
- Thermodynamic properties
- Temperature profiles
- Consideration of the flow above the wind turbine array concurrent with wake observations, momentum entrainment (Do large-scale motions above the wind farm affect systemwide performance by constraining or augmenting momentum flux?).

Turbine operational data/site characterization

- High-frequency SCADA data are important
- Surface characterization (soil moisture measurements are vital for large-eddy simulation)
- Quantification of (simple) terrain
- Consider which turbines are important (GE? Siemens?)
- Turbine-mounted instrumentation.

Logistics

- Duration: the longer the better; 3–5 years
- Plan for on-site maintenance
- Infrastructure (power and cell phone coverage)
- Webcams to monitor sensors
- Calibration plan.

Concurrent modeling

- Access to real-time model output for evaluating observations (as in WFIP2)
- Models that we want to evaluate include Simulator fOr Wind Farm Applications (SOWFA), Nalu-Wind, Weather Research and Forecasting Model (WRF), wake merging/meandering, constants or parameters and their stability dependence, inertial range of turbulence in wakes.)
- Consider low- or midfidelity models that can be used iteratively, or sweep over a larger experimental matrix, compare to high-fidelity modeling
- Tune observations to measure the quantities of interest for model validation.

B.3.2. Question #2 - What Should Be Removed from the Testing Plan?

Because the test plan has not yet come together into a final form, none of the small workgroups had suggestions for removing items from the plan.

B.3.3. Question #3 - What Instrumentation Should Be Added to Testing Plan and Why?

Meteorological masts

- Temperature profiles along with the wind speed measurements
- Sonic anemometers
- Pressure-temperature-humidity sensors
- Radiometers (calibrated before/after experiment)
- Thermal imaging, flux stations (thermal, momentum, moisture fluxes). Remote sensing
- Radiosondes
- Sodar and radio acoustic sounding systems (RASS)
- Ka-band and X-band radar
- A diversity of lidars, including:
  - Ground-based
  - Nacelle-mounted
  - Scanning
  - Profiling
  - All possible scan patterns.

- Sounding systems (difficult with the Federal Aviation Administration (FAA); a lot of labor)
- Field particle-image-velocimetry (PIV)-type systems have gone through some field testing, or possibly a lidar or multi-Doppler technique
- Satellite may be useful for cloudiness and crop measurements. Can use satellite cloudiness index for Nowcasting forecasting.

Other instruments

- Eddy covariance (surface flux stations) both upwind and throughout the wind plant
- Higher-frequency drone profiles for temperature, humidity, and wind that could include:
  - Multirotors for vertical profiling
  - Fixed wing for long transects
  - Tethered balloons for profiling winds and temperature
  - Ceilometer for planetary boundary layer heights.
- SCADA
  - Performance/operational data
  - Structural loads.

Other advice

- Plan for backup units
- Plan for calibration
- Plan for maintenance.

B.3.4. Question #4 - Any Recommendations for Test Site Location?

Sites

- Wind Forecast Improvement Project 2 (WFIP2) site (Columbia River Gorge): too complex terrain for this test, but instrumentation already exist
- Atmospheric Radiation Measurement (ARM) climate research facility
- Story County, Iowa: great towers

- North of Mason City, Iowa: there are three farms close together
- NextEra: pre/postconstruction test site
- Roscoe
- Loraine
- Champion
- ARM site: Oklahoma.

Considerations

- Keep site as simple as possible.
- Flat, simple terrain. As uniform as possible in roughness surrounding site.
- An area that already has atmospheric characterization should be prioritized.
- Atmospherically complex, topographically simple
- Consider a wind farm that operates with snow, if we wish to use particle image velocimetry (PIV)

B.3.5. Question #5 - Additional Thoughts on Test Planning?

Logistics

- Need to be flexible with landowner
  - Land use
  - Access
  - Incentives (?)
- 2020 seems like an ambitious timeline; more likely 2022.
- In Perdigao a lot of institutes could not commit until late, which then caused disruptions for others; need strict deadlines (but balance with flexibility)
- Do projects in phases
- Make sure safety rules are clear. Data access, archiving, and permissions
- Common architecture for data archive
- Create data management group
- Data transfer is important, often a bottleneck for partners
- Create a team responsible for permissions, access, nondisclosure agreements, and so on

- Infrastructure to support instrumentation, such as:
  - A data processing center
  - A data access center
  - An FTP server.

Instrumentation

- How will instruments be obtained? Some examples could include:
  - Subcontracts with universities
  - Lease from companies or international labs
  - Partnerships with other national organizations: National Oceanic and Atmospheric Administration, National Center for Atmospheric Research, and laboratories
- Power: deploying many lidars will require a reliable network for instrumentation
- Carnet issues for loaned equipment from Europe
- Careful calibration/alignment of the instrumentation.

Modeling component

- Should get the right suite of models, including:
  - High-fidelity models
  - Midfidelity and engineering models
  - Parameterized models
- These should be used in the design of the experiment, during the experimental campaign, and in data analysis.

B.3.6. Question #6 - Beyond Meeting Attendees, Any Potential Partners That Should Be Included?

Agencies

- United States Department of Agriculture
- International Energy Agency (IEA) Task 31/WakeBench, IEA Task (new).

Labs

- National Laboratory for Agriculture and the Environment
- National Center for Atmospheric Research
- National Renewable Energy Centre of Spain.

University partners/research groups

- ARM climate research facility
- Technical University of Denmark
- WindForS
- ForWind
- University of Hohenheim: ability to measure atmospheric properties, aerosols
- Wind tunnel researchers
- Sowento.

Industry partners

- Wind energy supply chain
- Lidar manufacturers
- Drone companies
- ECN wind energy facilities
- Someone that can provide operational forecasts to plan hours or a day ahead

C. Expert Elicitation Meeting Agenda

MEETING FOR AWAKEN - AMERICAN WAKE EXPERIMENT
University of Colorado
Sustainability, Energy and Environment Complex (SEEC) Boulder, Colorado, USA, March 21, 2018

8:00 am: Introductions
8:15 am: A2e Overview

- A2e Overview - Will Shaw, PNNL, DOE wind resource coordinator
- PIRT Development, Modeling needs and Validation and Verification plan - David Maniaci, Sandia National Laboratories
- AWAKEN Vision and Science Goals - Pat Moriarty, NREL

9:15 am: Breakout groups - Science goals feedback sessions #1

- Discussion of science goals, relevance, missing items etc.

9:45 am: Break - 20 minutes

10:05 am: Large Group - Summary of feedback session #1 and group discussion

10:35 am: Previous wake and atmospheric studies overview and future recommendations

- Perdigão overview - Jakob Mann, DTU
- SWiFT Testing - Tommy Herges, Sandia National Laboratories
- TWICS/CWEX 13 - Julie Lundquist, CU- Boulder

11:20 am: Breakout groups - Science goals feedback sessions #2

- From comments generated by groups in session #1, continue to modify science goals list and prioritize the science goals

11:50 am: Lunch

12:40 pm: Previous wake and atmospheric studies overview and future recommendations

- TTU radar studies/BEACON project - John Schroeder, TTU
- Perdigão wakes and others - Rebecca Barthelmie, Cornell University
- Iowa Wind Farm Observations - Daniel Rajewski, Iowa State University
- NextEra experience with wakes - Bob Conzemius, NextEra

1:40 pm: Large Group - Summary of feedback session #2 and group discussion

2:00 pm: Previous wake and atmospheric studies overview and future recommendations

- Lidar wake measurements - Peter Clive, Wood Group
- GE lidar, radar and validation - Stefan Kern, GE
- Texas Panhandle Lidar Study - Valerio Iungo, UT Dallas

2:45 pm: Initial testing plans for AWAKEN and instrumentation overview - Jeroen van Dam, NREL

3:00 pm: Breakout groups - Initial plan and instrumentation feedback session #3

- Discuss suggestion for improvement

3:30 pm: Break - 20 minutes

3:50 pm: Summary of feedback session #3 and group discussion

4:10 pm: Future testing plans with potential overlap

- WFIP II/III - Will Shaw, PNNL
- Overview of upcoming lidar wake experiments at DTU - Elliot Simon, DTU

4:40 pm: Partnering ideas and funding opportunities

- NCAR EOL Capabilities and Timeline for Involvement -Alison Rockwell, NCAR
- Group discussion - Pat Moriarty, NREL

5:15 pm: Next Steps and Wrap-up

5:30 pm: Adjourn

D. Expert Elicitation Meeting Attendees

Name	Affiliation
National Laboratories	
Pat Moriarty	National Renewable Energy Laboratory
Jeroen van Dam	
Jennifer Annoni	
Nicholas Hamilton	
Jason Fields	
Jason Roadman	
Andy Scholbrock	
Paul Fleming	
David Maniaci	Sandia National Laboratories
Tommy Herges	
Sonia Wharton	Lawrence Livermore National Laboratory
Will Shaw	Pacicifc Northwest National Laboratory
Yelena Pichugina	National Oceanic and Atmospheric Administration
Bob Banta	
Jim Wilczak	
Aditya Choukulkar	
Alison Rockwell	National Center for Atmospheric Research
Thomas Gerz	German Aerospace Center (DLR)
Norman Wildmann	
Universities	
Julie Lundquist	University of Colorado Boulder
Nicola Bodini	
Chris Heney	
Cory Dixon	
Jessica Tomaszewski	
Rochelle Worsnop	
Joseph Lee	
John Schroeder	Texas Tech University

Name	Affiliation
Jakob Mann	Technical University of Denmark
Robert Menke	
Elliot Simon	
Giacomo Valerio Iungo	The University of Texas at Dallas
Rebecca Barthelmie	Cornell University
Daniel Rajewski	Iowa State University
Tyler Bell	The University of Oklahoma
Joe Fernando	University of Notre Dame
Raghu Krishnamurthy	
Laura Leo	
Marc Calaf	The University of Utah
Jiarong Hong	University of Minnesota
Jose Lagniha Palma	University of Porto
José Carlos Matos	
Industry	
Cory Jog	EDF Renewables North America
Peter Clive	Wood Group
John Wang	
Bob Conzemius	WindLogics
Fatma Demet Ulker	Envision Energy
Ewan Machefaux	Vestas
Stefan Kern	General Electric
Lawrence Cheung	
Scott Schreck	Siemens Gamesa Renewable Energy
Jens Madsen	Suzlon Energy
Mark Stoelinga	Vaisala

References

Ahsbahs, T. T., M. Badger, P. Volker, K. S. Hansen, and C. B. Hasager. 2018. "Applications of satellite winds for the offshore wind farm site Anholt." *Wind Energy Science* 3 (2): 573–588.

Aitken, M. L., and J. K. Lundquist. 2014. "Utility-scale wind turbine wake characterization using nacelle-based long- range scanning lidar." *Journal of Atmospheric and Oceanic Technology* 31 (7): 1529–1539.

Aitken, M. L., R. M. Banta, Y. L. Pichugina, and J. K. Lundquist. 2014a. "Quantifying wind turbine wake character istics from scanning remote sensor data." *Journal of Atmospheric and Oceanic Technology* 31 (4): 765–787.

– . 2014b. "Quantifying wind turbine wake characteristics from scanning remote sensor data." *Journal of Atmospheric and Oceanic Technology* 31 (4): 765–787.

Banta, R. M., Y. L. Pichugina, W. A. Brewer, J. K. Lundquist, N. D. Kelley, S. P. Sandberg, R. J. Alvarez II, R. M. Hardesty, and A. M. Weickmann. 2015. "3D Volumetric Analysis of Wind Turbine Wake Properties in the Atmosphere Using High-Resolution Doppler Lidar." *Journal of Atmospheric and Oceanic Technology* 32 (5): 904–914. doi:10.1175/JTECH-D-14-00078.1.

Barthelmie, R. J., G. Larsen, S. Frandsen, L Folkerts, K Rados, S. Pryor, B Lange, and G Schepers. 2006. "Comparison of wake model simulations with offshore wind turbine wake profiles measured by sodar." *Journal of atmospheric and oceanic technology* 23 (7): 888–901.

Barthelmie, R. J., K. S. Hansen, S. T. Frandsen, O. Rathmann, J. Schepers, W Schlez, J Phillips, K Rados, A Zervos, E. Politis, et al., 2009. "Modelling and measuring flow and wind turbine wakes in large wind farms offshore." *Wind Energy* 12 (5): 431–444.

Barthelmie, R. J., S. C. Pryor, S. T. Frandsen, K. S. Hansen, J. Schepers, K Rados, W Schlez, A Neubert, L. Jensen, and S Neckelmann. 2010. "Quantifying the impact of wind turbine wakes on power output at offshore wind farms." *Journal of Atmospheric and Oceanic Technology* 27 (8): 1302–1317.

Barthelmie, R. J., et al., 2011. *Flow and wakes in large wind farms: Final report for UpWind WP8*. Tech. rep. Risø-R- 1765(EN). Danmarks Tekniske Universitet, Risø Nationallaboratoriet for Bæredygtig Energi.

Barthelmie, R., S. Pryor, N Wildmann, and R. Menke. 2018. "Wind turbine wake characterization in complex terrain via integrated Doppler lidar data from the Perdigão experiment." In *Journal of Physics: Conference Series*, 1037:052022. 5. IOP Publishing.

Bianco, L., K. Friedrich, J. M. Wilczak, D. Hazen, D. Wolfe, R. Delgado, S. P. Oncley, and J. K. Lundquist. 2017. "Assessing the accuracy of microwave radiometers and radio acoustic sounding systems for wind energy applications." *Atmospheric Measurement Techniques* 10 (5): 1707.

Bingöl, F., J. Mann, and G. C. Larsen. 2010. "Light detection and ranging measurements of wake dynamics part I: one-dimensional scanning." *Wind Energy* 13 (1): 51–61.

Board, O. S., E. National Academies of Sciences, Medicine, et al., 2018. *The Future of Atmospheric Boundary Layer Observing, Understanding, and Modeling: Proceedings of a Workshop*. National Academies Press.

Bodini, N., D. Zardi, and J. K. Lundquist. 2017. "Three-dimensional structure of wind turbine wakes as measured by scanning lidar." *Atmospheric Measurement Techniques* 10 (8).

Brock, F. V., K. C. Crawford, R. L. Elliott, G. W. Cuperus, S. J. Stadler, H. L. Johnson, and M. D. Eilts. 1995. "The Oklahoma Mesonet: a technical overview." *Journal of Atmospheric and Oceanic Technology* 12 (1): 5–19.

Bromm, M., A. Rott, H. Beck, L. Vollmer, G. Steinfeld, and M. KÃijhn. 2018. "Field investigation on the influence of yaw misalignment on the propagation of wind turbine wakes." *Wind Energy* 21 (11): 1011–1028. doi:10.1002/we. 2210.

Cal, R. B., J. Lebrón, L. Castillo, H. S. Kang, and C. Meneveau. 2010. "Experimental study of the horizontally averaged flow structure in a model wind-turbine array boundary layer." *Journal of Renewable and Sustainable Energy* 2 (1): 013106.

Chamorro, L. P., and F. Porté-Agel. 2010. "Effects of thermal stability and incoming boundary-layer flow characteris- tics on wind-turbine wakes: a wind-tunnel study." *Boundary-layer meteorology* 136 (3): 515–533.

Choukulkar, A., et al., 2017. "Evaluation of single and multiple Doppler lidar techniques to measure complex flow during the XPIA field campaign." *Atmospheric Measurement Techniques* 10 (1): 247–264. doi:10.5194/amt-10- 247-2017. https://www.atmos-meas-tech.net/10/247/2017/.

Christiansen, M. B., and C. B. Hasager. 2005. "Wake effects of large offshore wind farms identified from satellite SAR." *Remote Sensing of Environment* 98 (2-3): 251–268.

Cleijne, J. 1992. *Results of Sexbierum Wind Farm; Double Measurements. Tech. rep.* TNO Report 92-388. TNO (Netherlands).

– . 1993. *Results of Sexbierum Wind Farm; Single Wake Measurements. Tech. rep.* TNO Report 93-082. TNO (Netherlands).

Clifton, A., A. Smith, and M. Fields. 2016. *Wind plant preconstruction energy estimates. current practice and opportunities.* Tech. rep. NREL/TP-5000-64735. National Renewable Energy Lab. (NREL), Golden, CO (United States). Cline, J, S. Haupt, and W. Shaw. 2017. "Meteorology Research in DOE's Atmosphere to Electrons (A2e) Program."In *AGU Fall Meeting Abstracts*.

Debnath, M., G. V. Iungo, R. Ashton, W. A. Brewer, A. Choukulkar, R. Delgado, J. K. Lundquist, W. J. Shaw, J. M. Wilczak, and D. Wolfe. 2017. "Vertical profiles of the 3-D wind velocity retrieved from multiple wind lidars performing triple range-height-indicator scans." *Atmospheric Measurement Techniques* 10 (2): 431–444. doi:10.5194/ amt-10-431-2017.

Doubrawa, P., M. Debnath, P. J. Moriarty, E. Branlard, T. Herges, D. Maniaci, and B. Naughton. 2019. "Benchmarks for model validation based on lidar wake measurements." In *Journal of Physics: Conference Series*, 1256:012024. 1. IOP Publishing.

Emeis, S., S. Siedersleben, A. Lampert, A. Platis, J. Bange, B. Djath, J. Schulz-Stellenfleth, and T. Neumann. 2016. "Exploring the wakes of large offshore wind farms." In *Journal of Physics: Conference Series*, 753:092014. 9. IOP Publishing.

Fernando, H., J. Mann, J. Palma, J. Lundquist, R. J. Barthelmie, M Belo-Pereira, W. Brown, F. Chow, T Gerz, C. Hocut, et al., 2019. "The Perdigao: Peering into microscale details of mountain winds." *Bulletin of the American Meteorological Society* 100 (5): 799–819.

Fleming, P., et al., 2020. "Continued Results from a Field Campaign of Wake Steering Applied at a Commercial Wind Farm: Part 2." *Wind Energy Science Discussions* 2020:1–24. doi:10.5194/wes-2019-104. https://www.wind-energ- sci-discuss.net/wes-2019-104/.

Fleming, P., J. King, K. Dykes, E. Simley, J. Roadman, A. Scholbrock, P. Murphy, J. K. Lundquist, P. Moriarty, K. Fleming, et al., 2019. "Initial results from a field campaign of wake steering applied at a commercial wind farm–Part 1." *Wind Energy Science* 4 (2): 273–285.

Geerts, B., D. J. Raymond, V. Grubišic′, C. A. Davis, M. C. Barth, A. Detwiler, P. M. Klein, W.-C. Lee, P. M. Markowski, G. L. Mullendore, et al., 2018. "Recommendations for in situ and remote sensing capabilities in atmospheric convection and turbulence." *Bulletin of the American Meteorological Society* 99 (12): 2463–2470.

Hamilton, N., and R. B. Cal. 2015. "Anisotropy of the Reynolds stress tensor in the wakes of wind turbine arrays in Cartesian arrangements with counter-rotating rotors." *Physics of Fluids* 27 (1): 015102.

Hammond, S. W., M. A. Sprague, D. Womble, and M. Barone. 2015. *A2e High Fidelity Modeling: Strategic Plan- ning Meetings*. Tech. rep. NREL/TP-2C00-64697. National Renewable Energy Lab. (NREL), Golden, CO (United States).

Hansen, K. S., R. J. Barthelmie, L. E. Jensen, and A. Sommer. 2012. "The impact of turbulence intensity and atmo- spheric stability on power deficits due to wind turbine wakes at Horns Rev wind farm." *Wind Energy* 15 (1): 183– 196.

Hasager, C. B., A. N. Hahmann, T. Ahsbahs, I. Karagali, T. Sile, M. Badger, and J. Mann. 2019. "Europe's offshore winds assessed from SAR, ASCAT and WRF." *Wind Energy Science Discussions* 2019:1–27. doi:10.5194/wes- 2019-38.

Herges, T., D. C. Maniaci, B. T. Naughton, T. Mikkelsen, and M. Sjöholm. 2017. "High resolution wind turbine wake measurements with a scanning lidar." In *Journal of Physics: Conference Series*, 854:012021. 1. IOP Publishing.

Herges, T. G., D. Bossert, R. Schmitt, M. Johnson, D. C. Maniaci, C. Glen, and B. Naughton. 2016. "Field Demonstration of the Sandia Wake Imaging System Capabilities at the Scaled Wind Farm Technology Facility." In *34th Wind Energy Symposium*, 1524.

Hills, R. G., D. C. Maniaci, and J. W. Naughton. 2015. *V&V Framework*. Tech. rep. SAND2015-7455. Sandia National Laboratories. Albuquerque, NM (United States).

Hirth, B. D., J. L. Schroeder, W. S. Gunter, and J. G. Guynes. 2012. "Measuring a utility-scale turbine wake using the TTUKa mobile research radars." *Journal of Atmospheric and Oceanic Technology* 29 (6): 765–771.

Iungo, G. V., and F. Porté-Agel. 2014. "Volumetric lidar scanning of wind turbine wakes under convective and neutral atmospheric stability regimes." *Journal of Atmospheric and Oceanic Technology* 31 (10): 2035–2048.

Iungo, G. V., Y.-T. Wu, and F. Porté-Agel. 2013. "Field measurements of wind turbine wakes with lidars." *Journal of Atmospheric and Oceanic Technology* 30 (2): 274–287.

Jagdish, S. A. Kumar, A. Chakraborty, and R. Kumar. 2018. "Validation of wind speed retrieval from RISAT-1 SAR images of the North Indian Ocean." *Remote Sensing Letters* 9 (5): 421–428.

Käsler, Y., S. Rahm, R. Simmet, and M. Kühn. 2010. "Wake measurements of a multi-MW wind turbine with coherent long-range pulsed Doppler wind lidar." *Journal of Atmospheric and Oceanic Technology* 27 (9): 1529–1532.

Klein, P., et al., 2015. "LABLE: A Multi-Institutional, Student-Led, Atmospheric Boundary Layer Experiment." *Bul- letin of the American Meteorological Society* 96 (10): 1743–1764. doi:10.1175/BAMS-D-13-00267.1.

Kocer, G., M. Mansour, N. Chokani, R. S. Abhari, and M. Müller. 2011. "Full-scale wind turbine near-wake measure-ments using an instrumented uninhabited aerial vehicle." *Journal of Solar Energy Engineering* 133 (4).

Krishnamurthy, R., J. Reuder, B. Svardal, H. J. S. Fernando, and J. B. Jakobsen. 2017. "Offshore wind turbine wake characteristics using scanning doppler lidar." *Energy Procedia* 137:428–442.

Krogstad, P.-Å., and M. S. Adaramola. 2012. "Performance and near wake measurements of a model horizontal axis wind turbine." *Wind Energy* 15 (5): 743–756.

Kumer, V.-M., J. Reuder, M. Dorninger, R. Zauner, and V. Grubišic′. 2016. "Turbulent kinetic energy estimates from profiling wind LiDAR measurements and their potential for wind energy applications." *Renewable Energy* 99:898– 910.

Lawson, M. J., J. Melvin, S. Ananthan, K. M. Gruchalla, J. S. Rood, and M. A. Sprague. 2019. *Blade-Resolved, Single- Turbine Simulations Under Atmospheric Flow*. Tech. rep. NREL/TP-5000-72760. National Renewable Energy Lab. (NREL), Golden, CO (United States).

Lee, J. C., and J. K. Lundquist. 2017. "Evaluation of the wind farm parameterization in the Weather Research and Forecasting model (version 3.8. 1) with meteorological and turbine power data." *Geoscientific Model Development* 10 (11).

Lundquist, J., and L Bariteau. 2015. "Dissipation of Turbulence in the Wake of a Wind Turbine." *Boundary-Layer Meteorology* 154 (2): 229–241.

Lundquist, J., A. J. Clifton, S. Dana, T. Herges, A. Huskey, P. J. Moriarty, and J. J. Van Dam. 2019. *Wind Energy Instrumentation Atlas*. Tech. rep. NREL/TP-5000-68986. National Renewable Energy Lab. (NREL), Golden, CO (United States).

Lundquist, J. K., et al., 2017. "Assessing State-of-the-Art Capabilities for Probing the Atmospheric Boundary Layer: The XPIA Field Campaign." *Bulletin of the American Meteorological Society* 98 (2): 289–314. doi:10.1175/BAMS- D-15-00151.1.

Machefaux, E., G. C. Larsen, N. Troldborg, K. S. Hansen, N. Angelou, T. Mikkelsen, and J. Mann. 2016. "Investigation of wake interaction using full-scale lidar measurements and large eddy simulation." *Wind Energy* 19 (8): 1535–1551. Maniaci, D. C. 2016. *Overview of the V&V Framework*. Tech. rep. SAND2016-2862PE. Sandia National Laboratories. Albuquerque, NM (United States).

Maniaci, D. C., and J. Naughton. 2019. *V&V Integrated Program Planning for Wind Plant Performance*. Tech. rep. SAND2019-6888. Sandia National Laboratories. Albuquerque, NM (United States).

Maniaci, D. C., P. J. Moriarty, M. F. Barone, M. J. Churchfield, M. A. Sprague, and S. Arunajatesan. 2020. *Wind Energy High Fidelity Model Verification and Validation Plan*. Tech. rep. SAND2020-1332. Sandia National Laboratories. Albuquerque, NM (United States).

Mann, J., N. Angelou, J. Arnqvist, D. Callies, E Cantero, R. C. Arroyo, M. Courtney, J Cuxart, E. Dellwik, J. Gottschall, et al., 2017. "Complex terrain experiments in the New European Wind Atlas." *Philosophical Transactions of the Royal Society A: Mathematical, Physical and Engineering Sciences* 375 (2091): 20160101.

Menke, R., N. Vasiljevic′, K. S. Hansen, A. N. Hahmann, and J. Mann. 2018. "Does the wind turbine wake follow the topography? A multi-lidar study in complex terrain." *Wind Energy Science* 3 (2): 681–691.

Moriarty, P., et al., 2014. "IEA-task 31 wakebench: Towards a protocol for wind farm flow model evaluation. Part 2: Wind farm wake models." In *Journal of Physics: Conference Series*, 524:012185. 1. IOP Publishing.

Muñoz-Esparza, D., J. K. Lundquist, J. A. Sauer, B. Kosovic´, and R. R. Linn. 2017. "Coupled mesoscale-LES modeling of a diurnal cycle during the CWEX-13 field campaign: From weather to boundary-layer eddies." *Journal of Advances in Modeling Earth Systems* 9 (3): 1572–1594.

Murphy, P., J. K. Lundquist, and P. Fleming. 2019. "How wind speed shear and directional veer affect the power production of a megawatt-scale operational wind turbine." *Wind Energy Science Discussions* 2019:1–46. doi:10 . 5194/wes-2019-86.

Newsom, R. K., W. A. Brewer, J. M. Wilczak, D. E. Wolfe, S. P. Oncley, and J. K. Lundquist. 2017. "Validating precision estimates in horizontal wind measurements from a Doppler lidar." *Atmospheric Measurement Techniques* 10 (3): 1229.

Nygaard, N. G. 2014. "Wakes in very large wind farms and the effect of neighbouring wind farms." In *Journal of Physics: Conference Series*, 524:012162. 1. IOP Publishing.

Nygaard, N. G., and S. D. Hansen. 2016. "Wake effects between two neighbouring wind farms." In *Journal of Physics: Conference Series*, 753:032020. 3. IOP Publishing.

Nygaard, N. G., and A. C. Newcombe. 2018. "Wake behind an offshore wind farm observed with dual-Doppler radars." In *Journal of Physics: Conference Series*, 1037:072008. 7. IOP Publishing.

Ørsted. 2019. "Ørsted presents update on its long-term financial targets." *Ørsted Company Announcement List*. https://orsted.com/en/Company-Announcement-List/2019/10/1937002.

Peña, A., and J. Mann. 2019. "Turbulence Measurements with Dual-Doppler Scanning Lidars." *Remote Sensing* 11 (20). ISSN: 2072-4292. doi:10.3390/rs11202444.

Pichugina, Y. L., R. M. Banta, W. A. Brewer, S. P. Sandberg, and R. M. Hardesty. 2012. "Doppler Lidar-Based Wind- Profile Measurement System for Offshore Wind-Energy and Other Marine Boundary Layer Applications." *Journal of Applied Meteorology and Climatology* 51 (2): 327–349. doi:10.1175/JAMC-D-11-040.1.

Platis, A., et al., 2018. "First in situ evidence of wakes in the far field behind offshore wind farms." *Scientific Reports* 8 (1): 2163. doi:10.1038/s41598-018-20389-y.

Rados, K., G. Larsen, R. Barthelmie, W. Schlez, B. Lange, G. Schepers, T. Hegberg, and M. Magnisson. 2001. "Com- parison of wake models with data for offshore windfarms." *Wind Engineering* 25 (5): 271–280.

Rajewski, D., E. Takle, J. Lundquist, S Oncley, J. Prueger, T. Horst, M. Rhodes, R Pfeiffer, J. Hatfield, K. Spoth, et al., 2013. *Crop Wind Energy Experiment (CWEX): observations of surface-layer, boundary layer, and mesoscale interactions with a wind farm, B. Am. Meteorol. Soc., 94, 655–672.*

Rajewski, D. A., E. S. Takle, J. K. Lundquist, J. H. Prueger, R. L. Pfeiffer, J. L. Hatfield, K. K. Spoth, and R. K. Doorenbos. 2014. "Changes in fluxes of heat, H2O, and CO2 caused by a large wind farm." *Agricultural and forest meteorology* 194:175–187.

Reuder, J., and M. O. Jonassen. 2012. "First results of turbulence measurements in a wind park with the Small Unmanned Meteorological Observer SUMO." *Energy Procedia* 24:176–185.

Rhodes, M. E., and J. K. Lundquist. 2013. "The effect of wind-turbine wakes on summertime US midwest atmospheric wind profiles as observed with ground-based doppler lidar." *Boundary-layer meteorology* 149 (1): 85–103.

Sanchez Gomez, M., and J. K. Lundquist. 2020. "The effect of wind direction shear on turbine performance in a wind farm in central Iowa." *Wind Energy Science* 5 (1): 125–139. doi:10.5194/wes-5-125-2020.

Sanz Rodrigo, J., R. A. Chavez Arroyo, P. Moriarty, M. Churchfield, B. Kosovic´, P.-E. Réthoré, K. S. Hansen, A. Hahmann, J. D. Mirocha, and D. Rife. 2017. "Mesoscale to microscale wind farm flow modeling and evaluation." *Wiley Interdisciplinary Reviews: Energy and Environment* 6 (2): e214.

Schroeder, J. L., W. Burgett, K. Haynie, I Sonmez, G. Skwira, A. Doggett, and J. Lipe. 2005. "The West Texas mesonet: a technical overview." *Journal of Atmospheric and Oceanic Technology* 22 (2): 211–222.

Shaw, W. J., L. K. Berg, J. Cline, C. Draxl, I. Djalalova, E. P. Grimit, J. K. Lundquist, M. Marquis, J. McCaa, J. B. Olson, et al., 2019. "The Second Wind Forecast Improvement Project (WFIP2): General Overview." *Bulletin of the American Meteorological Society* 100 (9): 1687–1699.

Simley, E., N. Angelou, T. Mikkelsen, M. Sjöholm, J. Mann, and L. Y. Pao. 2016. "Characterization of wind velocities in the upstream induction zone of a wind turbine using scanning continuous-wave lidars." *Journal of Renewable and Sustainable Energy* 8 (1): 013301. doi:10.1063/1.4940025.

Sisterson, D., R. Peppler, T. Cress, P. Lamb, and D. Turner. 2016. "The ARM southern great plains (SGP) site."

Meteorological Monographs 57:6–1.

Sivaraman, C., E. G. Stephan, M. C. Macduff, and C. D. Hagler. 2014. *Data Archive and Portal Thrust Area Strategy Report*. Tech. rep. PNNL-23718. Pacific Northwest National Laboratory. (PNNL), Richland, WA (United States).

Smalikho, I., V. Banakh, Y. Pichugina, W. Brewer, R. Banta, J. Lundquist, and N. Kelley. 2013. "Lidar investigation of atmosphere effect on a wind turbine wake." *Journal of Atmospheric and Oceanic Technology* 30 (11): 2554–2570.

Stawiarski, C., K. Trümner, C. Knigge, and R. Calhoun. 2013. "Scopes and Challenges of Dual-Doppler Lidar Wind Measurements-An Error Analysis." *Journal of Atmospheric and Oceanic Technology* 30 (9): 2044–2062. doi:10. 1175/JTECH-D-12-00244.1.

Trujillo, J. J., J. K. Seifert, I. Würth, D. Schlipf, and M. Kühn. 2016. "Full-field assessment of wind turbine near-wake deviation in relation to yaw misalignment." *Wind Energy Science* 1 (1): 41–53. doi:10 . 5194 / wes - 1 - 41 - 2016. https://www.wind-energ-sci.net/1/41/2016/.

Trujillo, J.-J., F. Bingöl, G. C. Larsen, J. Mann, and M. Kühn. 2011. "Light detection and ranging measurements of wake dynamics. Part II: two-dimensional scanning." *Wind Energy* 14 (1): 61–75.

Tucker, S. C., C. S. Weimer, S. Baidar, and R. M. Hardesty. 2018. "The Optical Autocovariance Wind Lidar. Part I: OAWL Instrument Development and Demonstration." *Journal of Atmospheric and Oceanic Technology* 35 (10): 2079–2097. doi:10.1175/JTECH-D-18-0024.1.

Veers, P., et al., 2019. "Grand challenges in the science of wind energy." *Science* 366 (6464). doi:10.1126/science. aau2027.

Vollmer, L., M. van Dooren, D. Trabucchi, J. Schneemann, G. Steinfeld, B. Witha, J. Trujillo, and M. Kühn. 2015. "First comparison of LES of an offshore wind turbine wake with dual-Doppler lidar measurements in a German offshore wind farm." In *Journal of Physics: Conference Series*, 625:012001. 1. IOP Publishing.

Vömel, H., et al., 2018. *The NCAR/EOL Community Workshop On Unmanned Aircraft Systems For Atmospheric Re search*. UCAR/NCAR Earth Observing Laboratory. doi:10.5065/D6X9292S.

Wilczak, J., C. Finley, J. Freedman, J. Cline, L. Bianco, J. Olson, I. Djalalova, L. Sheridan, M. Ahlstrom, J. Manobianco, et al., 2015. "The Wind Forecast Improvement Project (WFIP): A public–private partnership addressing wind energy forecast needs." *Bulletin of the American Meteorological Society* 96 (10): 1699–1718.

Wilczak, J. M., M. Stoelinga, L. K. Berg, J. Sharp, C. Draxl, K. McCaffrey, R. M. Banta, L. Bianco, I. Djalalova, J. K. Lundquist, et al., 2019. "The Second Wind Forecast Improvement Project (WFIP2): Observational Field Campaign." *Bulletin of the American Meteorological Society*, no. 2019.

Wildmann, N., M. Hofsäß, F. Weimer, A. Joos, and J. Bange. 2014. "MASC-a small remotely piloted aircraft (RPA) for wind energy research." *Advances in Science and Research* 11:55.

Wildmann, N., S. Kigle, and T. Gerz. 2018. "Coplanar lidar measurement of a single wind energy converter wake in distinct atmospheric stability regimes at the Perdigão 2017 experiment." In *Journal of Physics: Conference Series*, 1037:052006. 5. IOP Publishing.

Wildmann, N., N. Vasiljevic, and T. Gerz. 2018. "Wind turbine wake measurements with automatically adjusting scanning trajectories in a multi-Doppler lidar setup." *Atmospheric Measurement Techniques Discussions (AMTD)* 2018:1–20.

Wilkinson, M. D., M. Dumontier, I. J. Aalbersberg, G. Appleton, M. Axton, A. Baak, N. Blomberg, J.-W. Boiten,

L. B. da Silva Santos, P. E. Bourne, et al., 2016. "The FAIR Guiding Principles for scientific data management and stewardship." *Scientific data* 3.

Chapter 7

The American Wake Experiment (AWAKEN): Groundbreaking International Field Research Addresses Wind Farm Atmosphere Interactions*

U.S. Department of Energy

Wind energy has become a low-cost renewable power source worldwide. To increase the energy output of wind farms, operators must minimize the effects of wakes—one of the least understood physical phenomena in wind plants today—which may result in a plant losing 10%–15% of its potential electric energy output.

As wind turbines rotate to extract energy and generate electricity from wind, they agitate the air around them, creating a wake like those formed behind a moving boat. Wakes reduce wind speeds, increase turbulence for other turbines downstream, reduce power output, and increase operation and maintenance costs. To reduce wake effects, operators need a better understanding of how air moves around turbines and how individual turbines interact with one another in the wind farm.

To help provide that understanding, the U.S. Department of Energy's (DOE) Wind Energy Technologies Office (WETO) funds the American WAKE experimeNt (AWAKEN). AWAKEN is an international, multi-institutional wind energy field campaign designed to answer the most pressing science questions about the impacts of wind turbine wake effects. From now through October 2023, AWAKEN researchers will use sensors, drones, and

---

* This is an edited, reformatted and augmented version of US depatment of Energy Office of Efficiency and Renewable Energy Publication No. DOE/GO-102021-5672, dated January, 2022.

In: Wind Energy
Editor: Mehmet Landrey
ISBN: 979-8-89113-528-4

aircraft to gather data about the atmosphere and wind around and within operational wind farms. Researchers will analyze and report the data through 2024, with the aim to help validate wind farm models and advance understanding of wind farm behavior.

AWAKEN is part of a larger DOE effort to understand and improve the efficiency of wind plants. The project leverages the expertise and resources of a variety of partners—from DOE national laboratories to international academia, research institutions, and industry.

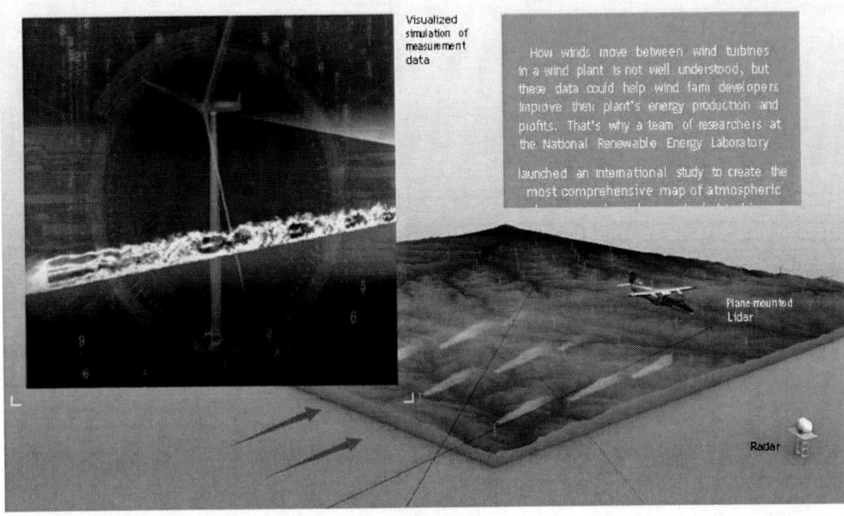

Field Research Benefits

With leadership from DOE national laboratories, AWAKEN project participants will organize, design, and execute the campaign, which is designed to:

- Produce a high-quality, comprehensive, globally available data set to advance the knowledge of how air moves within wind plants and between plants situated close to one another.
- Verify which of the tactics wind farms use to reduce wake losses and increase wind plant power production are most effective, reducing uncertainty for wind farm operators.

- Validate and improve existing wind plant simulation tools (e.g., ExaWind and FLORIS) that will create a more complete picture of how atmospheric changes affect wind turbines and wind farms.
- Inspire instrumentation innovation to enable better observations and enhance understanding of wake and other atmospheric interactions.

By collecting and sharing high-quality data and validated modeling techniques, AWAKEN's groundbreaking research will improve existing wind farm operations and develop new insights to enhance the capability of future wind farms.

The Wake Experiments

From 2022 to 2023, the AWAKEN research team will conduct field research in Oklahoma.

Designated Test Site: AWAKEN researchers chose a field campaign site that includes several wind farms in Oklahoma, between Ponca City and Enid. After equipment installation is completed in 2022, data collection will take place through 2023.

Data Collection Equipment: AWAKEN researchers will use scanning radars, lidars, and aircraft—each providing unique and complex measurements—to map the details of winds and turbulence around the wind turbines.

DOE's Wind Data Archive and Portal will host the data, allowing the AWAKEN research team and worldwide collaborators to paint a full picture of how wakes behave and to improve strategies to mitigate wakes and optimize power capture.

Data Analysis and Reporting: Collaborative efforts to analyze and report the data through various publications and dissemination efforts will conclude in 2024.

Project Outcomes

AWAKEN will result in a unique data set that will help:

- Significantly decrease the estimated 10%-15% loss of annual energy production caused by wake effects.
- Validate and improve wind farm models to predict future performance and wake impacts, ultimately reducing the uncertainty levels of current industry tools by half, if not more.
- Predict wakes, plant conditions, and turbine operation strategies to improve plant layout, power output, and turbine reliability.
- Improve understanding of how to design wind farms to optimize energy production and reduce operation and maintenance costs.

DOE National Laboratory AWAKEN Project Leads

- National Renewable Energy Laboratory
- Sandia National Laboratories
- Pacific Northwest National Laboratory
- Lawrence Livermore National Laboratory

For more information, visit: energy.gov/eere/wind

DOE/GO-102021-5672 • January 2022

Index

H

I

K

L

M

N

O

P

R

S

T

U

V

W